Professor Manning's *An Introduction to Animal Behaviour* is an established key text in its field. In this new edition, he has joined forces with Marian Stamp Dawkins, herself an experienced and accomplished writer, to present a fully revised and updated introduction to the subject. Distinguished by its broad biological approach, which links physiology, ethology and comparative psychology, and written in a clear and attractive style, this well-illustrated text will be of interest to all students of animal behaviour and psychology, as well as their teachers.

An Introduction to Animal Behaviour

An Introduction to Animal Behaviour

FOURTH EDITION

Aubrey Manning
Professor of Natural History,
University of Edinburgh

AND

Marian Stamp Dawkins
Department of Zoology,
University of Oxford

Published by the Press Syndicate of the University of Cambridge
The Pitt Building, Trumpington Street, Cambridge CB2 1RP
40 West 20th Street, New York, NY 10011-4211, USA
10 Stamford Road, Oakleigh, Victoria 3166, Australia

First, second and third editions first published by
Edward Arnold and © Edward Arnold 1967, 1972, 1979

This fourth edition first published by Cambridge University
Press 1992 and © Cambridge University Press 1992

Printed in Great Britain by The Bath Press, Avon

A catalogue record for this book is available from the British Library

Library of Congress cataloguing in publication data available

ISBN 0 521 41759 7 hardback
ISBN 0 521 42792 4 paperback

To Niko Tinbergen, FRS, Nobel Laureate,
1907–1988

CONTENTS

PREFACE TO THE FOURTH EDITION

Twelve years between editions is a long time for a subject moving as rapidly as animal behaviour. Aspects which formed just minor specialisms have become major fields in their own right, with their own textbooks and taught courses—sociobiology, behavioural ecology and neuroethology are obvious examples.

Nevertheless, we feel quite certain that there are dangers when everybody goes their own way, particularly for students coming to animal behaviour for the first time. The best understanding in one area will come from people who have a grounding that enables them to reflect on the whole field, with all the richness of interaction that this can bring. Sociobiologists must be aware of developmental problems, neuroethologists of evolutionary ones, and so on.

We have thus tried to keep the necessary breadth and, although obviously and extensively re-written, the original construction and approach of the earlier editions remains intact. Now, as then, the focus of the book is the organization and evolution of the behaviour of individuals.

We wish to express our thanks to Richard Andrew, Gabriel Horn and Richard Morris who helped us with preparation of some new sections, although all responsibility must remain with us! Joan Herrmann provided help and encouragement from the outset. Cathy Jones, Meg McRae, Gayle Stephens and Jackie Ward all helped greatly with the preparation of the text—not always a straightforward task.

Throughout our collaboration on this new edition we have found that Tinbergen's 'Four Questions for Ethology' has been the framework for our approach. Both of us acknowledge the enormous influence Tinbergen had over our own work and thinking. We dedicate this book to his memory.

Aubrey Manning
Marian Stamp Dawkins
June 1991

1 INTRODUCTION

Two polar bears rise upon their hind legs and joust; a flock of starlings sweeps in to roost; a mongoose swiftly and deftly bites its prey to death. The study of animal behaviour is about all these things and many others. It is about the chase of the hunter and the flight of the hunted. It is about the spinning of webs, the digging of burrows and the building of nests. It is about incubating eggs and suckling young. It is about the migration of a hundred thousand animals and the flick of a tail of one. It is about remaining motionless and concealed, and about leaping and flying. The study of animal behaviour is about all the static postures and active movements, all the noises and smells and the changes of colour and shape that characterize animal life.

Where in all this diversity do we start? What do people who study animal behaviour actually do and what do they want to find out?

There are two main approaches—the physiological and the 'whole animal'. Physiologists are mainly interested in how the body works, that is, in how the nerves, muscles and sense organs are coordinated to produce complex behaviour such as singing in a cricket or a bird. Those taking the 'whole animal' approach, on the other hand, although they are often also interested in the mechanisms of behaviour, study the behaviour of the intact animal and the factors that affect it. They may be interested in what it is in the environment of the cricket or bird that prompts them to sing at a particular time, or why they sing at all. Many such questions can only be studied by looking at animals in their natural environments and by studying the behaviour of a 'whole' animal without penetrating the skin, whereas a physiologist would tend to probe beneath the skin in the hope of finding the mechanisms that give rise to the behaviour of singing.

At their best, physiological and 'whole animal' approaches complement each other—behavioural studies sparking off the search for physiological mechanisms and, in turn, putting physiological studies into a functional perspective. One of our aims in this book is to give some idea of the extent to which this is now possible.

Within the 'whole animal' approach, a distinction is often made between psychologists and ethologists, both of whom could be described as interested in the behaviour of intact, functioning animals. Psychologists have traditionally been interested in learning and have tended to work in laboratories on the learning abilities of a restricted range of species, mainly rats and pigeons. Ethologists have been more concerned with the naturally occurring, unlearnt behaviour of animals, often in their wild habitats. Although this distinction still exists to some extent, there is now a fruitful coming together of the two. Ethologists have become interested in the role of learning in the lives of wild animals and psychologists are beginning to ask evolutionary questions about the learning abilities of a much broader range of species and to study their responses to more natural stimuli. Another of our aims is to show how much psychologists and ethologists are increasingly learning from each other, to mutual advantage.

But nobody—physiologist, ethologist or psychologist—can rely solely on one source of information. Physiologists sometimes like to emphasize that their methods are the more fundamental and it is true that ultimately it may be possible to explain behaviour in terms of the functioning of the basic units of the nervous system, the neurons. However, the main function of the nervous system is to produce behaviour and we must also investigate the end product in its own right. Even if we knew how every nerve cell operated in the performance of some pattern of behaviour, this would not remove the need for us to study it at a behavioural level also. Behaviour has its own organization and its own units that we must use for its study. Trying to describe the nest-building behaviour of a bird in terms of the actions of individual nerve cells would be like trying to read a page of a book with a high-powered microscope.

At the same time, ethologists recognize that what an animal does is only part of the picture. Phenomena such as learning or aggressive motivation are not directly observable from the outside and yet may have a profound effect on what an animal does. It is clear that the study of animal behaviour has to take into account processes taking place 'inside the nervous system'.

Because of the mutual dependence between the physiological and the 'whole animal' approaches, and because we will constantly refer to it throughout this book, we will start with a brief description of the way in which the nervous system works. Then we will look at ethological units of behaviour as recognized through studying the behaviour of whole animals 'from the outside'. We will find that many of the principles that operate at the neuronal level are also to be found when we look at reflexes and more complex behaviour.

UNITS OF THE NERVOUS SYSTEM

The fundamental unit of the nervous system is the nerve cell or neuron. Neurons are connected to each other in many complicated ways (Fig. 1.1), some individual cells being in contact with hundreds or even thousands of other cells. These connections are quite vital to the working of the nervous system because it is through them that nerve cells transmit information to each other. Each nerve cell has branching processes—dendrites—that receive information from other cells. It has a cell body or soma and, connected to this, a long tubular axon that may extend over a considerable distance and that transmits information to other cells (Fig. 1.2).

For the most part, information is transmitted

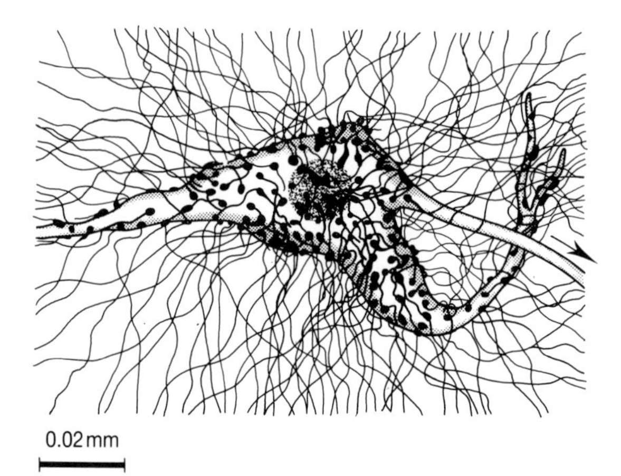

Fig. 1.1 Reconstruction of the cell body of a single motoneuron from the spinal cord of a cat. Parts of two dendrites can be seen (one branched) and the axon is indicated by the arrow. The cell and its dendrites are covered by synaptic knobs (boutons terminales) showing where the fine axons from hundreds of other neurons in the cord make contact with it (after Walsh).

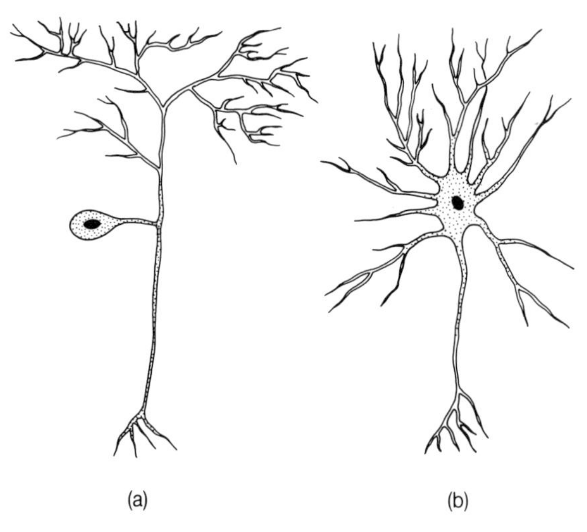

Fig. 1.2 Simplified drawings of (a) an invertebrate (arthropod) and (b) a vertebrate (mammalian) neuron. Real neurons have much more extensive branching than shown here.

through the nervous system by means of electrical impulses, although chemical transmission is usual at the junctions between cells—called synapses. The axon of the stimulating cell sends small packages of chemical—known as neurotransmitter—across the tiny gap that separates it from the next cell. When it arrives at the other side of the gap the neurotransmitter gives rise to new electrical activity in the receiver cell and the excitation is once again transmitted electrically until it gets to the end of that cell. (Some synapses do not employ this chemical package method and are completely electrical—two cells being close enough together for direct electrical propagation.)

Each nerve cell is electrically active all the time and, when not transmitting information, maintains a potential difference across its cell membrane (known as the membrane or resting potential); this means that the inside of the cell is slightly negative relative to the outside, i.e. it is polarized. Changes in the membrane potential—depolarization—away from its normal value of between 60 mV and 80 mV are the cells' way of transmitting information.

Being stimulated by another cell (through its dendrites) causes a change in the cell's membrane potential away from its resting value, just a very brief depolarization, but enough to pass on information if the depolarization is propagated along the cell axon. Now, all cells are poor conductors—an electrical impulse simply does not travel very far down the cell (certainly not as far as if the cell were made of copper wire). Sometimes, when one cell is stimulated by another, all that happens is that there is a tiny electrical signal in the receiver cell, which travels a few millimetres down a dendrite and then fades out because of the poor conductivity of the cell. The small induced signals are postsynaptic potentials, or PSPs, and, although they do not last very long or travel very far, they have a very important role in determining the kinds of operations that neurons can perform.

However, the axons of many neurons are much

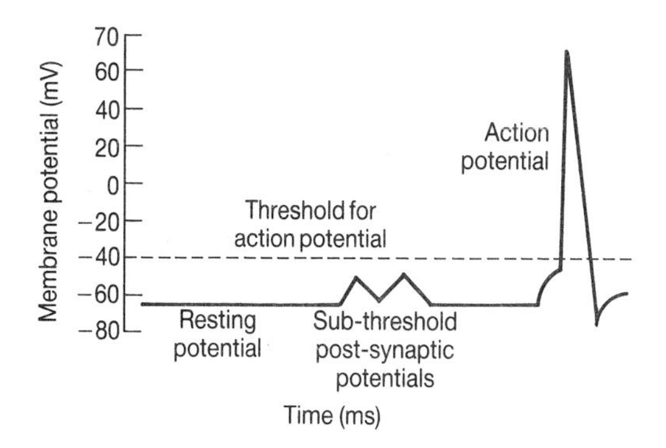

Fig. 1.3(a) Three examples of the electrical activity that can be recorded from neurons—the resting potential when the cell is at rest, post-synaptic potentials that do not lead to a 'nerve spike' or action potential and the massive action potential itself.

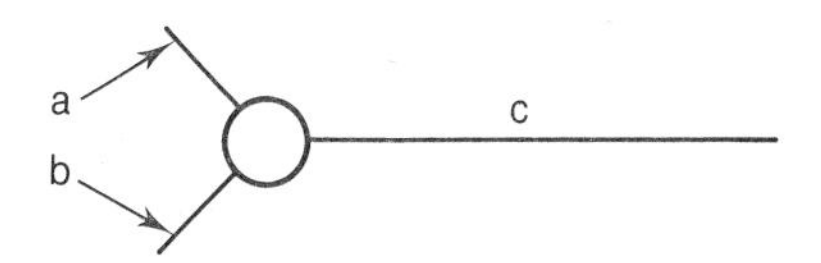

(i) A postsynaptic neuron, c, is shown diagrammatically with two presynaptic neurons, a and b.

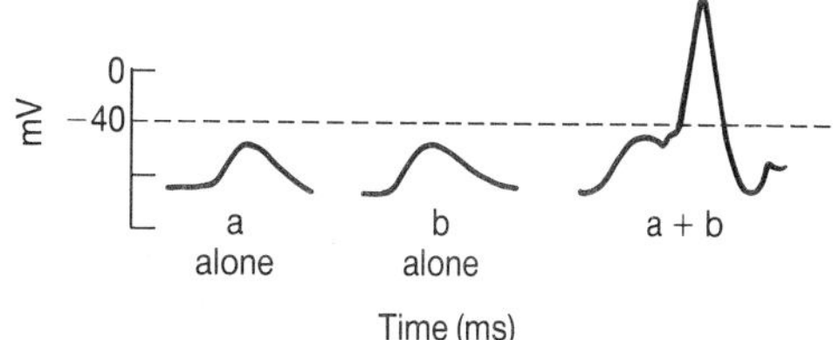

(ii) Spatial summation: the responses of presynaptic neurons a and b are summed by the postsynaptic neuron c.

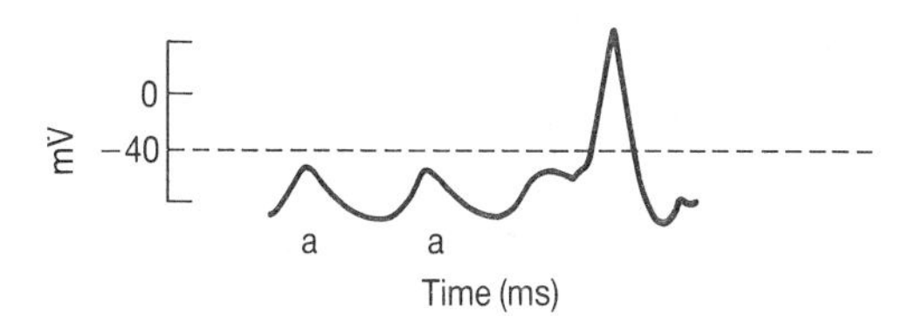

(iii) Temporal summation: the successive responses of one presynaptic neuron are summed.

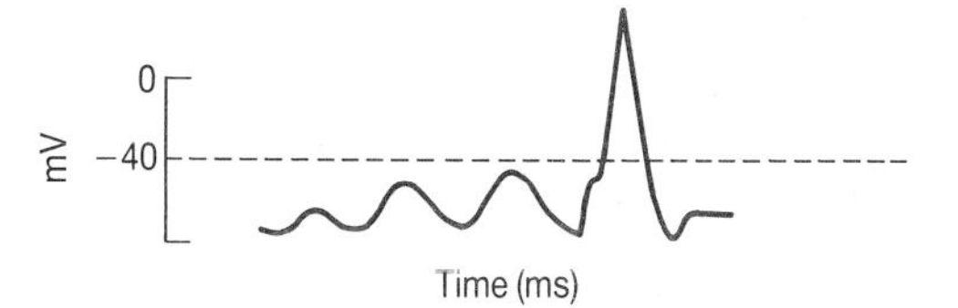

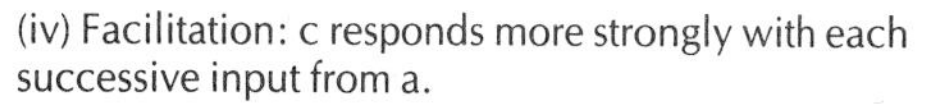

(iv) Facilitation: c responds more strongly with each successive input from a.

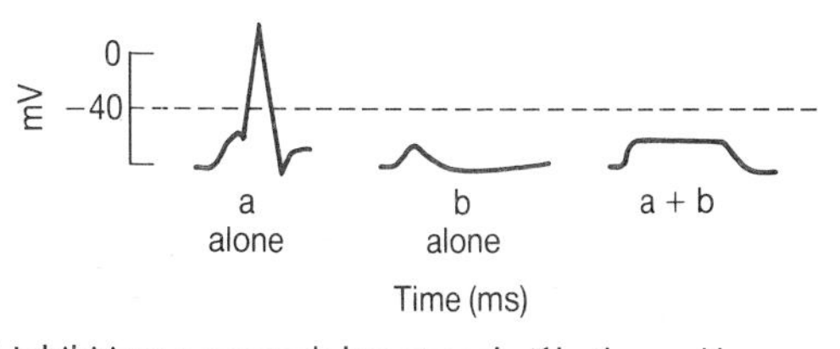

(v) Inhibition: c responds less strongly if both a and b are active.

Fig. 1.3(b) Some operations performed by neurons.

longer than the few millimetres that the tiny PSPs can travel and would be quite useless as transmitters of information over any greater distance if PSPs were their only response. Many neurons supplement PSPs with another type of signal altogether—a massive electrical change known as an action potential or nerve impulse (Fig. 1.3(a)). The nerve cell is capable of responding to a tiny PSP by generating an action potential that can travel along the entire length of a long axon without a decrease in its voltage. Sometimes one PSP alone is not enough to trigger a nerve impulse, but several PSPs coming rapidly over a short period of time, or from several different sources will give rise to a nerve impulse. In this way, the nerve cell shows **summation**: it adds together the effects of several subthreshold PSPs to give rise to the all-or-nothing response of the nerve impulse (Fig. 1.3(b)). At other times, the presynaptic neuron will deliver a barrage of several large action potentials to the receiver cell and each action potential results in more transmitter release than did the one before. Thus each PSP is larger than the one before, a phenomenon referred to as **synaptic facilitation**.

Interestingly, cells do not just excite one another. Some neurons also **inhibit** others. Their transmitters hyperpolarize the postsynaptic membrane, making it less likely that receiver neurons will respond. As we will see throughout this book, the idea of inhibition is of very great importance to the understanding of all animal behaviour.

Having looked at the basic attributes of nerve cells—excitation, inhibition, sub-threshold responses, summation and facilitation—we can now turn to more complex behaviour and see how some of the same phenomena are also to be found at 'higher' levels in the nervous system.

REFLEXES AND MORE COMPLEX BEHAVIOUR

Reflexes are often considered as among the simplest units of behaviour. The reflex that causes us to close our eyes when something flashes towards us, or to withdraw our foot when we step on a sharp object, is functionally very important. Yet it seems so very different in scale both from the action of individual nerve cells and from the kinds of behaviour with which most of this book will be concerned—building a nest, displaying to a mate, running through a maze to get food and so on. Such patterns are certainly more complex than reflexes but both are behaviour and we can learn some of the basic features of behavioural mechanisms from studying properties that both share with each other and also with individual nerve cells.

It is not possible to draw a firm line between reflexes and complex behaviour. Clearly complex behaviour can incorporate many reflexes; the swallowing reflex is the culmination of complex food-seeking behaviour and the reflexes controlling balance and walking are involved in almost everything animals do. What we choose to adopt as a 'unit' of complex behaviour will inevitably be somewhat arbitrary and dependent on the particular problem we are interested in. We have to make some decisions about what it is important to record and what can be safely ignored. We have to abstract and simplify; otherwise we would be overwhelmed with a mass of data.

For example, we could record every time an animal took a breath (breathing is certainly part of its behaviour) but we would probably decide that there was no need to record every breath if we were studying the courtship of sticklebacks or pheasants. However, if we were studying the courtship of male newts, where the need to take air from the surface can conflict with the need to display to a female under water (Chapter 4), then recording breathing might be quite critical if we were to understand what is going on.

Yet even when we have decided which behaviour we are going to include, it would remain impossible to describe, even for a moment, every twitch of every muscle or every small movement of a part of the body of an animal. What we have to do is to look for **patterns** in what we observe. Instead of describing the behaviour of an animal in terms of every small movement we can see (head does x, left front foot does y, tail does z, then after 1.5 s, head does x′, left front foot y′ . . . etc.), we hope to detect a recognizable pattern. We might notice that the head, the left foot and the tail always move in a coordinated sequence. There would then be no need to describe over and over again the exact way in which they moved because, like a repeated motif on a wallpaper, there are sequences of movement sufficiently similar from one occasion to the next to be given names like 'head-up display', 'facing away' or 'licking'. One of

Fig. 1.4 Male sage grouse at the peak of his 'strut' display, used in courtship.

the most spectacular examples of this is the courtship strut display of the male sage grouse (Fig. 1.4) in which the bird inflates a huge air sac under his neck and breast feathers and then suddenly deflates them with sharp snapping sounds that can be heard over a kilometre away.

A fairly full (although not down to the muscle twitch level) description of this display is given by Wiley (1973) as:

> The display begins from the strutting posture (tail fanned and cocked vertically, head raised, neck plumage erect) and consists in coordinated movements of the wings and the large oesophageal sac, together with associated movements of the legs, head and trunk. The oesophageal sac is inflated in the course of being twice lifted and dropped. Concurrently, the wings are extended forward and retracted twice. The culmination of the display follows immediately: a rapid compression and ballooning of the sac accompanied by a third excursion of the wings. Complex acoustic signals accompany this culminating action. These include two sharp snaps produced by the inflated air sac and an intervening whistle that rises and falls in pitch. The final 0.2 s second part sounds roughly like POINK. Preceding this sound are three low-pitched coos, probably produced by the syrinx, and two swishing sounds generated by the wings rubbing against the sides of the chest.

Now, male sage grouse will go through this whole sequence every 6–12 s and groups of males will cluster together in leks, or display grounds, each one displaying for hours at a time. For many kinds of investigation, it would be quite unnecessary, as well as logistically impossible, to give the whole of Wiley's description every time an observer saw that particular sequence. But the similarity of the sequence from male to male and from one occasion to the next means that the strut display can be recognized as a 'behaviour pattern' or unit of behaviour and different observers will know exactly what this means. Using slowed-down films of the behaviour, Wiley showed that for 45 consecutive struts by one male, the mean interval from the first wing swish to the peak of the first 'snap' sound was 1.55 s, with a standard deviation of 0.55 units (giving a coefficient of variation of less than 1 per cent).

Other behaviour patterns, such as the 'head-throw' courtship display of the golden-eye duck (Dane *et al.* 1959) and the strike of mantids, have also been shown to have a similar fixity in duration, indicating a clear-cut pattern. Theoretically, it would be possible to use film or videotape and to derive statistically each behaviour pattern from all animals from the raw material of component movements. In practice, this is so tedious to do that it is seldom done. The human eye is very good at detecting patterns of the order of a few seconds' duration and so most studies bypass the tedious analysis of film that would be involved in an objective detection of pattern and rely instead on the 'computer in the head' to give the basic units of behaviour.

While we can often describe the neural pathways and properties of a reflex in a rather exact manner, we can certainly not do this for complex behaviour such as strutting, where thousands if not millions of neurons are involved. Nevertheless, we should not be put off by quantitative differences, no matter how great. There is much to be learnt from comparing the properties of nerve cells, reflexes and more complex behaviour; such a comparison is a useful way emphasizing the unity of the study of behaviour.

In 1906, Charles Sherrington published *The Integrative Action of the Nervous System*. Sherrington, more than any other single person, can be regarded as the founder of modern neurophysiology. In his book he considered the way in which reflexes operate and how the central nervous system integrates them into adaptive behaviour, combining information gathered from different sources, arranging sequences of action and allocating priorities.

In the first few chapters of his book, Sherrington discusses some of the properties of reflexes and constrasts them with those of the same movements when elicited by direct stimulation of the nerves to the muscles concerned. We shall, in turn, use part of his classification to compare the properties of reflexes and more complex behaviour.

Latency

Reflexes and complex behaviour both show latency in response—there is a delay between giving a stimulus and seeing its effect. The latency between a dog encountering a painful stimulus with its leg and showing the flexion reflex by which it withdraws its leg usually lies between 60 and 200 msec. Only a small fraction of this delay is a result of the time taken for nerve impulses to be conducted along axons; most of it is caused by the delay at the synapses (a term we owe to Sherrington) between one neuron and the next. It is hardly surprising to find delays between stimulus and response in complex behaviour, for in the chain between receptors and effectors there are often dozens of synapses to cross.

Although it is often difficult to measure latencies for complex behaviour (it is often impossible to fix the time of stimulus onset precisely), nevertheless, results are sometimes vivid. Wells (1958) describes how, when a tiny shrimp is presented to a newly hatched cuttlefish there is no detectable response for perhaps as long as 2 min. Then the nearest eye of the cuttlefish turns to fixate on the shrimp. There is a further delay, but usually only a few seconds, before the cuttlefish turns towards the shrimp so that both of its eyes are brought to bear. Another brief delay follows and then it launches its attack and seizes the shrimp with its tentacles.

With reflexes, it is found that the stronger the stimulus, the shorter the latency (Fig. 1.5) and some evidence suggests that the same is true for certain complex behaviour. Hinde (1960) measured the latency between presenting various frightening stimuli to chaffinches and how soon they gave their first alarm calls. Just as with reflexes, the stimulus known on other grounds to be the strongest, produced the shortest latency.

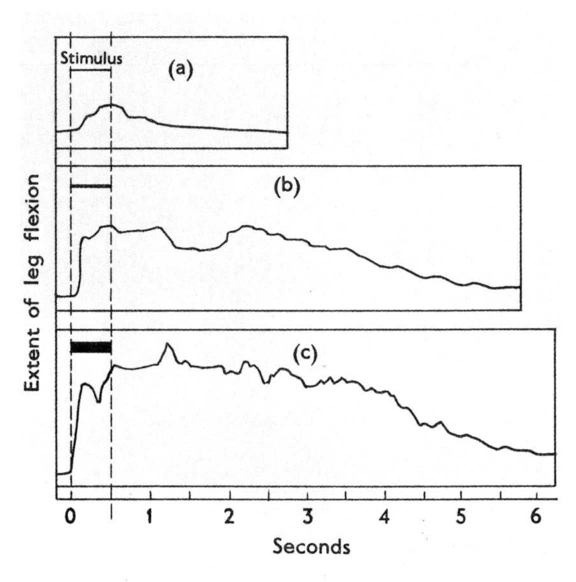

Fig. 1.5 The extent and persistence of the dog's flexion reflex with three strengths of stimulus, each of 0.5 s duration. The area enclosed between the line and the *x* axis gives a measure of the 'amount' of the response. Even with a weak stimulus (a) the after-discharge represents 75 per cent of the total 'amount'; with the strongest stimulus (c) it represents over 90 per cent. Note that the latency of the response decreases as the stimulus increases in strength (modified from Sherrington 1906).

Summation

We have seen that individual neurons are able to summate excitation coming at different times (temporal summation) and from different places (spatial summation). Sherrington gives several beautifully clear examples of summation at the level of reflexes. The scratch reflex of the dog is elicited by an irritating stimulus anywhere on a saddle-shaped area of its back. The hind leg on the same side is brought forward and rhythmically scratches at the spot. Weak stimuli—say, a series of 5 or 10 touches given in rapid succession—may not evoke any response, but after 20 or 30 scratching appears: the stimuli have been summed in time. Figure 1.6 shows the spatial summation of stimuli from two areas of skin 8 cm apart. Neither alone is strong enough to provoke scratching, but they are effective when given together.

Dethier (1953) studied the stimuli that cause blowflies to extend their proboscis before they drink. The flies can detect sugars and other food subst-

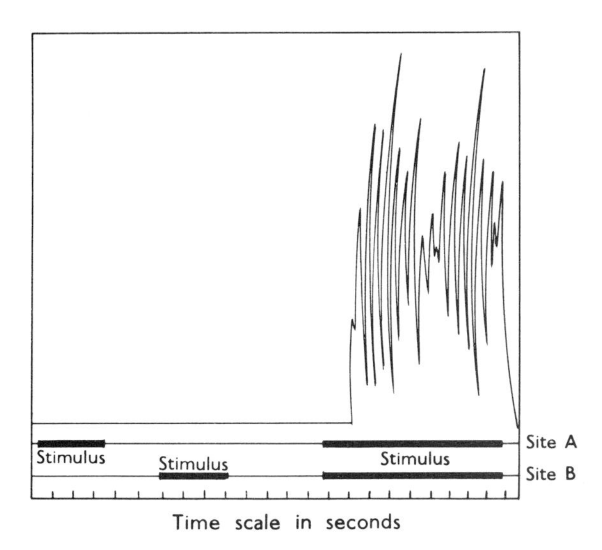

Fig. 1.6 Spatial summation leading to the appearance of the scratch reflex in the dog. The tracing represents the movements of the dog's leg when scratching. A and B are two points on the shoulder skin and weak stimuli, given singly first at A and then at B, do not evoke the reflex. When both points are stimulated simultaneously the reflex appears with a latency of about 1 s (modified from Sherrington 1906).

ances with sensory hairs on their fore tarsi. They search for food by running over a surface and extending the proboscis when the front legs encounter anything suitable. As measured by this proboscis extension, the flies can detect sugars at very low concentrations. Dethier found that when only one leg is dipped in a solution, the lowest concentration of sucrose to which 50 per cent of the flies respond is 0.0037 molar. However, if both legs are stimulated together there is summation and now 50 per cent of flies respond to only 0.0018 molar glucose.

With more complex behaviour, summation frequently occurs between stimuli of quite different types perceived by different sense organs. We all know how the sight and smell of food summate when we are hungry. Beach (1942) showed that male rats respond sexually to a combination of olfactory, visual and tactile stimuli from a receptive female. Young males do not respond unless two such sources are vailable—it does not matter which two. Mature males, with previous sexual experience, will respond to one type of stimulus alone.

'Warm-up' or facilitation

Sherrington found that some reflexes do not appear at full strength at first but, with no change to the stimulus, their intensity increases over a few seconds. Neurons, as we saw, show synaptic facilitation, each successive PSP being larger than the one before. At a behavioural level, Hinde (1954) found that chaffinches show a similar type of 'warm-up' effect when shown an owl. The bird's response to the owl is to give a mobbing call. Counting the number of calls given by a chaffinch in successive 10 s periods after the owl is shown to it indicates that it begins by calling at a relatively low rate and that the maximum calling rate is not reached for about 2.5 min, after which it gradually declines (Fig. 1.7).

Sherrington was able to show that 'warm-up' in some reflexes is due to summation of stimuli that come to evoke a response from more and more nerve fibres, producing a stronger contraction. He called this phenomenon 'motor recruitment'. Some analogous process probably occurs with complex behaviour, but what we commonly see is not only a change in the intensity of response but in the nature of the behaviour as well. Sherrington (1917) provides an excellent example from what he calls the cat's 'pinna reflex'. Repeated tactile stimulation to the cat's ear first causes it to be laid back. If stimulation persists, the ear is fluttered; thirdly the cat shakes its head and when all else fails to remove the irritation, it brings its hind leg up and scratches. Clearly there is more involved here than the recruit-

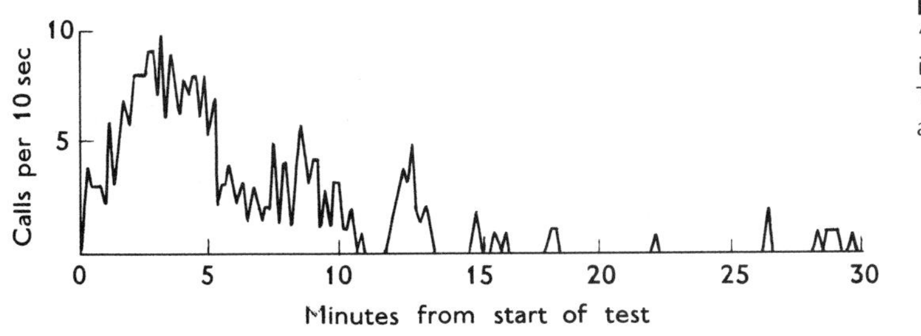

Fig. 1.7 The 'warm-up' and subsequent 'fatigue' of alarm calling when a chaffinch is presented with a stuffed owl in its cage. The maximum rate of calling occurs after about 2.5 min (from Hinde 1954).

ment of a few extra motor nerve fibres. Mechanisms that control patterns of movement such as ear-fluttering and head-shaking must be recruited. Perhaps all these mechanisms are activated in some way by stimuli to the ear but their thresholds are different. That for laying back the ear will have the lowest threshold, with successively higher ones for the other three patterns. Workers studying complex behaviour frequently rank the patterns they observe on a similar intensity scale of increasing thresholds. A similar threshold idea has been used by Bastock and Manning (1955) to explain the fact that a male fruit-fly switches from one courtship pattern to another when courting a female whose behaviour remains constant.

Inhibition

Inhibition operates at every level within the nervous system. As we have seen, nerve cells can actively inhibit each others' transmission of information. Similarly, the prevention of one activity's occurrence while another is in progress constitutes inhibition at the behavioural level. In many ways, inhibition is just as important for the coordination of behaviour as excitation and to see why we can again turn to Sherrington's work on reflexes.

Muscles are commonly arranged in antagonistic pairs, such that one flexes a portion of a limb and the other extends it. Clearly, it would be impossible to both extend and flex the same limb at the same time: Sherrington showed that excitation of one member of a muscle pair is accompanied by inhibition of its antagonist. Such inhibition is not absolute, and an inhibited muscle does not simply go limp. Once it is stretched by its antagonist then its own 'stretch reflex' (see pp. 9–10 for fuller description) will tend to make it contract. Although the antagonist muscle may override it, it will take up the slack, so to speak, in an active fashion. Much finer control of movement is possible if muscles can be made to work against one another in this way. Mutual inhibition allows them to take the lead in turn during limb movements, and to alternate flexion and extension of the limbs. Sherrington found that it is not only antagonists on the same limb that inhibit each other, but that muscles located on opposite limbs also have antagonistic effects during locomotion. When the flexors of one limb are contracting, the flexors of the opposite limb are inhibited. Reciprocal inhibition of this type is one of the basic integrating mechanisms for walking and without it coordination of the different limbs would be impossible.

The role of inhibition in complex behaviour is superficially less obvious than that of excitation. We stimulate an animal and the conspicuous result is that it makes a response. But in so doing it has made a swift transition that requires the inhibition of its behaviour prior to the stimulus and of other behaviour that it may be stimulated to perform at the same time. Sherrington saw reflexes as 'competing' for the final common pathway, i.e. the muscles whose action is common to several different reflexes. In an analogous way, we can see the different systems controlling patterns of complex behaviour like fighting, feeding and sleeping competing for the control of the animal's musculature. Such systems are obviously incompatible in the sense that only one behaviour can occur at a time. When, say, the animal starts feeding, other behaviour must be inhibited for the time being.

Sherrington found that when inhibition was removed from a reflex it returned at a higher intensity than it had previously. Figure 1.8 shows this phenomenon, which Sherrington called 'reflex rebound' for the scratch reflex. We commonly observe that when a particular type of complex behaviour, e.g. courtship, has not been elicited for some time, it has a lowered threshold and is performed with high intensity when it is, at last, evoked. Vestergaard (1980) found that if laying hens are kept on wire so that they have no substrate in which to dustbathe, when eventually given access to litter they start dustbathing quickly and dustbathe in very much longer bouts than hens kept all the time on litter. It is possible that the system controlling dustbathing shows something akin to reflex rebound. Kennedy (1965) interpreted some aspects of the behaviour of aphids along these lines. The behaviour of winged aphids alternates between periods of flight and periods of settling and feeding on leaves. If an aphid settles on an 'unattractive' surface—an old leaf, for example—it does not stay long and soon takes off but flies relatively weakly and soon settles again. Conversely, if it has settled on an attractive young shoot, it stays for a long period but, when it takes off, flies vigorously and for a long time.

In an elegant series of experiments, Kennedy was able to exclude any simple explanation for this relationship based on physical exhaustion during flight and recovery after resting and feeding on a young leaf. He suggested that there is mutual inhibition

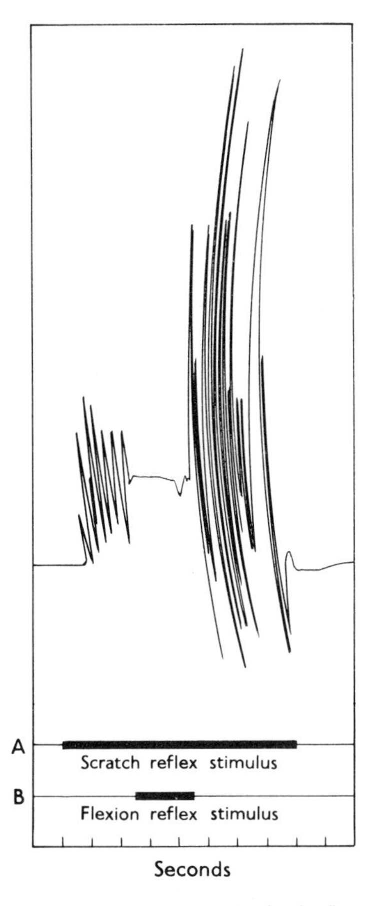

Fig. 1.8 Inhibition of the scratch reflex by the flexion reflex. The stimulus denoted on line A evokes the scratch reflex, but this response is inhibited when the stimulus on line B evokes the flexion reflex. The moment B is removed the scratch reflex returns, and much more vigorously than before—an instance of 'reflex-rebound' (modified from Sherrington 1906).

between the systems controlling flight behaviour and those controlling settling. As with reflexes, activation of the settling system may temporarily inhibit the expression of the flight system but, at the same time, gradually lower the threshold for flight. In Chapter 4, we will discuss other evidence that the systems controlling complex behaviour do inhibit one another and the various other interactions that occur between different control systems.

Feedback control

Very commonly, reflex or complex behaviour consists of a steady output of some activity that has to be held at a given level. When we 'stand at ease', our body is evenly balanced over the pelvic girdle and easily corrects for any slight jostling we may receive. To do so, the muscles of the legs and back must be held at a constant level of tension and, if shifted away from this level, they must correct to bring the body upright again. Analogously, under normal circumstances animals maintain a very constant body weight and eat and drink sufficient for their needs at regular intervals. If a surplus is available they do not overeat. In times of scarcity, they spend a higher proportion of their time in feeding and consume more when the chance arises to replace any deficit.

Both these examples show us behaviour acting as a **homeostatic** system ('homeostasis' means literally 'same state') and serving to preserve the *status quo*. In the first case, this was achieved through reflex systems controlling the leg and trunk muscles; in the second case it was achieved by a series of more complex systems regulating the search for food, feeding and satiation. In both cases the operation requires that the end result (posture and balance whilst standing; state of nutrition) is monitored in some way. When it deviates from a set value a signal is sent to the control mechanisms to correct the imbalance and bring the end result back to the set value again. This idea is shown diagrammatically in Figure 1.9 and its application to feeding and drinking behaviour is discussed in more detail in Chapter 4, page 87.

The homeostatic control of posture is understood quite thoroughly at a neurophysiological level. In some cases we know the paths of the neurons involved and can actually identify the structures that function as parts of the control system. One such is illustrated in Figure 1.10, which represents a typical muscle on the limb of a mammal, such as would be involved in maintaining posture. Motor neurons that have their cell bodies in the vental horn of the spinal cord run to the muscle and it is their activity that determines the tension developed by the muscle. In parallel with every skeletal muscle, and embedded within its fibres so that they contract and relax it, are muscle spindles. These are specialized sense organs for recording the degree of tension in the muscle. Their sensory nerves run back to the spinal cord and, entering through the dorsal root, synapse with the motor neurons to the muscle. Thus a loop that forms the basis of the stretch reflex is closed. When a muscle is stretched by the contraction of its antagonists, the muscle spindles are also stretched and their sensory fibres increase their rate of firing, stimulating the motor neurons so that the muscle contracts.

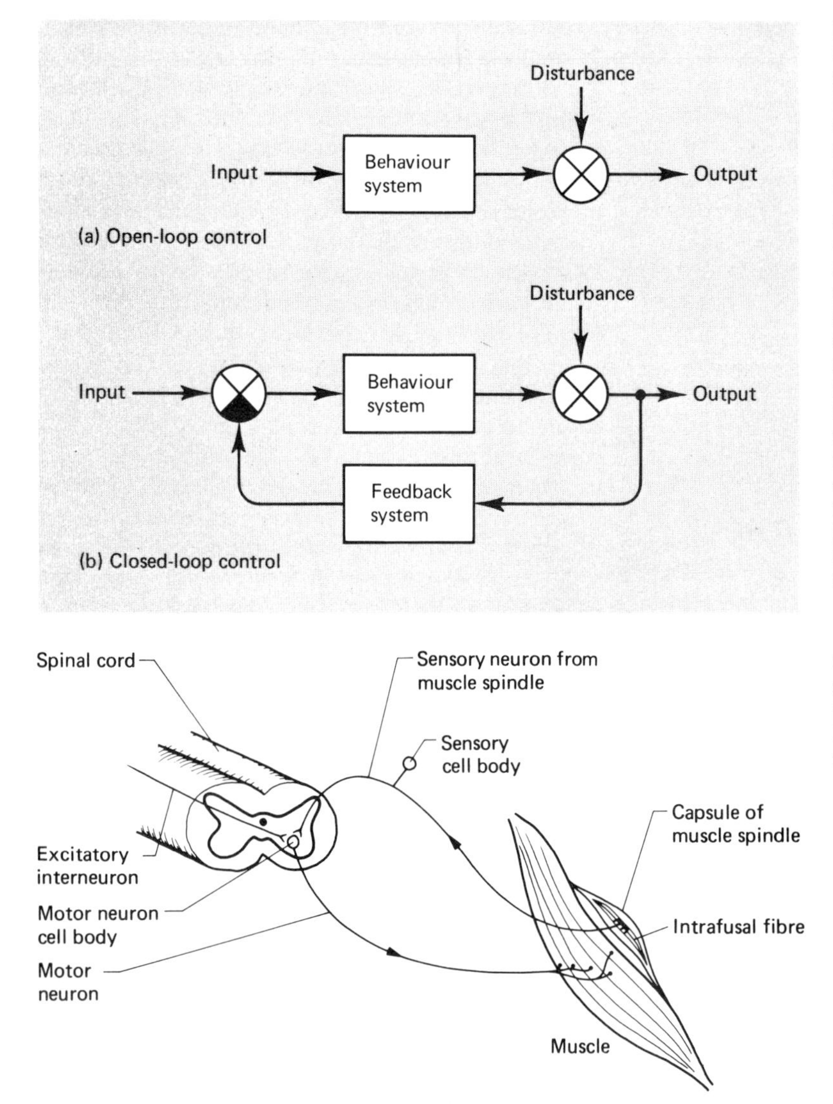

Fig. 1.9 Diagrams of simple open- and closed-loop control systems. The output from the behavioural system is affected by disturbance factors (the crossed circle represents interaction). In open-loop control (a), if the output is affected by disturbance no correction takes place. In the closed-loop arrangement (b) the results of the disturbance feed back to affect the input to the behavioural system. The dark segment of the crossed circle represents an interaction between the feedback mechanisms and the input, which tends to bring back the output to its original value. The feedback system produces its own output, which is proportional to the disturbance and changes the input to the behavioural system both to the right amount and in the right direction.

Fig. 1.10 The simplified diagram of some of the neural pathways involved in the stretch reflex. Further explanation is given in the text.

It is easy to equate the units of the close loop control system in Figure 1.9 with those of this real muscle mechanism. The output (state of tension in the muscle) is affected by a disturbance (being stretched by other muscles) and a feedback mechanism (muscle spindle) records the change and feeds back to change the input (motor nerve) and restore the original input. This is a simplified picture of the real situation, which in fact includes other regulatory mechanisms allowing for very fine graded control over the muscle contractions involved both in the maintenance of posture and in movements, but it serves to illustrate the reality of feedback control at a reflex level.

Not all behaviour involves feedback control. When a movement must be made very rapidly there is simply not time to modify the movement while it is in progress. The strike of the mantis is such a case. The mantis moves towards a fly and orientates its body slowly and precisely (operations that certainly involve feedback control) but, once aimed, the strike is an all-or-nothing movement. If the fly moves after the strike is initiated, this makes no difference to the form of the movement and the mantis strikes in the wrong place. Such behaviour, which occurs without feedback, is said to be under 'open-loop' control (as opposed to the 'closed-loop' control of a homeostatic system). The difference between open and closed systems is shown in Figure 1.10.

So, despite the many levels at which the behaviour of animals can be studied, we see that there are certain principles—excitation, inhibition, summation, facilitation and feedback control—that appear to be common to many different levels. Studying single neurons and studying the behaviour of whole animals may require very different techniques and, in many cases, very different concepts. Nevertheless, as we have seen, neurons, reflexes and more complex behaviour share many basic properties. It is often possible to break down complex behaviour patterns into smaller units, some of which are immediately equatable with reflexes. However, we cannot always explain behavioural observations using reflex terminology nor is there any point in trying to do so in many cases. There *are* differences in complexity and these often require different types of approach. Which we choose may well depend on what sort of questions about behaviour we are trying to answer. We must now turn to consider what these might be.

QUESTIONS ABOUT ANIMAL BEHAVIOUR

In a classic paper, which is as well worth reading today as when it was written, Tinbergen (1963) lists four possible kinds of questions that can be asked about behaviour:

1. Those about causation (by which he meant both external stimuli and internal mechanisms).
2. Those about survival value.
3. Those about evolution.
4. Those about ontogeny (development).

Because these are still the four major problems of animal behaviour, and indeed of biology as a whole, they will form a major theme of this book. To introduce them, we shall consider some examples of behavioural studies that illustrate each question in turn.

Causation

Anyone who has handled a cockroach will know that it behaves like greased lightning when an attempt is made to catch it. Toads, which are natural predators of cockroaches, seem to have almost as much difficulty as we do. A first possible kind of question that we could ask about the cockroach's behaviour is 'how does it avoid being eaten by a toad?' In other words, we would be asking about the mechanisms inside the cockroach and the stimuli impinging on it from the outside that enable it to detect that it is about to be attacked and to scuttle away so effectively.

The first clues about mechanism come from watching the behaviour of freely-moving cockroaches and toads, or rather, analysing film of them, because everything happens so quickly that the naked eye cannot follow it. Slowed-down film shows that just as the toad is about to flip out its tongue and strike, the cockroach will turn rapidly and then run. It seems, then, to have some way of detecting that a toad is going to strike and from which direction before it actually does so, because it always turns in the appropriate direction, away from the toad, about 16 ms before the toad's tongue appears. At this point, the toad is apparently already committed to striking in a particular direction (it thus shows open-loop control) and the tongue may strike in the wrong place.

The most important cue by which the cockroach

Fig. 1.11 The cockroach, *Periplaneta americana*. The cerci on the hind end are clearly visible. Photograph by Paul Embden.

detects when the toad is about to strike seems to be tiny gusts of wind produced by the toad's movements. Camhi *et al.* (1978) showed that these slight air movements are picked up by the cockroach through many tiny, wind-sensitive hairs on its cerci (tails) (Fig. 1.11); the cockroach can be made to make false escape movements if a small gust of air is blown at the hairs. The smallest gust of air that generates escape behaviour is 12 mm/s, with an acceleration of 600 mm/s; if the hairs are immobilized with glue, the cockroach is much less successful at escaping.

In behavioural experiments in which puffs of air were blown at cockroaches, it was shown that they started to show their escape behaviour just 44 ms after a puff started. Then, by measuring the wind generated by a toad when it strikes at prey, it was found that the critical (12 mm/s) gust of wind occurred on average 41 ms before the tongue emerged from the mouth—very close to the response latency of 44 ms shown by the cockroach to an artificial puff of wind. So it is the minute gusts of wind generated by a toad preparing to strike that sets off the escape turning behaviour. By waiting until the very last moment before turning, the cockroach evades the tongue that has already started to move and thus escapes because the toad cannot, at that late stage, change the direction of its strike.

So far, we can see that a great deal can be learned about the mechanism of the behaviour by studying the behaviour of whole animals—intact toads and intact cockroaches. Film records of strike and escape, experiments stimulating the sense organs of the cockroach and measurements of the wind stimulus produced by the toad give us a basic idea of the mechanism underlying escape behaviour at this level.

The next stage is to go 'inside the skin' and look at the same behaviour at the physiological level. Close examination shows that there are about 220 hairs on each of a cockroach's cerci and that each hair is hinged so that it can be moved most easily in just two directions at 180° to one another. They move less easily in directions at 90°. Different hairs have different preferred directions, which means that wind from certain directions moves some hairs and wind from other directions moves others. The basis for the cockroach being able to discriminate wind direction so accurately thus appears to lie in the mechanical construction of its hairs. By recording from the sensory nerve cells attached to the wind-sensitive hairs, Camhi *et al.* found that the nervous activity in these cells reflected the directionality of their hairs: the nerve cells responded much more to wind from some directions than from others.

Camhi *et al.* also found that there is a knot of nervous tissue—the terminal ganglion—at the hind end of the cockroach and that this contains a group of large cells known as giant interneurons (GIs). The GIs run up the nerve cord to the head, on the way passing through the thorax and linking to the motor nerves of the legs that do the rapid escape running. The GIs receive information from the sensory nerves in the cerci about wind direction and command the legs to run in the opposite direction (Fig. 1.12).

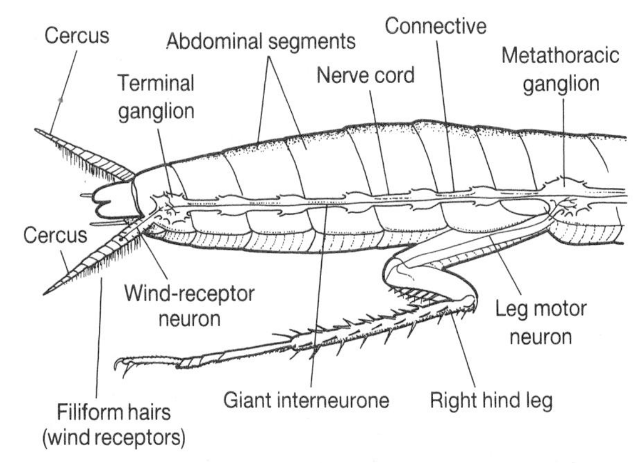

Fig. 1.12 Hind end of a cockroach showing the cerci with filiform hairs, which are the wind receptors (from Camhi 1984).

When the insect is stimulated with a wind puff the GIs become very active (as recorded by microelectrodes inserted into them), as do the motor neurons to the muscles of the hind leg. It is possible to inactivate some of the GIs by injecting them with an enzyme, leaving other GIs intact. This has the effect of making the cockroach turn in the wrong direction to a puff of a wind. So it appears that some sort of comparison between activity in different giant interneurons goes on in the normal cockroach, telling it which of its legs to move.

For relatively few behaviours, particularly in vertebrates, are the mechanisms understood in as much detail as this, right through from the detection of a stimulus to the command to the limbs. If we ask a comparable question about how the strut display of the male sage grouse is performed or the mechanism by which a flock of starlings coordinates its flight manoeuvres, our picture is very much less complete. Nevertheless, as we have seen, there seem to be a number of unifying principles underlying much animal behaviour, which gives us hope that the task of understanding mechanism is not an impossible one, even for complex behaviour. Indeed, Chapters 3 and 4 will show how these principles apply when we look in more detail at how animals respond to the stimuli around them and how they make 'decisions' about what to do.

Function or the adaptive significance of behaviour

The question 'What causes this behaviour to happen?' is in one sense—the immediate or proximate mechanism sense—the same as asking 'Why does the animal do this or that at this moment?' But we can also ask 'Why?' in a more functional or ultimate sense, meaning 'How does the behaviour contribute to an animal's survival and reproductive success?'

We have seen something of the mechanisms underlying the escape response of the cockroach and we could ask 'Why does the animal turn first and then run?' or 'Why does it only start its escape response when the toad's tongue is emerging from its mouth?' We could look at a group of male sage grouse, bobbing and cracking as each performed the strut display over and over again and ask 'Why do they do it?' Why should a male bird spend so much time and energy inflating his air sac and deflating it again, making himself vulnerable to predators all the while? And if, as it turns out, females are attracted by the display, that only raises another series of 'Why' questions. Why should females be attracted to males inflating and deflating air sacs? And why should males inflate their sacs only when in groups and not on their own?

In some cases, the answers to these 'Why' questions are already understood. The cockroach that turns first before running appears to stand a good chance of escaping from a predator whose method of attack is an open-loop strike with the tongue. Once committed to a particular direction, the toad's tongue continues in its path and so misses an insect that is quick enough to change direction after the strike has started.

In other cases, the sage grouse display being one of them, the answers to the 'Why' questions are very much less clear, and are in fact still the object of debate and research. Why sage grouse males should have evolved throat sacs while male peacocks evolved enormous tails; why group display is used and why it should be important for females to choose males that display the longest on the display ground are all questions that, as we shall see in later chapters, have many possible answers but few definite ones.

Davies and Houston's (1981) study of the feeding behaviour of pied wagtails near the River Thames shows just how complex the answers to such 'Why' questions may be. During the winter, wagtails spend over 90 per cent of the daylight hours feeding and they have to find a prey item on average once every 4 s in order to achieve energy balance. Davies and Houston found that some of the birds defended territories along the riverbank and fed on the insects that were washed up on the shoreline (Fig. 1.13). The owners walked around their territories in very regular circuits, taking about 40 min to get round. The reason 'why' they did this is that it is the most efficient way of harvesting food from the territory. The birds picked up the insects from one section of riverbank and so depleted the food supply there and then moved on to the next stretch. After about 40 min, the original section of the bank had collected a new lot of insects and the bird could harvest it again.

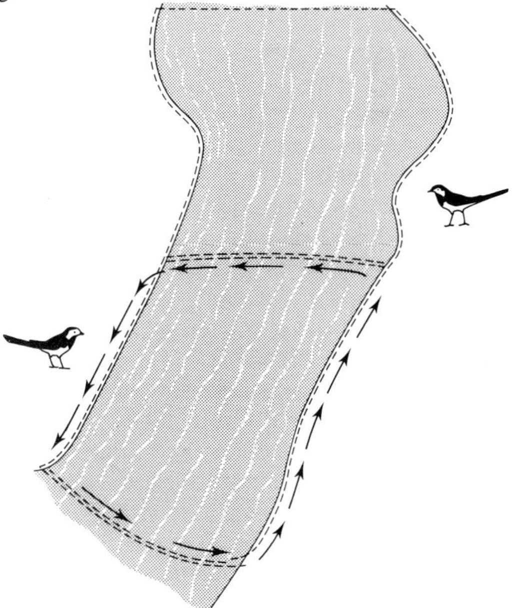

Fig. 1.13 Pied wagtails on riverside territories. The birds patrol around their territories collecting food washed up by the river.

All this holds good only as long as another wagtail or other insect-eating species of bird does not enter the territory and deplete the food before the owner can get round to it. Wagtails vigorously defend their territories against such intruders and behave most aggressively towards intruders that land close by in front of them on their patrol circuit, where of course, the greatest amount of food is waiting to be picked up. Intruders that land in a part of the territory that the owner has just been over often leave of their own accord without having to be evicted by the owner. The intruder's feeding rate is so low anyway that it is not worth its while to stay, while the owner, who knows which parts of the territory contain food and which ones have just been depleted, can feed at a much higher rate. The territory is therefore more valuable to the owner than to the intruder, which goes some way to explaining why the owners defend territories and why intruders may leave without a fight.

It is somewhat surprising, then, to find that some territory owners tolerate another bird in their territory for long periods of time. Such 'satellites' also walk around the territory, following the same path as the owner, but half a circuit behind so that each section of riverbank is left for only 20 min before being revisited by a bird. The owners would appear to be collecting only half the food they could collect if they occupied territories on their own and did not allow the satellites to feed too. The satellites do take food but they also help with defence of the territory.

Davies and Houston found that when an owner was alone on the territory, 60 per cent of intruders were spotted immediately. The remaining 40 per cent took some food from the territory before being chased. When a satellite was present, however, 85 per cent of intruders were spotted and chased immediately and only 15 per cent took any food. When food was sufficiently abundant, owners could increase their own feeding rate by up to 33 per cent by having a satellite to help them with defence, which more than compensated for the food eaten by the satellite. When food became scarce, however, and there was not enough food for two birds, the owners became aggressive to the satellites and evicted them from their territories. So the question of why wagtails sometimes attack other birds and sometimes tolerate them as semi-permanent residents appears to be related closely to the net feeding rate the owner can attain with and without another bird present.

But this does not explain why territory-owning birds sometimes leave their territories and join a flock of wagtails feeding elsewhere. If short term feeding rate was the only reason for this, we should expect that they would only go back to their territories when they could obtain a higher feeding rate there than in the flock, as is true for satellites. This was not the case for owners. Owners still spent some time on their territories even when, in the short term, they would have obtained more food by spending the whole day with the flock.

Davies and Houston argue that wagtail territory owners defend their patch of ground as a long term investment and the periodic (temporarily unprofitable) visits to a territory are to prevent intruders that might settle in the owner's absence from taking over. In thinking about 'why' animals behave in certain ways, we have to think not just about immediate short term gain, even when the gain is as important as food in winter to a small bird, but about the long term effects on the animal's survival and lifetime ability to reproduce. As we will see subsequently, understanding why animals behave in certain ways means understanding why natural selection favoured one kind of animal over another and the true reasons for this may not be apparent over a day, a week or even a year. We may need an understanding of the animal's whole life span and its ability to contribute to the next generation before the full explanation becomes clear.

Ontogeny (development)

A third sort of question that can be asked about behaviour is 'How does the behaviour develop during an individual animal's lifetime?' Does the animal have to learn or can it behave appropriately without any previous experience? If experience is necessary, what sort of experience? Does the animal have to be at a particular age or stage of its life cycle for learning to be effective?

As with questions about mechanism, questions about development can be asked at many different levels. Does a newly hatched cockroach escape as effectively from a toad as an adult or does it improve with practice? How do the connections between the sensory nerves and the giant interneurons form so that escape behaviour can be accurate? Do male sage grouse have to learn how to do the strut display? Does the display change with experience?

In a way, there is a simple and direct answer to all such questions: it is always true that an animal's genetic makeup and the environment in which it grows up both contribute to the final behaviour it shows. Neither genes nor environment on their own would produce a fully functional, behaving animal. But the way in which genetic factors and environmental ones interact may be very complex. For example, a cockroach that has just hatched and has had no previous experience of wind, turns away from a puff of air just as accurately as an adult cockroach (Dagan and Volman 1982). With the very first puff, then, the escape behaviour is fully formed and very accurate, even though the hatchling has only four sensory hairs on its cerci instead of adult's 440. Nevertheless, this 'innate' behaviour can still be modified later in life. If the adult cockroach loses one of its cerci, its escape behaviour becomes at first very inaccurate but then gradually improves. Immediately after the loss, the cockroach erroneously turns towards a wind source instead of away from it. However, after 30 days, with or without practice, it improves markedly and turns consistently away from the wind source, despite still having only one cercus and no extra sensory hairs. Its nervous system has gradually altered so that the giant interneurons on the side of the body from which the cercus has been removed start becoming responsive to sensory cells on the other side of the body. By some mechanism not yet fully understood, the giant interneurons on both sides of the body come to respond to sensory input from the one intact cercus and the cockroach is able to adapt to its changed sensory picture of the world and escape in the correct direction. So, even when a behaviour is fully functional at hatching, modification and improvement are still possible. The genes and the environment continue to interact throughout life.

Evolution

The fourth question that can be asked about behaviour is how, in evolutionary history, it came to be the way it is. Just as we might ask how a horse's hoof evolved or what a bat's wing evolved from, so we might also ask how a behaviour pattern evolved. Here answers have to be somewhat speculative as we have few fossils and can never have a re-run of evolutionary events that took place thousands or millions of years ago. If the distant ancestors of present-day animals had no hooves or wings and their not-so-distant ancestors had hoof-like or wing-like structures, we might use the fossil record to find out what happened in between and what has happened since to give rise to modern structures. In the case of behaviour, the fossil record (apart from a few tracks and footprints) is of little help. Nevertheless, it is sometimes possible to see, in closely related species that are still alive today, how a behaviour pattern may have evolved.

Morris (1959) describes such an evolutionary series in the courtship of tropical grass finches, illustrated in Figure 1.14. Male zebra finches often perform beak-wiping on the perch when they are courting a female (a). In the related striated finch (b) and spice finch (c), the male performs a bow and holds himself with head lowered close to the perch. In neither of these two species case does the male actually wipe his beak on the perch, but the similarity of the bow posture to bill-wiping in the zebra finch is

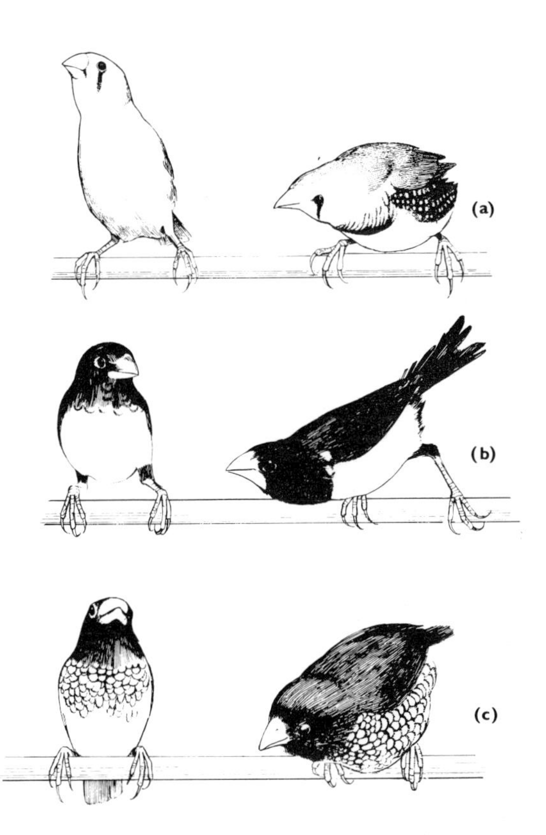

Fig. 1.14 Beak-wiping in the courtship of three grass finches: (a) the zebra finch, in which the male is just about to wipe his bill across the perch, just as he would if he were cleaning his beak; (b) the striated finch and (c) the spice finch, in both of which species the male remains stationary in this position for some seconds and the 'beak-wiping' now looks rather like a bow. In all three cases the bird on the left is a female (from Morris 1959).

quite striking. Morris suggests that the bow of the striated and spice finches has evolved from an ancestral finch that beak-wiped during courtship as the present-day zebra finch still does. Without being able to compare the three living species, it would not have been possible to attempt to reconstruct evolutionary history in this way.

DIVERSITY AND UNITY IN THE STUDY OF BEHAVIOUR

It will be clear by now that the study of animal behaviour presents us with a great diversity of subject matter, levels of analysis and questions to be answered. At one end, neuroethology merges into cellular physiology and biochemistry. At the other, the study of animal groups moves into evolutionary theory and ecology. Psychologists, ethologists, physiologists and behavioural ecologists all contribute to this diversity, so much so that it may be tempting to think that there is no sense in which there is a single 'study of animal behaviour' at all.

There are, however, two very positive reasons for seeing it as a single subject of rich diversity rather than as a collection of isolated disciplines. The first is that the same questions, particularly those of mechanism, survival value and ontogeny, can be asked at all levels. It is as important to ask 'Why' (in the survival value sense) an animal has a nervous system with sensory, motor and interneurons connected in particular ways as it is to ask why male sage grouse display. It is as important to ask about causation ('How does it work?') when looking at the way in which a bird assesses the amount of food in its territory as it is when recording from a single nerve with a microelectrode. Studying adaptation without looking at mechanism tends to become sterile because it is essential to understand the constraints imposed by what the animal's body can do before we can understand the evolutionary significance of what it does. Studying mechanism without asking about adaptation tends to give lists of facts with no context. Evolution through natural selection is the unifying theme of all biology and the study of animal behaviour, at all levels, is at its best when questions about mechanism and questions about evolutionary significance are asked in parallel.

The second reason for not compartmentalizing the subject and trying to see it as a whole is that the different levels of analysis, although distinct, are not totally separate. Many of the same concepts that have been used and tested at one level have been found to be useful at other levels. The behaviour of animals is, ultimately, due to the workings of nerves, muscles, hormones and sense organs. Equally, however well we understand the workings of those parts of the body at a physiological level, our understanding is incomplete unless we understand how they interact to produce complex behaviour. The most fruitful studies have been those in which the boundaries between the different levels have been breached and in which 'outside the skin' and 'inside the skin' have not been the demarcation lines of a particular investigation.

2 THE DEVELOPMENT OF BEHAVIOUR

One of the most remarkable features of living organisms—plants or animals—is the way in which a single-celled zygote, the fertilized egg is transformed by cell division, cell differentiation and cell movement into the adult form, sometimes many millions of times larger and far more complex. Development (often called embryology when it describes the progress from egg to a young but free-living stage) remains one of the most challenging problems of all science.

The zygote contains all the information necessary to build a new organism provided that it can develop in and interact with a suitable environment. When we study the development of behaviour, we must obviously concern ourselves with some aspects of embryology, e.g. the way in which the basic framework of a nervous system is laid down, but we shall need to go far beyond this. A young animal's behaviour may continue to develop long after it is independent and it is perfectly reasonable to argue that, in some animals, behavioural development continues throughout life. Thus, learning (on which we shall concentrate in Chapter 6) might well be regarded as a form of development. However, here we shall concentrate on the early part of life, where behaviour changes most rapidly and dramatically.

At the outset we must recognize that young animals are not simply partly formed creatures, inadequate stages on the path to adulthood; they have at all times to be fully functional animals capable of behaving effectively in their own world. During their early stages of growth some animals may be protected inside an eggshell or a uterus or watched over by attentive parents, but others are free-living, having to look after themselves entirely. Young animals may emerge as miniature adults, gradually growing in size, but their behavioural responses must also change to keep pace. Young cuttlefish begin and remain as carnivores, but at first they can kill only tiny crustacea, which are ignored as prey when the cuttlefish become larger. They move on to larger and larger food and the behaviour patterns employed for detecting and catching prey have to change accordingly with growth towards the adult size. The behavioural and morphological changes can be even more dramatic, for some young animals have a life totally different from that of the adults. Tadpoles swim and breathe like fishes and are herbivores before metamorphosing into land-living, carnivorous frogs or toads. Aquatic filter-feeding rat-tailed maggots (so called because they breathe air through a long snorkel tube at their rear end) metamorphose into flower-feeding hover-flies. In such life histories young and adult each require an almost totally separate repertoire of behaviour.

Such changes mean that development often has to generate patterns that operate for only part of an animal's life and then disappear. Provine (1976) has described the particular synchronized movements by which cockroaches break out of their individual eggshells and the cocoon that packages a batch of eggs together. They involve a series of reversed waves of contraction along the body from tail to head and they are seen only on this one occasion. They appear, precisely timed at the close of early development in the egg, and serve to launch the young cockroach nymph into its next stage of growth; this done, they are never elicited again. The feeding behaviour patterns of the aquatic larvae of hoverflies, just mentioned, must have a rather longer life than this because they have to carry the larva through its whole growth period until it pupates but then, like the cockroach hatching movements, they disappear. Animals with complete metamorphosis have, in effect, to be capable of organizing two different lives, carrying two sets of genes that organize two developmental programmes, interacting with two different environments. Just sometimes the two stages of life can combine. There is a remarkable example of 'cooperation' between young and adults within colonies of weaver ants. Using adult behaviour patterns the workers bring together the two opposing edges of a leaf to form a sheltered nest space within it. They seal the edges together by carrying up larvae and inducing them to spin silk across the gap—a behaviour pattern the larvae have evolved for spinning their cocoons prior to pupation. Oppenheim (1981) provides an excellent review of this whole topic.

We shall be describing some of the developmental processes in the nervous system that underlie this dual behaviour system later (see p. 24). Here we may note that specialized infantile behaviour patterns do not always disappear but may return in a slightly different context. Baby meerkats, an African mongoose, go limp and behave passively when their mother seizes them by the scruff of the neck. This reflex facilitates their being moved without injury. Adult female meerkats similarly relax when seized in a neck bite by the male during copulation. The neural basis for the reflex remains beyond infancy and we have some examples that reveal that even if they are not used again, some juvenile reflexes remain. In a series of elegant experiments, Bekoff and Kauer (1984) showed that the unique movements made by the chick to break out of the eggshell (beating the beak against the shell and strong thrusting movements of the legs, which serve to rotate the body) will be performed again by young chickens several weeks after hatching if they are gently pushed back into a huge artificial eggshell and then placed in the posture of the chick at hatching. Probably one factor that contributes to the retention of the neural circuitry underlying this pattern is that elements of it are involved in adult locomotor behaviour. We shall discuss other examples of this kind of 'neural recycling' from insects later (see p. 26).

The progression, sudden or gradual, that marks the transition from young to adult is not always made along a single pathway. In this chapter we shall be considering a number of ways in which behavioural development can proceed and how far young animals can compensate for deviations from the normal path. However, few animals have one single end-point, one fixed pattern of adult behaviour. Some animals become males, others females and their behavioural repertoires may be very different. Honey-bees develop into queens or workers, whilst young grasshoppers of certain species may continue development into solitary grasshoppers or alternatively turn into gregarious, swarming locusts. In these cases we know that there are clear-cut genetic or environmental triggers that send subsequent development along one of two or more alternative paths. The presence or absence of a Y chromosome in the zygote of mammals determines the sexual pathway, although as we shall see, subsequent 'sign posts' along the path are not genetic (p. 28). The length of time that royal jelly is provided to the growing larval honey-bee determines whether it develops into a queen or a worker (p. 155). The degree of crowding—how many other young hoppers are in the vicinity—switches the young grasshopper into solitary or gregarious phases, with such astonishingly different appearance and patterns of behaviour.

Not all triggers are clear-cut. Males of some vertebrates fight to defend territories to which females are attracted and then mate with them. Other males may not fight and hold a territory of their own but rather hang around submissively in another male's territory as so-called 'satellite' males, not very dissimilar to the satellites in the winter feeding territories of pied wagtails described in Chapter 1. Such behaviour in the breeding season is found in animals as diverse as ruffs (van Rhijn, 1973) and white rhinoceros (Owen-Smith, 1971). Although satellites may appear to be unsuccessful, close observations show that they quite often manage to mate with females—often when a territory holder's attention is concentrated on fighting off other territorial males. In other words, becoming a satellite may be a viable alternative strategy for adult life. Identifying the events, genetic or environmental, that bias such development one way or the other, may not be easy. We must be aware of ecological factors, as well as genetic or physiological ones, that may affect these life-history 'decisions'. Nor will they necessarily be permanent decisions: further change and development may be possible as animals get older and individual experience enables them to adapt.

NATURE AND NURTURE

Whether we are observing young animals or adults, one of the most compelling aspects of behaviour is its adaptiveness. Animals do make mistakes and may appear clumsy at times, particularly when they are put into unnatural situations, but for the most part their behaviour is beautifully matched to their way of life. They respond appropriately to the features of their world and thereby feed themselves, find shelter, mate and produce offspring.

How can behaviour develop so accurately as to match an animal's way of life? This question has fascinated human beings for centuries, because we have always been observers of animal behaviour. Walker (1983) and Sparks (1982) discuss the history of our thinking about animal behaviour. The manner in which different ages and cultures have explained behavioural adaptation has depended to a large extent on how they viewed their relationship to animals and the world around them; in other words, it has been treated as a metaphysical question, not just as a biological one. One of the most influential views in Western culture was stated most forcibly by the French philosopher, René Descartes, in the seventeenth century. As a devout Christian of his time he viewed human beings as set apart from brute creation, yet he had to recognize that the brutes did cope remarkably well with the problems of their lives. How could they do this, lacking souls or the power of reasoning that enabled humans to survive and prosper? Descartes supposed that animals were able to respond adaptively because they operated using **instincts** endowed by the Creator, which automatically provided them with the correct response, requiring neither experience nor thought. We choose the term 'automatically' because Descartes viewed animals as little more than automata, blindly driven along pre-set but adaptive paths.

The concept of instinct or instinctive behaviour (the terms 'innate behaviour' or 'inborn behaviour' are synonymous) is still a familiar one. It is often thought of as a pattern of inherited, pre-set behaviour responses that develops with the developing nervous system and can evolve gradually over the generations, just like structural features, matching an animal's behaviour to its environment. We shall need to examine these ideas more critically and, as we shall see, they are not without difficulties. For the moment, let us define instinctive behaviour in a negative way, as that behaviour that does not require learning or practice, which appears appropriately the first time it is needed. This negative definition immediately suggests the converse and the other familiar way in which behaviour can become adapted to the environment. Animals may be able to modify their behaviour in the light of their individual experience. They can learn how to behave and perhaps practise or even copy from others to produce the most adaptive responses.

Expressed in these terms, we appear to have come up with a clear dichotomy—rather as Descartes intended—**instinct** (or nature) in which adaptation occurs over generations by selection, or **learning** (nurture) where adaptation occurs within the lifetime of an individual. As we shall see, studies of the way in which behaviour develops do not support such a distinction, and suggest that the categories themselves are inadequate. Nevertheless, it is useful to look briefly at the sources of adaptiveness across the animal kingdom, with its huge diversity of form and behaviour, for some discernible pattern.

INSTINCT AND LEARNING IN THEIR BIOLOGICAL SETTING

Pre-set behaviour that requires no learning is obviously going to be advantageous for animals with short life spans and no parental care. Some of the insects lead almost totally solitary lives, with no contact between the generations. Mason wasps of the genus *Monobia* construct a series of cells inside the hollow stems of plants. A female wasp emerges from its cell and has a brief moment of social contact when she mates with a male. Thereafter she is on her own. She selects a hollow stem and builds a partition of mud mixed with her saliva at the inner end. She then lays an egg, which is attached to the roof of the stem, close to this partition. Next she hunts for caterpillars, which she lightly paralyses with her sting, and provisions the cell with between five and eight of them; these are the future food supply for the larva when it hatches. This done, she builds another partition sealing off the egg with its food supply and, laying a second egg beyond this, provisions the second cell, seals that off and so on. In this way eight to ten cells may be constructed in line along the cavity of the stem until the female, reaching the outside end, plugs this with mud. She then moves off to seek another stem and constructs more cells (Fig. 2.1).

The female wasp lives only a few weeks and carries out this elaborate series of behaviour patterns in total isolation. She could not possibly achieve this tight schedule if she had to acquire everything from scratch by trial and error; she has to rely on pre-set, unlearnt responses. This conclusion is given further emphasis by the fascinating observations made by Cooper (1957) on this wasp. If you examine a stem in which the larvae have pupated prior to emerging as adults, they have all done so with their heads facing the open end. Making the correct choice of end is a matter of life or death for, although sometimes the emerging adult could turn round in the narrow stem, they do not do so but move on ahead breaking through the partitions. Normally the outermost pupae (although they derive from the later eggs of the series of cells) emerge first leaving a clear passage, as it were, for their siblings from deeper in the stem. Adults emerging from artificially reversed pupae struggle on inwards through the deeper cells and accumulate at the blind end (Fig. 2.1).

How can a larva, about to pupate, make the correct decision? Cooper's experiments showed clearly that there was no possibility that they detected light, or used gravity or oxygen concentration as a cue. They rely on information left behind by their mother. As the female wasp retreats outwards building the partitions between the cells, the inner side of each is, of necessity, left as rough mud whilst the outer sides she smooths into a concave form. Cooper showed that it is the characteristics of concavity and smoothness versus those of roughness that the larva uses; it pupates with its head towards the latter (Fig. 2.1). Information is thus passed from one generation of wasps to its offspring and must be encoded genetically in a way that allows the larvae and the adult female to develop appropriate behaviour. Again we must recognize that neither the mother wasp's actions nor the response of the larvae can rely on experience.

For the most extreme of contrasts we may compare the mason wasps' life history with that of the African elephant. The elephant's life span is comparable to our own and the key element of elephant society is the matriarchal group led by a mature female with her daughters and their offspring. A baby is born into a group where every individual knows all the others from long experience in their company. It is nourished and closely protected by its mother for several years. Slowly it acquires the adult repertoire of feeding behaviour, learning how to select food and how the group migrates around its home range to match the seasonal changes of vegetation and water supply. Females do not become sexually mature until about 20 and puberty is even

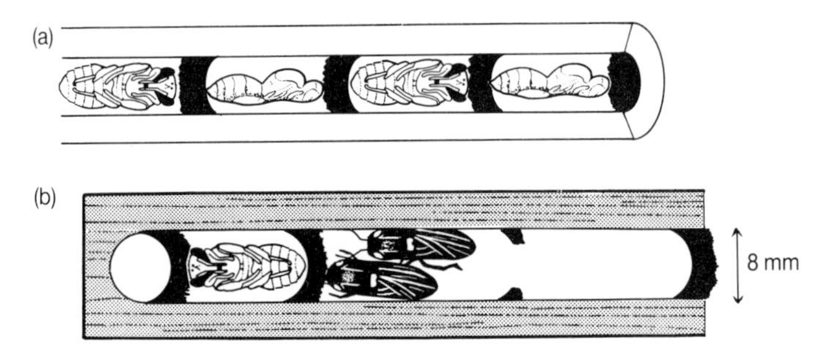

Fig. 2.1 (a) Cells of the mason wasp, *Manobia quadridens* in a hollow stem, note the form of the mud partitions, rough and convex on the 'outer' side, smooth and concave on the 'inner'. The larvae have all pupated with their heads towards the rough wall. (b) An artificial stem of hollowed out dowelling made by Cooper (1957). The partitions were reversed in orientation and, as a result, emerging adults have headed 'inwards'. They gather at the blind end where one wasp still remains in the pupal stage—also headed inwards.

later for males, who leave the group when mature to lead more solitary lives. The behaviour of individuals and of groups varies considerably according to their history. For instance, Douglas-Hamilton and Douglas-Hamilton (1975) record how one group of elephants at Addo, in South Africa, are abnormally nocturnal in their habits and, whilst fearful, are also unusually aggressive towards humans. This behaviour can be traced back to an attempt to annihilate them by shooting in 1919! Very few, if any of the elephants alive at that time can still be there, but their descendants have acquired and transmitted the behaviour that enabled a few to survive over 70 years ago.

The mason wasp, which must rely on pre-set instinctive behaviour, and the elephant, which can learn at relative leisure and even transmit information from one generation to the next, represent two extremes on the behavioural scale. In fact, our descriptions are greatly over-simplified, because the wasp can and must learn many things during its brief life—the exact locality of each of its nest stems, for example, so that it can return to them after its hunting trips. The young elephant possesses some instinctive tendencies, such as those for feeding and reproduction, even though it may have to learn how to direct them. Certainly, all animals beyond the annelid worm level show both types of behaviour and each has its own special advantages. One advantage of learning over instinct is its greater potential for changing behaviour to meet individual changing circumstances. Such a consideration is obviously more important to a long-lived animal than to an insect that lives only a few weeks. A further relevant factor may be body size, because highly developed learning ability requires a relatively large amount of brain tissue, insupportable in a very small animal. Body size and life span are usually positively correlated to some extent, and large animals live longer than small ones.

Apart from these physical constraints it is clear that natural selection can operate to match different degrees of learning ability to a species' life history. The two most advanced orders of the insects, the Hymenoptera (ants, bees and wasps) and the Diptera (two-winged flies), are comparable in size and life span. The Hymenoptera, in addition to a rich instinctive behaviour repertoire, show an extraordinary facility for learning, albeit of a specialized type, and this plays an important role in their lives. During her brief 3 weeks of foraging a worker honey-bee will learn the precise location of her hive and the locations of the series of flower crops on which she feeds. She may move from one to another of these during the course of a day's foraging because she also learns at what time of day each is secreting the most nectar. After three visits to a food dish marked by a particular colour, a honey-bee worker retains her memory of this colour for the rest of her foraging life. Even after one visit it is 5 or 6 days before her responses to the colour are back to the neutral level (see Menzel 1985 and further discussion in Chapter 6). Even flies can learn. Hover-flies learn something of the position of the flowers they visit and house-flies tend to return to the same place to settle in a room. It has proved possible, using some ingenuity, to get fruit-flies to learn some simple discriminations. But for the most part the dipteran memory is brief and their learning powers very limited. Unlike the Hymenoptera they do not reproduce in fixed nests to which they must return regularly. Their short lives are largely governed, with complete success, by inherited responses to stimuli that signal the presence of food, shelter and a mate.

The characteristics of instinct and learning

We mentioned earlier that instinct and learning are inadequate labels to describe or help us understand what actually happens during the development of behaviour. Superficial considerations may suggest that whereas learning results in flexible patterns of response, instinctive behaviour is characterized by rigid, stereotyped responses and patterns of movement. Yet common observation will tell us that learnt patterns can be just as stereotyped as instinctive ones. In his delightful book, *King Solomon's Ring*, Lorenz (1952) describes how water shrews learn the geography of their environment in amazing detail. If at one point on a trail they have to jump over a small log, this movement is learnt with such fixity that they continue to make the jump in precisely the same fashion long after the obstacle is removed. All mason wasps build the partitions between cells in the same way and all kittiwake gulls show the same 'choking' display (Fig. 2.2) on their cliff nest-sites. Neither of these patterns require learning for their development. But all chimpanzees of one community in Tanzania show one specialized mutual grooming pattern (Fig. 2.3), which is not seen in other communities and is certainly copied afresh by each

Fig. 2.2 The 'choking' display of the kittiwake. The bill is wide open revealing the brilliant reddish-orange gape and tongue. As the body inclines forward the bird makes a rapid series of jerky up and down movements of the head and neck (after Tinbergen 1959).

Fig. 2.3 Chimpanzees of the Kasoge community in western Tanzania raise arms and clasp hands for mutual grooming of each others underarms (photograph by W.C. McGrew and C. Tutin).

generation of young animals. All male chaffinches end their song with a 'terminal flourish' (see p. 41), which we know is learnt by the young birds as a result of listening to adults singing. There are many routes along which animals may acquire adaptive behaviour and hence all behaviour presents us with the problem of how it develops. So far we have used the absence of learning or practice for a kind of working description of instinctive behaviour, but such a negative definition is of no help in studying development. Instincts cannot develop *in vacuo* and learning or practice do not cover all the kinds of influence that may come to bear on a young animals as it develops. It is certainly not adequate to separate off instinctive behaviour as that which requires no experience although this feature was certainly emphasized by some workers (see, for example, Lorenz 1966) who laid great stress on the importance of the 'isolation' or 'Kaspar Hauser* experiment'.

This attempts to keep animals solitarily, out of all contact with others from as early an age as possible. When mature, their responses to a variety of stimuli are tested and compared with those of animals reared normally. The life histories of many insects are, as we have seen, effectively natural isolation experiments. Few other animals have been tested under very rigorous conditions, but some fish and birds have been shown to perform various feeding, sexual and alarm patterns of behaviour quite normally following isolation. These are interesting data but, however easily isolation experiments enable us to eliminate the possibility that animals learn how to do something, they offer only a very restricted view of what happens during development. Isolation can tell us only what factors are *not* important for the development of behaviour, it tells us nothing of what is involved and some factors may not be at all obvious. We ought to begin our experiments by asking from what, exactly, are the young animals being isolated?

* Kaspar Hauser was a youth who appeared on the streets of Nuremberg, Germany on 26 May 1828 with confused stories of his origin and upbringing. He was obviously mentally disturbed and displayed bizarre behaviour. He died, in 1833, of wounds that were almost certainly self-inflicted. Kaspar Hauser attracted the attention of various philosophers and occultists of the day and was sometimes regarded (for no fathomable reason) as an example of the totally untutored human being. He was claimed to be an isolate, revealing native human characteristics unchanged by experience. This usage of his name has been picked up by some German ethologists as a convenient label for isolation experiments with animals.

Gottlieb (1971) has carried out a series of investigations into the way in which young ducklings of various species come to respond correctly to the calls of their mother. One might regard the young bird, developing inside its eggshell, as another natural isolation experiment but this is too simple a conclusion. Young mallard (*Anas platyrhynchos*), for example, hatched from an incubator respond preferentially to the call notes of mallard ducks over other related calls the first time that they hear them. They clearly show preference for their own species' call —we would call this instinctive. But experiments reveal that one factor necessary for this instinct to develop is an embryo duckling's ability to hear the calls—quite unlike those of the mother—that itself and other ducklings make whilst still inside the eggshell. Even less expected is Gottlieb's (1983) finding that young wood duck, which hatch with an equivalent preference for the characteristic descending calls of the mother, subsequently lose this preference unless they continue to hear the calls of other ducklings around them; their own calls are not sufficient.

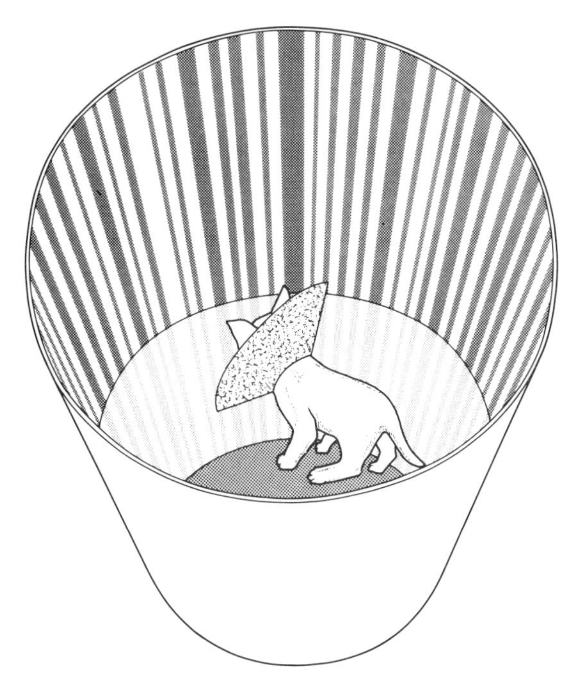

Fig. 2.4 Apparatus used by Blakemore and Cooper to restrict the visual environment of kittens. The kittens wore a black collar so that they couldn't see their bodies and stood on a glass plate inside a high cylinder whose walls carried vertical or horizontal stripes of varying thickness (after Blakemore and Cooper 1970).

The auditory system of the ducklings requires experience—perhaps one might call it 'priming' —with certain types of sound if it is to develop normally and mediate the response to the mother's calling. We know of many comparable examples in the development of sensory systems. Mammalian eyes develop all their basic structure in total darkness, but they do not develop their normal functioning unless 'primed' with certain types of visual experience.

Blakemore and Cooper (1970) reared kittens in complete darkness except for brief periods in a very visual environment (Fig. 2.4). From 2 weeks of age, when their eyes opened, they spent an average of 5 h per day looking at a pattern of either horizontal or vertical stripes. When first brought out into the normal world at 5 months the kittens were, understandably enough, disorientated. However, they quite quickly managed to cope, although their vision remained abnormal. This was revealed dramatically when a 'horizontal' and a 'vertical' kitten were put together. If the experimenters presented them with a rod held vertically and shaken the 'vertical' kitten moved forward to play with it, but the 'horizontal' kitten ignored it. When the rod was turned to the horizontal, the 'horizontal' kitten immediately took interest and approached whilst the 'vertical' kitten no longer responded—it behaved as if the rod had suddenly disappeared.

Subsequently, Blakemore and Cooper used neurophysiological techniques to look at the functioning of neurons in the kitten's visual cortex. (This is the part of the brain that receives, in coded form, the visual information from the cat's retina.) Each of these neurons has a 'preferred orientation', which means that they fire when lines or edges at that orientation are presented to the eye. In a normal cat the orientations are distributed all around the clock, but vertically- or horizontally-reared kittens showed a very strong bias towards the corresponding orientation. Thus horizontal kittens had no units responding to vertical or to lines or edges within 20° either side of vertical; they literally cannot see a rod when it is held vertically.

Such experiments reveal clearly how the early environment, sometimes before hatching or birth, sometimes later, shapes the development of fully functioning behaviour. Major problems arise if one tries to force behaviour into categories, such as instinct or learning. There is a temptation to do this because instinctive behaviour is such a constant feature of an animal species—unless their environment is grossly changed, they all develop the same patterns of behaviour and perform them in the same way. The so-called 'fixed action patterns' typical of courtship and aggression, which were discussed as units of behaviour in the previous chapter, are a typical example. It seems reasonable to call such patterns innate or inherited but, unfortunately, such observations do not allow us to conclude anything certain about the operation of genes. Common sense tells us that genes must be involved when we observe instinctive behaviour emerging fully fledged, as it were, at the first performance in every individual of a population. But such constancy of development gives us no way of knowing if and how genes are acting. We can take an analogy from morphology. It is very easy to work out how genes affect the colour of eyes in fruit-flies or in human beings. Eye colour varies and the inheritance of the variants leaves no doubt that they are under simple genetic control. What eye colour you develop usually depends entirely on what genes you are carrying. This means we can readily proceed to study how the genes affect the synthesis of pigments, at what stage in the development of eye colour they act and so on. However, if we are interested not in the development of eye colour, but of eyes themselves, we have no such opportunities. Apart from certain gross abnormalities, all fruit-flies and all humans develop eyes. Genes must be involved at all stages in the complex growth and differentiation that lead to the formation of the eye cup and the retina, but unless we can isolate suitable mutations we can say nothing about how they act.

So it must remain with most instinctive behaviour, and it serves little purpose to equate instinct with genetic control. In any case, the complexities of some of the developmental pathways we have already described show how difficult it is to press behaviour into hard and fast categories—learnt versus unlearnt, for example. The open-minded study of behavioural development itself is the only way by which we can come to understand how genes and environment interact to produce adaptive behaviour. This is a very broad topic and it is deservedly attracting a lot of attention at present. Here we can take only a few examples to illustrate some of the complex range of influences which must be considered.

MATURATION—DEVELOPMENT INVOLVING GROWTH

It is logical to begin with behavioural development that occurs in parallel with normal growth processes both of the nervous system and the rest of the body. Thus the emergence of sexual behaviour in vertebrates depends on the growth of the gonads, which begin to secret hormones. Very often the nervous and muscular systems of young animals have to go through further growth or differentiation before they reach adulthood and the term 'maturation' is commonly used to describe behavioural changes that can be linked to such growth.

For example, young birds can often be seen making vigorous flapping movements with their wings whilst still in the nest and it is commonly supposed that they are practising flying. Human parents often support young babies on their legs and encourage them to 'practise' walking. In fact there is no evidence that the early development of bird flight, or human walking are affected in any way by such activities. Over a century ago Spalding (1873) showed that young swallows reared in cages so small that they could not stretch their wings flew just as well when released as normally reared birds. By the age of 18 months, when the majority of children are walking, their skills are very similar whether they walked first at 10 months or 15. In both cases it is the maturation of the central nervous system and its coordination with muscular development that count. Practice, of course, eventually adds all the finer points of skill—young fledglings are notoriously clumsy fliers—but the basic pattern matures without it.

The development of pecking in newly hatched chicks provides another example of the interaction between maturation and practice. Young chicks have an inherited tendency to peck at objects that contrast with the background but their aim is at first rather poor, and various workers have studied how it improves. One of the most complete studies was made by Cruze (1935). He hand-fed chicks in the dark on powdered food for periods of up to 5 days before testing the accuracy of their pecking. Whilst in the dark they are inactive and have no chance to practise the movement. Cruze measured accuracy by putting the chicks individually into a small arena with a black floor, on to which he scattered two or three grains of millet. Each chick was allowed 25 pecks scored for miss or hit and grains were replaced if the chick swallowed them. After accuracy tests the chicks were allowed to feed naturally in the light and the effects of practice on their accuracy measured again after 12 h.

Table 2.1 shows the results of one experiment. There is a steady improvement with age (heavy type) but, at any age, 12 h of practice greatly improves accuracy (light type). Much of this improvement must result from maturation, although we must bear in mind that mechanisms other than those specifically controlling pecking will influence accuracy. The chicks' legs grow stronger and perhaps their stability improves, which would help their aim.

Age (h)	Practice (h)	Average misses (25 pecks)
24	0	**6.04**
48	12	1.96
48	0	**4.32**
72	12	1.76
72	0	**3.00**
96	12	0.76
96	0	**1.88**
120	12	0.16
120	0	**1.00**

Table 2.1 The pecking accuracy of chicks at different ages before and after 12 h of practice. Each figure represents an average from 25 chicks (modified from Cruze 1935)

Because the developing embryos of amphibia and birds are relatively accessible it has been possible to investigate more closely the nature of maturational changes to their nervous system and behaviour. As development proceeds, the increasing complexity of an embryo's structure is paralleled by an increasing repertoire of behaviour, both spontaneous and in response to external stimuli. Oppenheim (1974) reviews work on chicks that shows how movements appear quite suddenly when the requisite connections between growing nerve fibres are made. The sensory fibres that grow into the spinal cord from the muscle spindles and provide information on the state of contraction of the muscle (Chapter 1, p. 10) must form their central connections before the muscle can take part in coordinated movements with other muscles.

The gradual nature of development in crickets enabled Bentley and Hoy (1970) to study one type of maturation in great detail. Crickets go through a series of 9–11 larval or nymphal stages (instars),

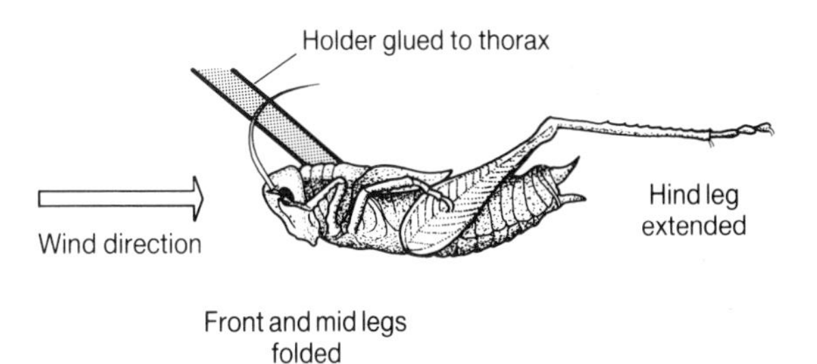

Fig. 2.5 A 7th instar nymph of the field cricket, *Teleogryllus commodus*, adopts the flying position when suspended in a current of air, although it has no wings. The antennae point ahead, the fore- and mid-legs are drawn close to the body and the hind legs are held straight back parallel to the body. All the features are identical to the position of the adult in suspended flight (drawn from photographs in Bentley and Hoy 1970).

which become increasingly similar to the adult with each successive moult. The final moult usually involves the biggest changes, with the full development of wings and adult genitalia. Since they have no wings cricket nymphs do not fly, but some elements of the flight pattern can be detected as early as the 7th instar. Nymphs at this stage will adopt the typical flying posture (Fig. 2.5) when suspended in a wind tunnel. All the muscles necessary for flight are present from an early stage, albeit reduced in size. Bentley and Hoy inserted fine wire electrodes into some of these muscles of the thorax and recorded action potentials from them. Since the relationship between the input from motor neurons and the output from the muscles is relatively simple in insects, the muscle activity gives a very accurate picture of the nerve activity responsible for it. Bentley and Hoy could detect signs of the rhythmic nerve impulse characteristics of flight from 7th instar nymphs, but it was incomplete and not sustained. It became more complete with each successive instar —new elements could be identified as becoming active, presumably as functioning synaptic contacts were established. Although timing varied between individuals, the units always developed in the same sequence until, by the last instar before the final moult, the entire pattern was complete. At this stage, then, the maturation of the nymph's nervous system is finished and it awaits only the development of wings. Substantially the same story is true for the cricket's song, another activity that requires the wings of an adult (they are rubbed together rapidly to produce sound) for proper expression. For both flight and singing, Bentley and Hoy found that they could not elicit neural activity from the thoracic centres of nymphs unless they made lesions in the brain. This presumably removed inhibition, which normally prevents the nymphal muscles from being stimulated into useless activity until the rest of the body's development has caught up.

The crickets and their relatives show gradual development towards the adult form and this provides a gradually growing morphological framework in which to build up the neural basis of adult behaviour. Insects with complete metamorphosis have to reorganize their nervous system quite radically during the pupal stage. In experiments similar to those on crickets, Kammer and Rheuben (1976) could detect patterns of activity in the developing muscles of insect pupae of some large moths, which indicated that the neural basis for both flight and warming-up (shivering) movements have developed by less than half-way through the 21 days of the pupal stage. Elsner (1981) reviews these and other studies in more detail.

Studies on the changes to the nervous system of insects during larval growth and metamorphosis have revealed some remarkable parsimony in their development (Levine 1986). Insects sometimes recycle the same neurons for use in both larva and adult forms. Levine and Truman (1985) have shown that, in the hawkmoth *Manduca sexta*, all the motoneurons supplying the abdominal muscles are recycled. Different as the two body forms are, the abdominal movement of the caterpillar and those of the adult moth may not be totally dissimilar. Some of the motoneurons can stay innervating the same muscles, but others lose their specifically caterpillar connections at pupation and grow new processes to innervate newly formed adult muscles (Fig. 2.6). Most larval sensory neurons supplying sensory hairs die away at pupation but one group come to innervate specialized trigger hairs on the abdomen of the pupa. They trigger a reflex, unique to the pupa, which causes it to flex suddenly in a movement that defends against predators.

And so behaviour patterns mature, flourish and disappear, correlated with growth, modification and death of the underlying neurons. It all looks very much like the running of a pre-set programme of development, but this is a description, not an explanation. Close examination of neuroembryology can help us to understand what 'pre-set' implies and this is certainly an example where studying the neurobiology is an aid to studying behavioural development itself.

Again, insects are excellent material for studying the way in which a nervous system is built up during development. For example, in grasshoppers the axons of sensory neurons from the developing legs grow out from cells in the sensory hairs and make connections with the central nervous system. Bentley and Keshishian (1982) have shown that the first axons to grow this way—the pioneers, as they call them—use special cells situated at intervals along the route as 'guide posts'. As the limb grows and

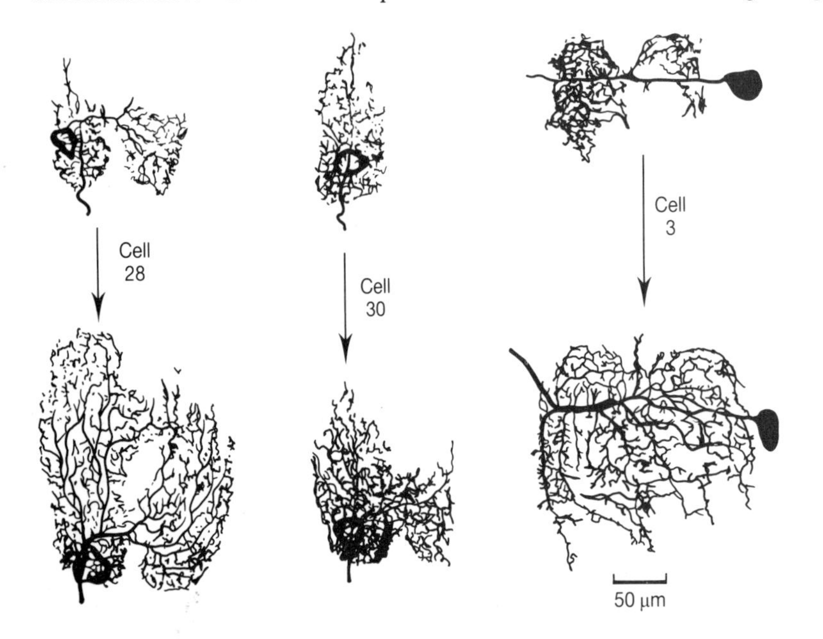

Fig. 2.6 How three motoneurons of the moth *Manduca* are 'remodelled' during metamorphosis. Individual cells can be identified regularly in all moths and injected with a cobalt dye, which reveals the rich arborization of the dendrites. The top row shows the relatively simple dendritic pattern of each cell in the larva—cell body and axon to the larval muscle are clear. After metamorphosis into the adult each cell has a greatly enlarged dendritic field, which has grown into areas not penetrated in the larval state and the axons have, in fact, lost their target larval muscles and now innervate new muscles in the adult moth (redrawn from Levine 1986).

lengthens, the guide posts are stretched apart and later axons do not use them but grow along the line already established by the pioneers. Thus, although the final result looks uniform—sensory axons running smoothly into the central nerve cord—the actual developmental histories of those neurons have not been uniform. There is also considerable flexibility in the way detailed synaptic connections are formed as sensory axons reach their destination. Which cells they connect with and how they do it depends upon what other cells are in the vicinity. If cells are missing then those that would normally have synapsed with them are able to compensate, at least to some extent. There must be a genetically based programme for development, although this is more likely to lay down ground rules rather than a specific point-to-point blueprint. Murphey (1985, 1986) and Stent (1980) provide an interesting discussion of such evidence and its implications, which are certainly relevant to the development of behaviour itself and its genetic basis. We shall meet similar concepts when we come to discuss the development of bird song (p. 41) and the evolution of specialized types of learning ability (p. 132).

HORMONES AND EARLY DEVELOPMENT

It is appropriate to discuss some of the fascinating discoveries about early sexual development here, because they involve clear maturational changes to the nervous system. Sexual development, as we mentioned earlier, is a clear case of alternative pathways, along one or other of which an animal's development (both morphological and behavioural) is switched by an early 'triggering event'. Because being male or female is such a pervasive difference in ourselves and the animals we are most familiar with, it is important to keep some broader biological perspective and remember that very many invertebrates are hermaphrodites and have only one developmental pathway to becoming male and female. Even more strange in some ways, there are a number of fish that retain the capacity to switch paths. For example, in the coral reef fish, *Anthias*, all individuals develop into females, although some transform later into males. They normally live in mixed groups and, if a male dies, one of the females changes sex. This change is very rapid—it takes only a few days for the previously female fish to develop male features and begin making sperm in its gonads rather than eggs. But even earlier, within hours in some cases, its behaviour begins to change and it is treated like a male by other females in the group (Shapiro 1979; Fig. 2.7).

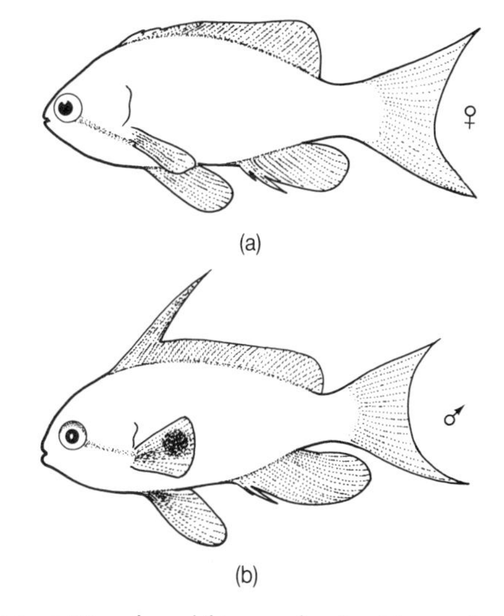

Fig. 2.7 (a) Female and (b) normal male of the coral reef fish *Anthias squamipinnis*. The males are brightly coloured and have a conspicuous dark spot on the pectoral fin. The females are a uniform orange–gold colour; they begin to show colour changes towards the male type within 3–6 days of the removal of the male from a group (drawn from photographs in Shapiro 1979).

For most vertebrates the switch is permanent. In some reptiles it appears to be controlled by the temperature at which the eggs develop. Bull (1980), investigating Mississippi alligators, found that shaded nests, where temperatures were slightly lower and incubation slightly longer, had a high proportion of males, while nests in the open sunlight of the riverside produce mostly females. Birds and mammals do not rely on such environmental triggers to achieve a balanced sex ratio. With them the switch is a genetic one, depending on the segregation of one pair of sex-determining chromosomes in the gametes. The fish that can change sex demonstrate that they inherit the potential for both types of sexual behaviour. The same must be true for mammals and birds, where the two sexes share almost all their genes. The 'problem' for natural selection has been how to make them develop differently. In morphology we can see that sometimes it has not been worth differentiating between the sexes. Male mammals develop functionless mammary glands and nipples. For those aspects of morphology and behaviour where sexual differentiation is vital, the switch is a hormonal one. Chromosomes determine whether the gonad will develop into an ovary or a testis and thereafter the pattern of hormones is also determined. Short (1982) and Birke (1989) give complete reviews of the complex story of sexual differentiation, which has many interesting variants. Here we must summarize the main events as they affect behaviour.

The differentiation of the genitalia and of those neural mechanisms responsible for initiating sexual behaviour in mammals is not determined at fertilization, but much later in development. Both males and females inherit a brain that can mediate both masculine and feminine behaviour; which pattern becomes dominant depends on whether the brain receives a pulse of hormone. The embryonic or newborn testis, unlike the ovary, has a brief period of activity and hormone is picked up by particular regions of the hypothalamus. This is a small region of the brain much involved in regulatory activities of all kinds, and which controls the functioning of the pituitary gland, which lies attached to and immediately below it. (The anatomy of these structures, crucial for the control of sexual behaviour, is described in more detail in Chapter 4, p. 88.)

If the hypothalamus picks up hormone at a certain period its development is switched along masculine lines; if not it becomes feminine. If females are given tiny quantities of hormone at the right stage their behaviour is masculinized. It is interesting to note that either the male or female sex hormones—testosterone or oestrogen, respectively—will serve to masculinize male or female embryos—they are both steroid hormones and are closely related in chemical terms. Denying male embryos their pulse of hormone at the normal time, removing the testes before they secrete or injecting a chemical that inhibits the action of testosterone, results in demasculinization. We may note that this is a clear case of a critical period of development (see p. 36), because hormone treatment earlier or later has no effect. The die is cast once the period is over and, for example, a male demasculinized by early castration will not have his masculinity restored by large doses of testosterone given later in life. When adult, masculinized females show changed sexual behaviour, sometimes dramatically so. Thus, in some rodents such as rats and mice, masculinized females will mount and thrust on receptive females and go on to perform the whole pattern of ejaculation; they also show male-like aggressive behaviour. In addition, such females are 'defeminized' in that their vagina fails to open, they lack oestrus cycles and are consequently sterile. Female primates can readily be masculinized and, as with the rodents, it is not only specifically sexual behaviour that is altered. They are more aggressive and as infants show more of the 'rough and tumble' patterns of play that are characteristic of males. However, unlike rats and mice, such females show little sign of defeminization—they remain fertile and can reproduce. Actually, it is rather difficult to compare how early hormone treatment affects behaviour in different species because the extent and nature of the critical period varies. In sheep, for example, it extends for some weeks during midgestation and there are different critical periods for masculination of genital development and for sexual behaviour itself. In primates it is also the middle period of gestation that is critical but in rats and mice it is very late in development and in fact can be extended until a day or two after birth.

Careful observations have shown that the masculinization of females is not simply an artificially induced condition in animal behaviour laboratories. Female mice produce 8–10 pups after a 3-week pregnancy in which the fetuses have been developing in a row along each horn of the uterus (Fig. 2.8).

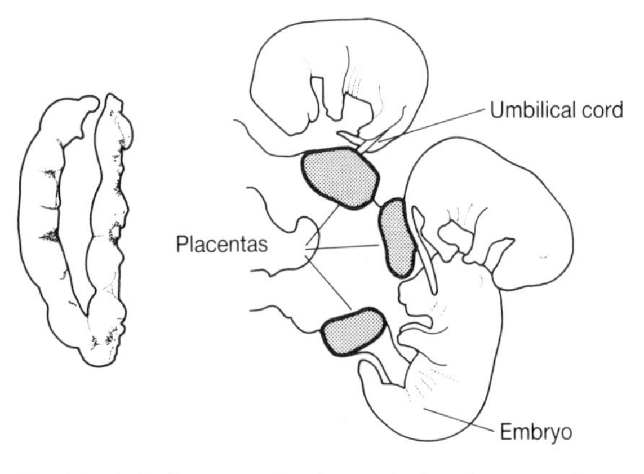

Fig. 2.8 Left: the two uterine horns of a female mouse about two-thirds of her way through pregnancy. The rows of developing fetuses can be clearly made out, strung out like peas in a pod. Right: mouse fetuses dissected out to show each with a separate, disc-shaped placenta attached to the wall of the uterus. No matter how separate they appear, some diffusion of testosterone must occur between adjacent fetuses (see the text for further explanation).

Each has its own placenta but it appears that some mingling of secretions must occur between adjacent fetuses. Vom Saal and Bronson (1980) found that, at birth, male fetuses had three times the level of circulating testosterone in their blood as females —this is presumably a result of the pulse of secretion by the testis, which we know occurs around birth. Having delivered mouse pups at full term by caesarean section, they noted the position each female had occupied in the uterus. They compared females that had developed between two male fetuses (2M females) and those who had had two female neighbours (0M females) and found a number of significant differences. Most interestingly, they found significantly more testosterone in the blood of 2M females than in their 0M sisters and they could correlate this effect—presumably due to leakage between adjacent blood systems—with a number of morphological and behavioural differences that extended into adult life. 2M had slightly masculinized genitalia, had longer and less regular oestrus cycles, showed more aggression towards other females and in defence of their young and were less attractive to males than 0M in a choice situation. This does not necessarily mean they are at a disadvantage. Vom Saal and Bronson suggest that under crowded conditions, the increased aggression shown by 2M females may improve their survival and that of their young. It is probably a case of swings and roundabouts in fluctuating environmental conditions. Such naturally arising variation within a litter may, in fact, be advantageous for the parents. Consequently, natural selection may not act so as to better insulate female fetuses from their brothers' hormone.

And so it appears that some of the normal variation we expect to find between females may have its origin in the chances of fetal life. In human beings too there are clearly detectable effects when female embryos are exposed to male hormones during pregnancy. This happens when the fetus has a rare genetic abnormality that leads to the so-called 'adrenogenital syndrome' (AGS), a condition known as congenital adrenal hyperplasia (CAH). The adrenal glands of the fetus function abnormally and release a masculinizing hormone into the bloodstream. When CAH girls are born their genitalia are often somewhat masculinized but this can be corrected with surgery and they can grow up to be fully functioning females. However, it is fascinating that the effects of masculinization can be detected in their behaviour as children. Ehrhardt and Baker (1974) have shown that CAH girls are more 'tomboyish'. For example, they indulge in more rough play, prefer boys to girls as playmates and tend to reject dolls as playthings. Now certainly we know that such preferences can be greatly influenced by parents and the particular culture in which children grow up (Money and Ehrhardt 1973). However, it seems very probable that the CAH girls were treated as girls by their parents and that the changes to the behaviour were spontaneous.

Erhhardt and Baker are cautious in their interpretation of how early hormones may act. It might bias infants towards specifically male patterns of behaviour, but in humans these would always be interpreted by a developing child in ways that reflect the 'male role' as presented by its culture. Of course, it may not be a specific effect at all. Tiefer (1978) has suggested that early androgen may do no more than increase metabolic rate and hence energy levels. An especially active girl infant will be most likely to end up playing with other active infants and they will be boys, the other male-orientated patterns will follow. We may note that the behaviour of AGS boys appears qualitatively unchanged, but they do have raised energy expenditure.

The early pulse of hormone is picked up by the hypothalamic region of the brain. Here the preoptic nucleus (so called because it is a discrete cluster of neurons situated just anterior to the point where the optic nerves enter the brain) is a particular centre of hormone concentration. The preoptic is far larger in male rats than in female and this difference is due to the hormone received in early life. Oestrogen or testosterone added to cultures of brain cells from the hypothalamus cause a great surge of growth but they have no such effect on cells from other regions of the brain. Presumably the differentiation of the preoptic nucleus is one of the maturational changes leading to the organization of male or female sexual behaviour. Certainly, this region also concentrates hormone during periods of sexual activity when the animal is mature (see Goy and McEwan 1980, for a full review of this topic).

We think of such maturational effects as part of growth, which ends with the attainment of maturity. With a few exceptions, such as the extensive powers of regeneration in urodeles, the nervous system shows no further growth or cell division in adults. However, recent work with canaries reveals remarkable seasonal changes in those parts of the brain responsible for the control of singing. Nottebohm

(1989; see also Bottjer and Arnold 1984) has shown that at the end of each breeding season some brain nuclei diminish greatly in size. Early in spring, new growth, almost certainly involving the differentiation of new neurons, begins again under the influence of testosterone. These changes are associated with the annual cycle of singing and the learning of new song types (see p. 43). This growth is quite distinct from, for example, the differentiation of the preoptic nucleus in rodents because not only is it reversible and continues into adult life, but it can be shown by female, as well as male, canaries. Testosterone injections lead to the growth of the appropriate brain regions of adult females and they begin to sing. (Canaries may be rather exceptional in this because in some other birds adult females cannot be induced to sing in this way.) These remarkable results illustrate the value of a comparative approach that makes use of the diverse specializations that can be found if we look beyond the conventional set of laboratory animals. The work with canaries is forcing a new look at mammalian brain growth and is a good example of the value of combining behavioural, anatomical and physiological studies—real neuroethology.

EARLY EXPERIENCE

With maturation in its diverse aspects, we have been concentrating on behavioural development that is proceeding along guidelines parallel to and influenced by the normal growth processes of the young animal. In the course of maturation the young animal will be interacting with its environment, both animate and inanimate, but not all individuals will undergo identical experiences. We must now turn to examine some results of such early experience.

A great deal of work on a variety of animals—birds and mammals especially—has shown how sensitive they are to events when young. We have already discussed Gottlieb's work on the effects produced by the experience of sounds heard by young ducklings whilst still in the egg. Such responsiveness may have surprising results. Vince (1969) has shown that embryo quails also respond to the clicking calls made by the other members of the clutch. Remarkably, this communication between embryos serves to synchronize their hatching. Slow developers are accelerated by hearing the calls characteristic of advanced embryos and, to a lesser extent, advanced embryos are slowed down by the calls from less advanced neighbours. In nature this synchronization may be important, because it will reduce the dangerous period when there are both chicks and unhatched eggs in the nest before the mother quail can lead the whole brood away.

In mammals, the embryo is more insulated from the external world but, of course, more directly dependent on its mother's physiological state. If female rats are kept under stress during pregnancy the behaviour of their offspring is affected (Joffe, 1965). The type of change produced concerns the degree of fearfulness that they show when first put into a strange environment. Similar effects can be produced by a whole variety of treatments to young animals after birth. Mild electric shock, cooling, or even the simple act of lifting them from the nest and then replacing them can affect their subsequent behaviour as adults. Rats that have been handled during infancy mature more quickly—their eyes open sooner and they are heavier at a given age than unhandled controls. In addition, they appear less 'emotional' in frightening situations. By this we mean, for example, that when placed in a large, brightly lit arena they move around and explore more than unhandled rats, which spend more time crouching and cling to the edges of the arena if they do move.

The nature of such effects and the way in which early experience produces them is not at all clear, nor is it a very easy subject to follow because, as with many complex questions, the results obtained are not always consistent. The nature of emotionality and its relationship to fearfulness and other stress-related responses is certainly not straightforward. Archer (1973, 1979) provides a critical review of the various measures of emotionality used with rats and mice. Probably the reason we can detect effects from such apparently trivial manipulations is because we are measuring them against the background of the laboratory-reared animal, which is certainly deprived of much stimulation. Rather than emphasizing how early handling 'adds' something we ought, perhaps, to recognize that laboratory rearing takes something away. In the wild, for example, it is common for a mother rodent to abandon her nest and shift the litter to a new site, sometimes more than once. Several mothers may share a nest at times and the environment around the nest, when the young begin to emerge at about 12–15 days of age, is usually far more complex than a laboratory cage. A number of studies have found that rats and mice that grow up in 'enriched' environments, i.e. those more complex than the standard laboratory cage, show improved learning in a variety of situations and their brains actually come to contain higher concentrations of enzymes concerned with neural functioning (Rosenzweig 1984).

The 'extra' stimulation of young animals may, in part, act on them directly so as to change their subsequent development but it will also affect them through their mother. What such experiments do, in effect, is to break into the normal intense and almost continuous communication system that operates between a mother rat or mouse and her growing litter. The pups provide a variety of stimuli and in particular have a range of calls, the most important of which are ultrasonic and far above the limits of our hearing. Disturbing the pups in any way leads to an increased rate of calling and the mother responds (see Cohen-Salmon *et al.* 1985 for an introduction to such work). A litter of pups may get scattered and if, for example, some towards the outside of the nest get chilled, their outburst of calling will accelerate the mother's retrieving them back to the nest cup. As the pups get older their normal rate of calling decreases, they move around voluntarily and the responses of the mother are less intense. The behaviour of both mother and developing young are reasonably balanced so as to promote successful weaning. Elwood (1983) provides an excellent source book of work on parent–offspring interactions in rodents.

Whatever the influence of parents on young, we must remember how much variation there is in the amount of contact between them even within a single group like the mammals. The details of the developmental pathways towards behavioural independence must be equally diverse. Marsupial infants are born after a very brief gestation period and they are at first nothing more than externalized fetuses (Fig. 2.9(a)), fused to the mother's nipples. Their sensory capacities are so restricted at this stage that mother–infant interactions can only be through the milk supply, but even when they become more independent they are constantly with the mother, carried on her body and often in a pouch (Fig. 2.9(b); Tyndale-Biscoe 1973). This represents a far higher degree of contact than

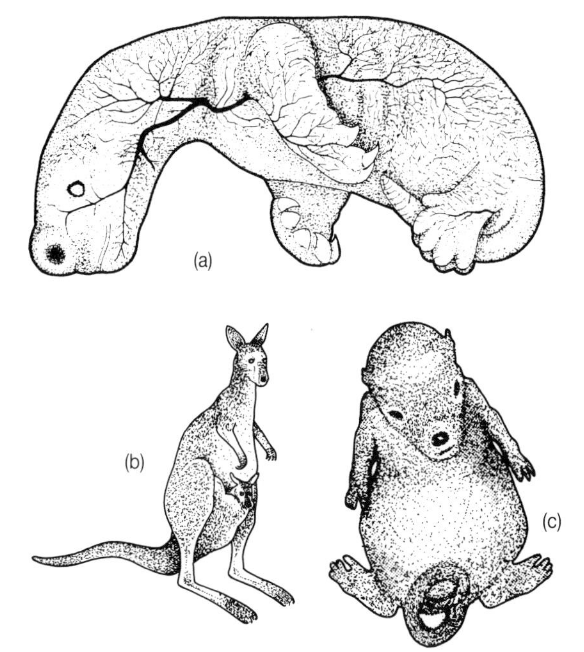

Fig. 2.9 Contrasting mammalian young and contrasted maternal styles. (a) A newly-born kangaroo, minute in relation to the size of its mother, is effectively still an embryo. The forelimbs haul it into its mother's pouch where it attaches to a nipple. The prominent blood vessels run just beneath a moist skin and at this stage the young kangaroo may well breathe through its skin as much as through its embryonic lungs. Many months later as a large subadult it still retreats into the pouch when alarmed (b). (c) A newly-born tree shrew, its abdomen distended with milk pumped immediately after birth by its mother who will now leave it alone for 48 hours with its other litter mate. Note how advanced it is physically with light fur, eyes and ears open and well developed limbs and paws—the litter mates groom one another in the nest (from Tyndale-Biscoe 1973; D'Souza and Martin 1974).

the part-time attention received by rats and mice. At the other end of the scale are rabbits, where Zarrow *et al.* (1965) have shown that the mother visits her litter only once per day and then for only a few minutes. Milk is pumped into the babies, they are briefly groomed and then the nest is covered and left. Presumably this behaviour helps to minimize the chances that predators will find the nest by picking up the mother's scent. A similar explanation is suggested for the yet more remarkable maternal behaviour of tree shrews—small insectivore-like mammals probably quite similar to the early ancestors of the primates. Martin (1968; see also D'Souza and Martin 1974) observed pairs breeding in captivity and found that they build two separate nests, in one of which the adults sleep whilst the female produces her litter, usually of two, in the other. The young are visited briefly on every second day; they suckle rapidly and become distended with milk (Fig. 2.9(c)). The mother then leaves them with the very minimum of grooming and typically does not return for 48 h. Martin discovered that the young are capable of grooming and licking themselves from the day of birth. They stay alone in the nest until they emerge at 33 days of age. This is a very remarkable type of development for a mammal and forces us to recognize that natural selection can produce a range of specializations in the pattern of development which are matched to other aspects of a species' life history. The influence of the mother on the earliest stages of behavioural development must be very different in the marsupial, rat, rabbit and tree shrew. Even close relatives can vary enormously in their stage of development and independence at birth: we can contrast the helpless blind naked young of rats and rabbits with the active young—fully furred and with eyes open—born to guinea pigs and hares.

It is within our closest relatives, the primates, that we find some of the longest and closest associations between parents and offspring, extending to a matter of several years in the apes and ourselves. The relationship between a primate mother and her offspring is absolutely crucial in almost every aspect of behavioural development. Separation, particularly in the early months, has very severe effects on the infants. In the experiments of Harlow and his group (1965) young rhesus monkeys were isolated at birth and reared artificially. Although they grew well enough they were behaviourally crippled and when subsequently put with other monkeys, showed almost none of the normal social responses. In particular they showed totally inadequate sexual and parental behaviour.

This is another illustration of the subtlety of behavioural development. If we watch a young monkey with its mother, it is not easy to identify in their behaviour those factors that are contributing crucially to the infant's social development, yet we know they are there. The work of Hinde's group, well described in his book (1974) has followed in great detail the normal development of rhesus monkeys reared by their mothers living in small groups. They could trace the gradual growth of independence due in part to the infant and in part to the mother. At first the infant is scarcely ever out of contact. The mother rarely allows her infant to move beyond arm's reach and even when it leaves its mother to explore it returns to her frequently, using her as a secure base from which to investigate new objects. Gradually, in parallel with the infant's increasing independence, the mother becomes less solicitous and even begins to reject some of the infant's approaches (Fig. 2.10), a point we return to in Chapter 5, p. 116.

Knowing the normal course of development, Hinde and his co-workers then went on to study how it was disturbed by forced separation of mother and infant. They studied the effects of levels of deprivation far less drastic than those imposed by Harlow. In a typical set of observations a baby rhesus and its mother, living in a group, were watched regularly for some weeks. Then, when the baby was 6 months old—well able to feed itself—its mother was removed from the group for a few days. Far from being isolated, the baby was usually 'adopted' by other females of the group and given a great deal of attention. Nevertheless, its behaviour showed a marked change—its distress calls increased, it moved around less and spent long periods in a very characteristic hunched posture. When the mother was returned there was usually an instant reunion and the infant spent much more time clinging to her than just before separation. The pattern of their relationship was different from that normal with a 6-month infant and it took several weeks to recover.

Several fascinating conclusions, with obvious human parallels, have emerged from systematic studies of this type. For instance, the infants worst affected by the brief separation are those whose relationship with their mother before separation appeared to be less good. From our point of view, in this discussion of early experience and development, we may note how profoundly the normal smooth adjustment of mother–infant relations was affected by even a brief interruption. The effects on the young monkey were both general and persistent, because Hinde could detect even years later that monkeys that had been separated were more fearful in strange environments.

Just as within the rodents, observations on primates in the wild have revealed striking differences between species in what we would call child-rearing practice. Thus, mother langurs are quite permissive in that they allow other females to take their babies from time to time. Rhesus macaques are far more restrictive, scarcely allowing their babies to leave their side for several weeks and controlling outside contact for much longer. However, a close relative of

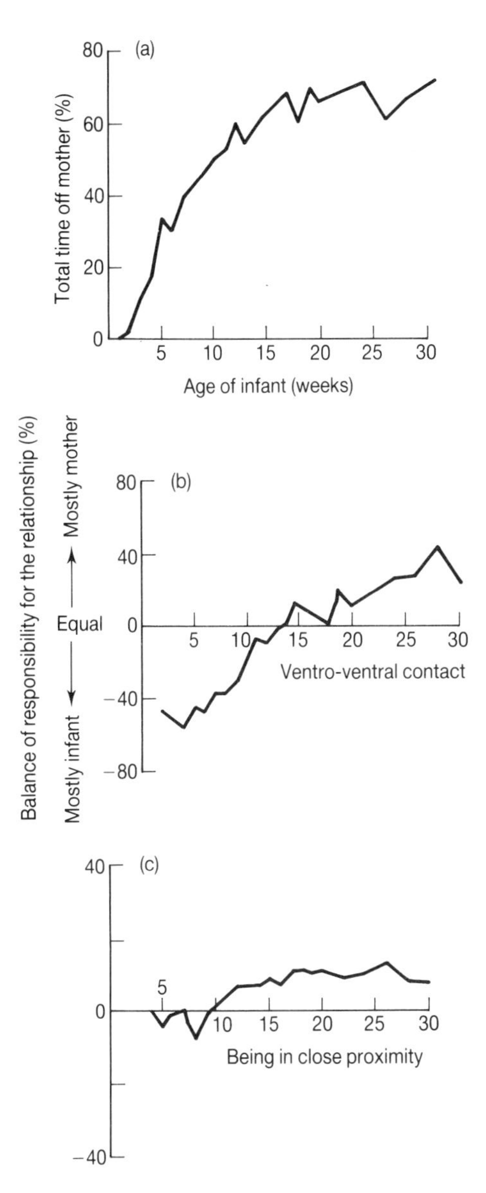

Fig. 2.10 Three facets of the changing mother/infant relationship in rhesus monkeys taken from the work of Hinde's group. (a) The increasing proportion of time infants spend off the mother (i.e. not in contact with her body, although it is often very close) as they grow. (b) and (c) are more complex measures; (b) is the ratio of responsibility for ventroventral contact, i.e. whether mother or infant initiates or terminates bouts of clinging, (c) is a similar measure for close proximity between them—being within arm's length effectively. Further details are given in Hinde 1974.

the rhesus, the Barbary macaque, behaves quite differently. Mothers allow other individuals, males as well as females, to hold their babies even within a few days of birth. Deag and Crook (1971) describe how young babies can be seen moving about in the group not obviously under the care of any particular adult.

Even within a single species mothering styles can vary substantially, depending on circumstances. Altmann's (1980) long term study of baboons describes the behaviour of young adults and suggests how differences in the social, exploratory and feeding behaviour of young animals relate to the type of mothering they have received. Are there then 'good' and 'bad' mothers and ought not natural selection to eliminate the latter? Altmann's conclusion is that, in a varying and uncertain environment, restrictive or permissive mothers may sometimes be 'good' and sometimes not. Perhaps natural variations in the pathways of behavioural development are not, over a lifetime of reproduction, disadvantageous but quite the opposite, a point we have already made (p. 29) when considering variations in the development of females within a mouse litter.

PLAY

For anybody who has enjoyed watching kittens or puppies grow up, the most conspicuous aspect of their behaviour is play (Fig. 2.11). They spend long periods playing with each other and with objects but, although adults may continue to play intermittently, it does tend to fade out with age. It is commonly suggested that play has a role in the development of adult behaviour. Certainly, there is an enormous literature on the importance of play in infancy and childhood for the normal development of human beings (Bruner *et al.* 1976).

In animals we can try to argue from analogy, but with care. If we observe that they spend long periods engaged in a certain activity, we would, as a first assumption, conclude that it is important for their survival. A comparative survey across the group reveals that, apart from a few fragmented anecdotes, there is no evidence of play outside the mammals and a few bird species. Even here it is not universal: for example, young rats show chasing and wrestling behaviour, reasonably described as play, but for mice there are few records. It is among the ungulates, carnivores and primates that play becomes so conspicuous. We observe chases, mock fights, stalking, leaping and all aspects of prey-catching behaviour. Adults may join in and there are sometimes special postures that indicate that an individual

Fig. 2.11 Kittens face and threaten one another in play (drawing by Melissa Bateson).

wants to engage in play (p. 62). Such descriptions give the impression that play is easily recognized and distinguished from 'real' behaviour, but this is certainly not true. It is often impossible to identify play with certainty, particularly as animals get older and we have all observed that animals themselves may have the same problem, because mock fights may change rapidly into real fighting as one partner really gets hurt.

When discussing parental care we had to accept that, although there could be little doubt that the interchange between mother and offspring contributed to the healthy development of the latter's behaviour, it was difficult to identify exactly how it operated. The same problem is met, even more forcibly, in play. As we watch mock fights among kittens or the amazing high speed chases and wrestling bouts of young monkeys, we feel convinced that such behaviour must play a part in the development of skills that are going to become vital for survival in adult life. Apart from the obvious physical skills, numerous other functions have been suggested for play—gaining knowledge of the social group, exploration of the environment, and so on. Yet whatever common sense may seem to tell us, it is extremely difficult to demonstrate convincingly any effects of play itself on subsequent adult behaviour.

Although we may not understand how parental care operates, we can at least demonstrate the effects of its deprivation, as in some of the work just described. However, it is almost impossible to deprive a young animal of the opportunity to play without depriving it of many other things, e.g. social contact, which we already know have drastic results. One natural experiment comes from Lee's (1983) work on vervet monkeys in East Africa. Her observations extended over several seasons and included periods of severe drought. She noted that the play that was such a conspicuous feature of the infants and juveniles during normal seasons, and that occupied a great deal of their time, virtually disappeared during drought. All animals, including the young, managed to survive only by constant searching for food. When Lee compared young adults who had grown up during the dry period—and hence had been deprived of play opportunities—with the rest, she could detect no differences in their behaviour. Baldwin and Baldwin (1973) found a rather similar situation in some troops of squirrel monkeys where there was great variation in the amount of play; in some troops almost no play was ever seen. Nevertheless it was hard to identify corresponding differences in their social behaviour as adults. We shall need many more such observations to obtain a reasonable assessment of the role of play and it is bound to be different in different animals. Fagen (1981) and Martin and Caro (1985) provide good reviews of this perplexing but fascinating field of research.

IMPRINTING

So far, the effects of early experience that we have been considering have been fairly general in their nature, affecting a wide range of behaviour. Sometimes particular early experiences can be shown to have rather clearly defined results. Of course, whether we call the results of an experience 'general' or 'specific' depends to a great extent on how closely we look for effects. It is very easy to label an effect as specific if only a few aspects of behaviour are studied. Nevertheless, some examples are quite striking: playing a tape recording of song to a young bird may affect the song it sings when it becomes sexually mature months later (p. 44) but so far as we know nothing else in its behaviour is changed by this experience. Some of the results of 'imprinting' appear to be equally specific.

Imprinting cannot be precisely defined. As we shall use the term here, it refers to various behavioural changes whereby a young animal becomes attached to a 'mother figure' and/or a future mating partner. It was Lorenz who introduced the topic to most behaviour workers by experiments with geese, in which he got broods of goslings to follow him and treat him as their mother figure. A good deal of the work on imprinting has been carried out using birds like geese, ducks, pheasants and chickens, which have precocial young (i.e. those that can walk at hatching and do not stay in the nest). Imprinting may be measured by the amount of attention paid to the mother, time spent close, latency to approach, time spent following if she moves, and so on. This type of response to the mother figure is usually called 'filial imprinting' to contrast with 'sexual imprinting', by which early experience affects the subsequent choice of sexual partners when mature.

Imprinting usually takes place soon after hatching or birth and often results in a very fixed attachment, difficult to change. Lorenz described it as a unique form of learning that, unlike other forms, was irreversible and restricted to a brief 'sensitive period' just after hatching. It clearly does involve a learnt association between a particular stimulus and a response such as we shall be considering in more detail in Chapter 6, on learning. The context in which we observe it, so early in life, makes the results of imprinting very distinct, but few would now wish to consider the actual process of imprinting as special and different from other types of associative learning. It seems more useful to concentrate on the role that it plays in development. Because imprinting occurs before anything else has been acquired by learning it is often a very clear and identifiable event in development and this has made it extremely useful for studies of the neural basis of learning and memory (Chapter 6, p. 148).

As Lorenz suggested, the range of objects that elicits approach and attachment in young birds is very large. His original observations were most dramatic because he got broods of greylag goslings to imprint on himself. There seems to be no limit to the range of visual stimuli for imprinting. Birds have been imprinted upon large canvas 'hides' inside which a man can move, down through cardboard cubes and toy balloons to matchboxes. Colour and shape seem to be equally immaterial (Fig. 2.12). Movement helps to catch attention but it is not essential—a stationary object will attract young birds provided it contrasts with its background, so will flashing lights. Bateson and Reese (1969) have shown that within a few minutes, day-old chicks and ducklings will learn to stand on a pedal in order to switch on a flashing light, which they then approach. The rapidity with which they acquire this response suggests that the light has reinforcing properties even before the young birds are imprinted to it. They

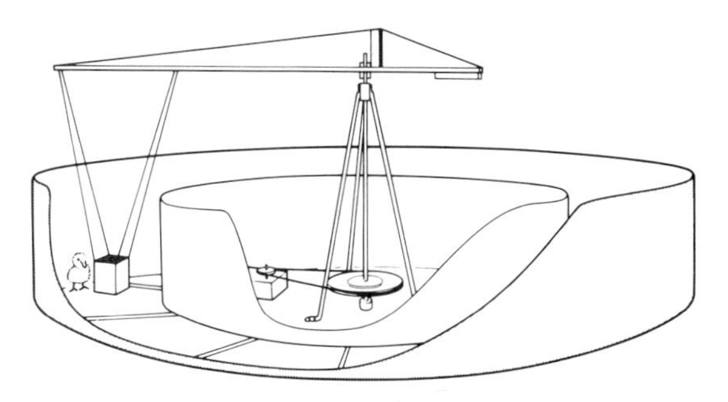

Fig. 2.12 Apparatus used to study the following responses of young birds. Different models can be attached to the long arm, which rotates slowly, moving them around the circular runway (after Hinde 1974).

actively seek such stimulation and such attraction forms the basis for the imprinting attachment.

The young animal comes prepared or biased to respond to conspicuous objects, to approach them, to learn their characteristics, visual, auditory or olfactory, and to form an attachment. This sensitivity can be affected by its early environment. Bateson (1976) discusses a number of experiments showing that exposing chicks to light just after hatching markedly increases their responsiveness to conspicuous objects in comparison to chicks kept in the dark. Just 18 min of constant light before returning them to the dark had effects that could be detected up to 12 h later. This effect is not just due to the light arousing the chicks in a general way, because playing them sounds, or stroking them in the dark, did not make them more responsive to visual stimuli—if anything, the reverse.

This specific arousal of the visual system reminds us of Gottlieb's similar experiments, described on page 23, where the auditory responsiveness of ducklings was heightened by playing them sounds. Auditory stimuli are themselves attractive to young birds and will often enhance imprinting to objects close to their source. Sound is almost essential to induce following in mallard ducklings, for example. Wood-ducks nest in holes in trees, and the young usually hear their mother calling from the water outside the nest hole before they have ever seen her.

Imprinting-like phenomena are clearly involved in the social development of mammals. It is common knowledge that the younger we take and rear a litter of wild mammals the easier it is to tame them. Orphan lambs reared by humans follow them about and often show little attraction towards other sheep. This is not just a filial attachment; a form of sexual imprinting must also occur and zoo authorities know to their cost that hand-reared animals are often useless for breeding when they are mature. Whilst visual and auditory stimuli are dominant in birds, the behaviour of mammals is very much dominated by their sense of smell and it is not surprising to find that their early olfactory experience often affects their choice of a mate. In guinea-pigs, mature animals are more attracted by others whose scent matches that which was present in the nest during the time that they were being reared (Carter and Marr 1970). With mice, the situation is more complex and is greatly affected by the genotype of the animals, but experience of scent is always a key factor on which they base their choice (D'Udine and Alleva 1983; Lenington and Egid 1989).

Imprinting has often been cited as one of the best examples of a 'sensitive' or 'critical' period in development. Our perception of the sensitive period depends to a considerable extent on how exacting are the criteria we use. Taking filial imprinting in birds as an example, if the criterion 'follows a moving object at first exposure' is used, then many precocial birds will satisfy it for 10 days or more after hatching. However, if one takes as a criterion 'the formation of a lasting attachment following a single exposure to an object', then the sensitive period may appear much more limited. Figure 2.13 illustrates some results using both types of criteria. Ramsay and Hess (1954) measured how well mallard ducklings discriminate and stay close to a model 5–70 h after a single 30-min exposure to it. Boyd and Fabricius (1965) simply scored birds that followed a model on their initial exposure.

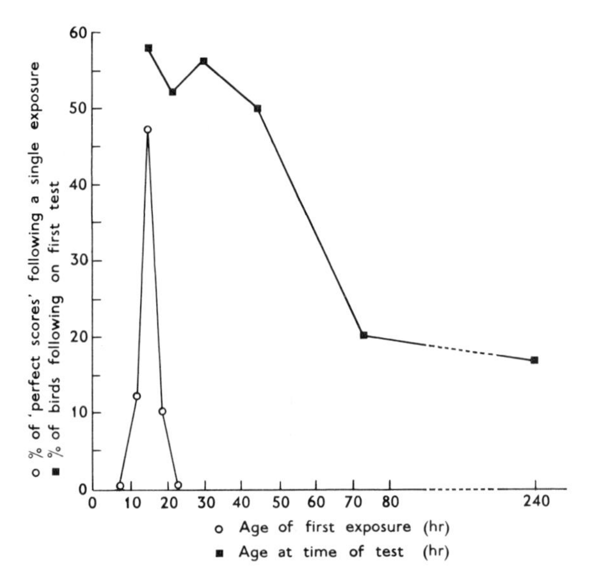

Fig. 2.13 Two ways of expressing the sensitive period for filial imprinting. The line joining the open circles shows the very sharp peak obtained when mallard ducklings were tested for following and discrimination of a moving object after a single exposure to it at different ages. The line joining the black squares shows the much broader peak obtained by scoring the percentage of duckings that follow a moving object in their first exposure to it (data from Ramsay and Hess 1954 and Boyd and Fabricius 1965).

Guiton (1959) proved that the length of the sensitive period was not independent of external factors because, whilst young chicks kept in groups ceased to follow moving objects 3 days after hatching, chicks reared in isolation remained responsive much longer. He could also show that the socially reared chicks did not remain inert and ignore their surroundings until exposed to a moving object: they became imprinted upon one another. This phenomenon may be important in the natural situation. Boyd and Fabricius (1965) point out that mallard ducklings do not normally leave the nest until the second day after hatching, well past the peak of the experimentally demonstrated sensitive period. By this time some of them may be imprinted upon each other. Even if only a few of the brood actively approach and follow the mother bird as she leads them to water, the brood will act as a group and stay together because the rest will follow the maternally imprinted ducklings.

Social rearing restricts the sensitive period and it may be that the imprinting process itself plays a part in ending this period. Bateson (1964) reared chicks in small pens whose walls had conspicuous patterns of horizontal or vertical stripes in black and white or red and yellow. After 3 days of isolation in a pen the chicks were tested in a plain runway with moving objects of similar striped patterns. They preferred to follow the model whose pattern matched that of their rearing pen. In a rather similar experiment, Taylor *et al.* (1969) found that exposure of young chicks to coloured walls in their pen shifted their subsequent choice of colours towards that which was familiar, even if the initial exposure was only 15 min. Such experiments strongly suggest that imprinting involves the young animal learning to discriminate the familiar from the unfamiliar and approaching the former. In the absence of any conspicuous moving object the chick, in effect, becomes imprinted on its surroundings and its responsiveness to new objects declines.

Other workers had laid more emphasis on the growth of fear itself as the cause of waning responsiveness. There is no doubt that escape responses do increase from the first day; newly hatched chicks or ducklings will approach new objects, but by the third day they will flee and crouch. The point at issue is whether the escape tendency grows only when the mother figure becomes sufficiently familiar through imprinting to make other environmental stimuli recognizably unfamiliar, or whether it grows anyway, independently of such experience, as a result of maturation.

Evidence relevant to this issue comes from Moltz and Stettner (1961), who reduced the visual experience of ducklings by rearing them with translucent hoods, which meant that their eyes received light but no pattern. We know from the 'visual priming' experiments of Bateson, described on page 36, that such exposure to light will keep responsiveness high. When tested at 2 or 3 days of age without their hoods, such birds followed a moving box more than control birds, but they also showed less fear. This certainly suggests that previous visual experience of the surroundings does affect the growth of fear and the weight of the evidence now suggests that young birds became afraid once they have formed an attachment to the object that serves as a reference point from which they can discriminate against the unfamiliar. The sensitive period ends when this particular developmental stage has been achieved.

The development of social responses in mammals is also highly dependent on early experience and once again the concept of sensitive periods has been invoked to explain the timing of certain crucial events. Scott and Fuller (1965) summarize the extensive work of their group on dogs (see also Scott 1962). They have found that there is a period from about 3–10 weeks of age during which a puppy is forming normal social contacts. If isolated beyond the age of 14 weeks they no longer respond and their behaviour is very abnormal. A very short exposure at the height of this sensitive period is sufficient for them to form a normal relationship with human beings. Dogs, like some sexually imprinted birds, seem perfectly capable of accepting both humans and their own species as social partners.

Imprinting has always attracted the attention of psychiatrists because there is no doubt that human infants are extremely sensitive to early experience of many kinds. In particular, Bowlby (1969, 1973) developed a theory of the attachment of a baby to its mother, which drew extensively from animal research. In earlier work he suggested that the period from 18 months to 3 years was especially important and that separation from, or lack of an adequate mother figure at this time led to a greatly increased risk of psychological disturbance in adolescence and later life. The idea of a restricted sensitive period in human development with its implicit suggestion of irreversibility has been widely criticized (Clarke and Clarke 1976; Rutter 1979). Nobody would deny that maternal separation in childhood may have long term effects on human children but inevitably the situation is not as clear-cut as with simpler animals.

Separating a human baby from its mother will involve a whole series of changes to its life, almost always deleterious, and it is difficult to link cause and effect very closely. The whole topic is well discussed by Hinde (1974) in his Chapter 13.

There is also considerable argument over the recognition of sensitive periods in the social development of other mammals and papers relating to this dispute are introduced and collected together in McGill (1965; see also Denenberg 1964; Bateson 1979). As we have seen when discussing sensitive periods for imprinting in young birds, it is often very difficult to distinguish between a change in responsiveness that occurs as a result of previous experience and a change involving some kind of maturation. We can certainly point to clear sensitive periods at particular turning points in the behaviour of adult mammals. The time around birth requires a crucial switch in the responses of the mother, who must accept her newborn and start lactation. Rosenblatt and Siegel (1983) discuss the behavioural and physiological events surrounding parturition in rodents. They point out that the high arousal of the female facilitates the development of interaction between her and the newborn. A brief contact at this stage has been found to change the behaviour of females for weeks subsequently, even though there is no further opportunity to interact.

The argument about sensitive periods is not merely academic because it relates to contrasting views of behavioural development. Is it a totally continuous process with interactions between the growing animal and its environment possible at every stage? Or is it more like embryonic development, where we know that certain events must take place within a critical period if they are to take place at all? Once the critical period is past embryonic cells may no longer be competent to respond. The development of behaviour probably exhibits both sets of characteristics and much will depend on a particular animal's life history. With short-lived animals, which must grow rapidly and are sexually mature at a few months of age, behavioural development is highly compressed and events like imprinting come to take on an all-or-nothing, irreversible appearance. (When we come to discuss the development of bird song (p. 41) we shall certainly find evidence of sensitive periods beyond which experience has no effect.) But the infancy of some mammals is very prolonged and development is far more a continuous process—we have only to think of the work with rhesus monkeys that was discussed earlier (p. 32). The behaviour of the young monkey gradually develops over months or even years. Experience has effects at all stages, some transient, some permanent, but rarely appearing as irreversible because there is usually time for later experiences to change the direction of development.

Sexual imprinting

Lorenz found that the early experience of his young geese and ducks affected their choice of sexual partner when they were mature. Subsequent work with a wide variety of birds and mammals has confirmed this (Bateson 1976). As with filial imprinting there seems to be little limit to the range of objects which can provoke attachment. Male and female turkeys have been sexually imprinted on human beings, cockerels imprinted on to cardboard boxes, which they courted and attempted to mount, whilst we have already referred to the sexual imprinting of many species of mammal kept in zoos and artificially reared by human foster parents.

Some of the most complete experimental work has used Estrildine finches, in particular the zebra finch and the Bengalese finch. Immelmann (1972) carried out a range of experiments to investigate the effects of cross-fostering between these two species. He placed a single zebra finch egg in a clutch of the Bengalese finch and allowed the Bengalese parents to rear the whole brood. Subsequently, cross-fostered zebra finch males were isolated until they were sexually mature. Immelmann then gave them a choice between a zebra finch female and a Bengalese finch female. The results were unequivocal: the zebra finch male directed his courtship towards the Bengalese finch female. This preference was all the more striking because when a zebra finch male was put in with the two females the zebra finch female usually responded at once with all the usual conspecific greeting calls. The Bengalese finch female was, at best, neutral and usually showed avoidance as he approached her.

Immelmann found that if cross-fostered males were obligatorily paired with females of their own species most would eventually mate with them and raise broods of young. Astonishingly, such experience did nothing to alter the preference described above. When once again given a choice, the males ignored conspecific females and courted the foster species. Sonnemann and Sjolander (1977) studied female zebra finches similarly reared by Bengalese. Although the results are not so striking as with the males, the females do show a definite preference for males of the foster species and choose to perch close to Bengalese males rather than males of their own species. The strength of imprinting in the finches was not affected by their experience after leaving the nest—as we have seen, no amount of subsequent contact with their own species modified their preference. It is unlikely that sexual imprinting is so strong in all birds, but we lack much comparative evidence. Imprinting of this type certainly occurs in pigeons but note that it cannot occur in the European cuckoo and other parasitic birds, whose young are always reared by foster species.

Immelmann's experiments, in which he varied the exposure of developing zebra finch males to siblings and foster parents, suggest that they imprint more rapidly on zebra finch parents than on foster parents. This may be because they have an inherited responsiveness towards their own species (albeit one that can readily be overridden) and this innate bias has also been suggested for other species. However, Ten Cate (1982) made a close examination of the parental behaviour of both species. He finds that zebra finches interact more actively with their offspring than Bengalese finches and suggests that this increased attention may account for the more rapid imprinting to zebra finch parents. Further, the young themselves are affected by the other young they are reared with. Thus, a single zebra finch foster chick in a Bengalese finch brood is more likely to develop the very strongest preference for a Bengalese sexual partner. One reared with both zebra and Bengalese finch siblings will show a less marked preference (Ten Cate *et al.* 1984). Experiments of this type help to unravel the interactions between inherited tendencies and social experience operating during development, other examples of which we shall discuss further in the section on bird song.

All the evidence suggests that the sensitive period of sexual imprinting is both later and more extended than for filial imprinting. Immelmann found that sexual imprinting is not complete until about 33 days of age, when the fledglings are just able to look after themselves. Keeping males with the foster species until 94 days, i.e. until just before sexual maturity, had no additional effect. Vidal (1980) exposed male domestic chicks to a moving model for a period of 15 days at one of three stages of the young bird's early life—0–15 days, 16–30 days or 31–45 days. They were then kept either with a hen or isolated until full sexual maturity at 150 days, when they were given a series of choice tests. The strongest sexual preference for the object over a normal hen was shown by those birds that had been isolated after exposure from 31 to 45 days, the group that showed the least following or signs of filial imprinting.

Bateson (1979) has argued that it makes sense to delay the onset of sexual imprinting until young birds are beginning to develop their adult plumage because siblings will then become better models. In chickens, Vidal's result fits in with this idea and the timing of the sensitive period in quail and in mallard also agrees. Bateson (1980) has good evidence that a rather detailed and intense form of learning is taking place, which may require a longer period to become established than filial imprinting. He suggests that, during this period, the young bird is learning the detailed characteristics of its mother and siblings as a basis for its subsequent choice of sexual partner. The bird chooses some individual that is neither too similar (and therefore likely to be a close relative) nor too dissimilar. He calls this a choice for 'optimal discrepancy' and he has tested this hypothesis with domestic quail using the apparatus illustrated in Fig. 2.14. Quail show great variation in the details of their head-markings, which has been shown to be genetically based and therefore a good indicator of relationship. Birds are reared in groups and then given a choice between three birds of the opposite sex:

1. A close relative with very similar head pattern to the test subject.
2. A bird with different markings but of the same basic colour.
3. A bird with white plumage.

The choice of which bird to approach was consistently for the second type. Choice for optimal discrepancy could lead to a reasonable balance between avoiding inbreeding and avoiding outcrossing too distantly, with the risk that a genetic constitution that has been selected to match local conditions is broken up (see p. 44 for a further discussion of this point in relation to bird song dialects).

Imprinting is not the only means whereby identification of a sexual partner can be achieved. We have already noted that brood parasites like the cuckoo must take a different developmental path. Yet Bateson's hypothesis does help to explain how

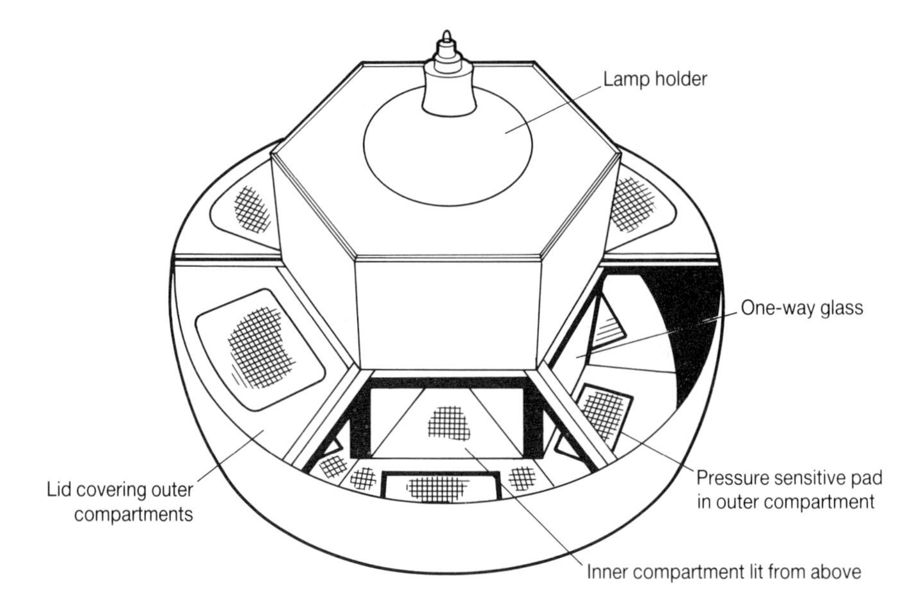

Fig. 2.14 Apparatus used for testing courtship preferences of adult Japanese quail. Stimulus birds of various types are put singly into the inner compartments, which are brightly lit from above. Their outward-facing windows are fitted with one-way screens, so that although the stimulus birds cannot see into the unlit outer part of the apparatus, the bird making its choice can see through into each compartment. A sensitive platform in front of each window records automatically when a bird is standing on it. Those familiar with that city will understand why Bateson called this 'The Amsterdam Apparatus' (after Bateson 1983).

sexual imprinting, with its gradual learning of species and individual characteristics, may have evolved. Imprinting also removes any constraints on the rapid change of plumage characters because the learning system can immediately follow, something that would be very difficult for a system that involved inherited responsiveness to one plumage type. Filial imprinting is more rapid and probably less exact, although we lack much hard evidence of this. It is adequate to keep the young animal close to its mother and siblings, something that is essential immediately after birth, because most animals discriminate against and kill strange young that approach them (Bateson 1979).

THE INTERACTION OF INHERITED TENDENCIES WITH LEARNING

Imprinting has just provided us with an example of how an inborn tendency to approach is linked with learning of the characteristics of the conspicuous object. We have also mentioned previously how a relatively crude basic pattern such as bird flight may mature in the absence of practice but the finer skills of flying are added later with practice.

Eibl-Eibesfeldt (1970) describes a number of similar examples from his studies of young mammals. He reared young polecats in isolation, with no opportunity to catch live prey. Such animals showed varying degrees of interest when first presented with a live rat, but no hint of attack unless the rat ran away from them. If it did so, a polecat would instantly pursue and seize the rat in its mouth, usually shaking it rapidly in a characteristic way. Their bites were at first badly orientated, but after a few trials they were seizing prey by the nape of the neck and killing with a single bite. Clearly there are inherited components to the killing pattern that are completed by learning. Eibl-Eibesfeldt suggests that normally polecats pick up the necessary practice during play sessions with their litter mates. In a similar series of observations with hand-reared squirrels, he has shown that, although they respond to nuts and try to open them on the first exposure, their efforts are uncoordinated. They have to learn to direct their gnawing to the thinnest portion of the shell and to confine it to this one area.

Some of the most beautiful examples of the

dove-tailing of inherited and learnt components during development come from studies of bird song. Song is just like any other behaviour pattern—a controlled sequence of muscular activity which, in this case, we perceive as sound. Detailed analysis of song is in fact easier than for most motor patterns because it can be recorded and then played into a sound spectrograph, which yields a chart on which marks of different densities show how much energy was emitted at the various sound frequencies over time. Some examples of sound spectrographs of bird song are shown in Figure 2.15. With practice they can be interpreted rather like a musical score and they show up variations in sound pattern that are hard to detect by ear alone.

Bird vocalizations exhibit an astonishing variety from the briefest of call notes through to elaborate songs lasting several minutes. Song, as we popularly understand it, is characteristic of one of the most abundant and advanced groups of birds, the Oscine families of the order Passeriformes or perching birds (see Campbell and Lack (1985) for a concise account of this group). This order includes some of our most familiar birds, the finches, warblers (both Old World and New World) and thrushes. The ways in which they acquire their songs have proved to be a rich source of material for investigation. They show great diversity and provide us with opportunities to study some important functional and evolutionary aspects of behavioural development. Reviews by Catchpole (1980) and Slater (1983, 1989) provide an overview of this rapidly growing field.

The songs that have proved most useful in developmental studies are those of intermediate complexity, such as those of the chaffinch illustrated. Chaffinches produce a 2–2.5 s burst of sound consisting of several sequences of repeated notes of different types. As we listen to chaffinches singing we notice immediately that individuals sing several variants of the song—all clearly recognizable as belonging to that species but differing in detail. Neighbouring males often have several songs in common but each bird will usually have its own individual mix of song types (Fig. 2.16 illustrates

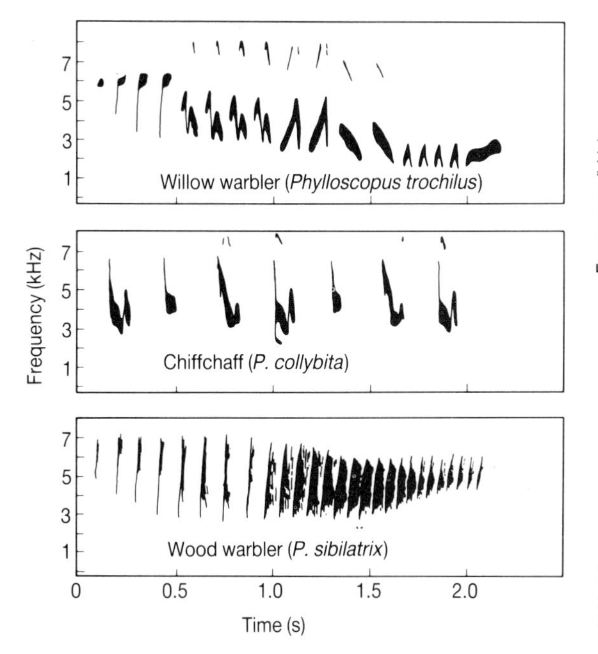

Fig. 2.15 Sound spectrographs of three closely related species of European warbler. The three are very similar in plumage and share the same general habitat, hence the songs probably serve, in part, as a mechanism of sexual isolation, see page 118. These warblers were first clearly distinguished as separate species by the eighteenth century English naturalist Gilbert White, who published *The Natural History of Selborne* in 1788. He separated them by their songs.

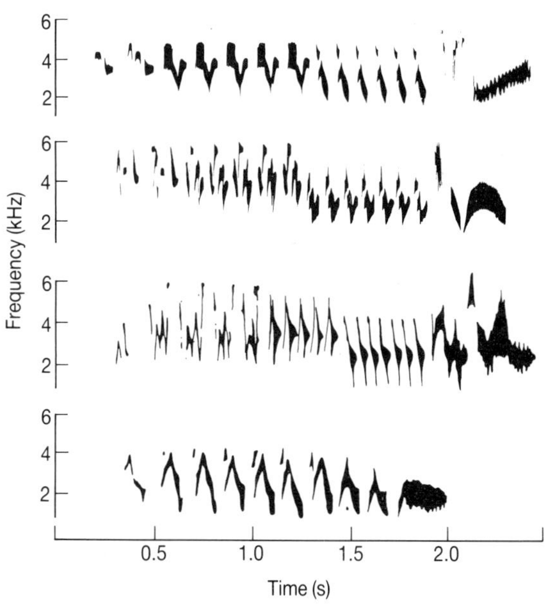

Fig. 2.16 At the top, three examples of chaffinch song. They share the same basic pattern but are very different in detail although they are all unmistakably chaffinch to the human—and certainly to the chaffinch—ear. One male will commonly have several such variants in his repertoire and share some of them with birds in neighbouring territories. The lowermost song is that sung by a chaffinch brought up from the chick stage in complete auditory isolation from others. It has chaffinch timbre and rhythm but is obviously far simpler in form and, in particular, lacks the 'terminal flourish' so typical of the normal songs (from Ince *et al.* 1980 and Thorpe 1961).

some of these variants from a study by Slater and Ince 1979).

Such variability raises intriguing developmental questions; what are its origins? It has been known for some time that chickens and pigeons produce all their characteristic calls when isolated and even if deafened. Could the song of chaffinches be similarly resistant to environmental influences; could the variants even be genetically based?

To answer these questions Thorpe and his colleagues (Thorpe 1961) took young birds from the nest immediately after hatching and reared them alone in sound-proof chambers. The following spring the males began singing in isolation; their song was recognizably chaffinch but was much simpler in form and showed little variation, either within or between individuals (Fig. 2.16). Following our discussion of the validity of isolation experiments on page 22, we may note that here such an experiment gives a very clear answer to a first stage question. The different song variants are not genetic but due to some influence from outside. It was an obvious guess that this external source was the song of adult birds that the young bird heard both in the nest and early in its first spring, and this proved to be the case.

Work by Marler and Tamura (1964) with a relative of the chaffinch, the American white-crowned sparrow, allows us to go into more detail. This small finch has a wide range on the Pacific coast and, like the chaffinch, its songs vary. But they vary in a more systematic way in that birds in a given geographical area tend to sing similar song variants so that there are clear local 'dialects' (Fig. 2.17). If young males are taken immediately after hatching and isolated, as the chaffinches were, then no matter which region they come from, they all eventually sing very similar and simplified versions of the normal song. Obviously they must pick up the local dialect by listening to adult birds and modifying their own simple song pattern accordingly. Marler and Tamura found that this learning process usually takes place during the first 50 days of life—before the bird has ever sung itself. Males captured in their first autumn and reared alone begin to sing for the first time with a recognizable version of the local dialect. There are clear indications of a sensitive period for learning dialect characteristics. Up to 50 days of age, isolated males can be 'trained' to sing their own or other dialects by playing to them tape-recorded songs, although the results of such training do not show until the birds begin to sing themselves some months later. Beyond about 50 days the birds are unreceptive to any further tape training and their songs, when they begin singing, are not affected. Here, then, we appear to have a simple, inherited song pattern that is sensitive to modification by

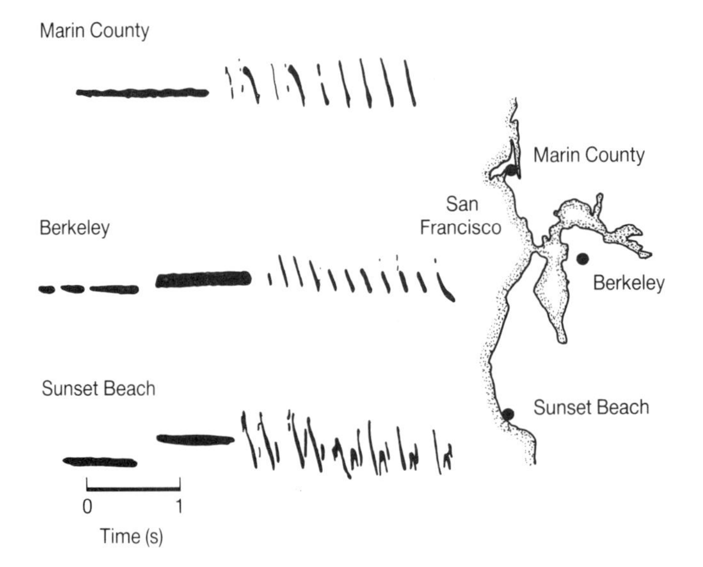

Fig. 2.17 The song 'dialects' of the white-crowned sparrow around San Francisco Bay, USA. The examples given are typical of their area and, unlike the chaffinches of Figure 2.16, there is much less song variation within geographical areas than between them. The use of the term 'dialect', as in human geographically-based dialects, seems justified (after Marler and Tamura 1964).

learning but only during early life. The young birds 'carry' the memory of the songs they hear and reproduce them when they first sing.

Experiments by Konishi (1965) take the analysis further and enable us to qualify the conclusion that simple song is inherited. If a young fledgling is deafened just as it leaves the nest it will subsequently sing, but it produces only a series of disconnected notes. These are quite unlike the song of isolated birds which, although simple in form, would still be recognizable as 'white-crowned sparrow' to an ornithologist. The bird has to be able to hear itself in order to produce this simple song pattern. In other words, it would be more accurate to say that it is not the capacity to produce the simple song that is inherited, but rather some kind of neural template representing this song, against which the bird matches the notes that it produces and adjusts them to fit. It requires auditory feedback if it is to realize this inherited potential.

In some birds, such as the chaffinch, the way the song develops within an individual perhaps reveals this feedback control in actual operation. They begin the breeding season singing a 'subsong', a soft rambling pattern of notes varying in pitch and length (Fig. 2.18). As this is repeated the notes become louder and less variable, always moving towards the final pattern, which presumably corresponds to that held in the template.

Konishi found that if he deafened young white-crowned sparrows after they had been 'trained' with normal song, but before they had themselves sung, their subsequent song resembled that of birds deafened as fledglings. They need to hear themselves in order to match the song they produce with that stored in their memories. Presumably the songs they have heard as fledglings modify their inherited template so that it conforms to the more complex characteristics of normal adult song. Once the birds have matched their own output with this, and sung the adult song, they can go on singing normally, even when deafened. At this stage song development comes to an end in the white-crowned sparrow; after its first spring the bird is no longer susceptible to further experience and keeps much the same song pattern for the rest of its life.

The results of these experiments are summarized in Figure 2.19. One final experiment illustrated there needs to be mentioned. Playing tape recordings of other bird species to the young white-crowns during their sensitive period has no effect, and the

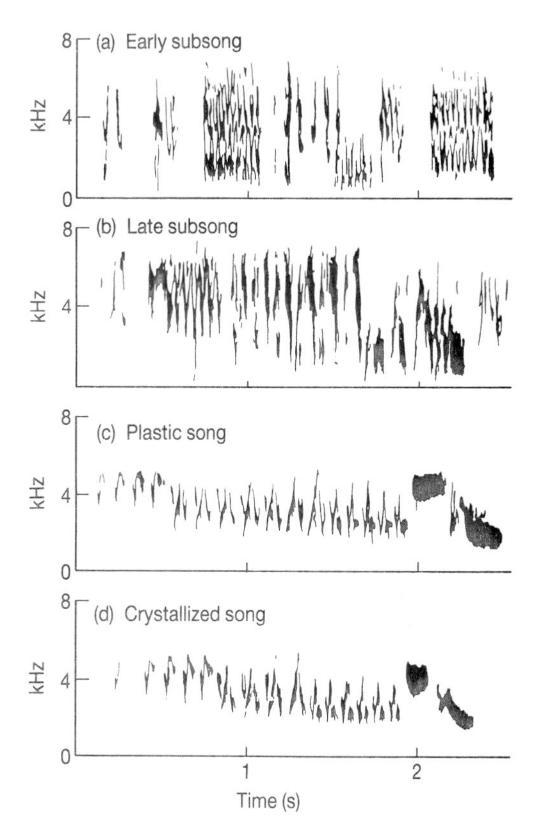

Fig. 2.18 The development of song within a young chaffinch in his first breeding season. Subsong is quiet and rambling, often quite lengthy. As the bird continues, some signs of phrasing are heard and louder and more structured songs begin—so-called plastic song. It justifies this name because it precedes a final version, which is completely structured with fixed length and pattern. After this stage, no further change takes place and the song is said to be 'crystalized'. Each male usually develops two or three different crystalized songs (Fig. 2.16) and, after its first season, does not change this repertoire, although early in spring it will run rapidly through these preliminary stages before crystalizing again into the familiar forms. Note how subsong is completely different from the song of isolated birds (Fig. 2.16, bottom). This latter is preceded by subsong during development and is, itself, a crystalized song (from Catchpole 1980).

birds' subsequent songs sound like those of isolated males. It appears that the young birds are highly selective in what they will learn. Young chaffinches react similarly and Thorpe (1961) points out that this can scarcely be a result of limitations in its sound-producing organ, or syrinx. The related bullfinch and greenfinch have syrinxes of almost identical structure. They have only a poorly developed natural song but are good mimics and will learn to reproduce the songs of many other birds. Some clue to the chaffinch's limitations is provided

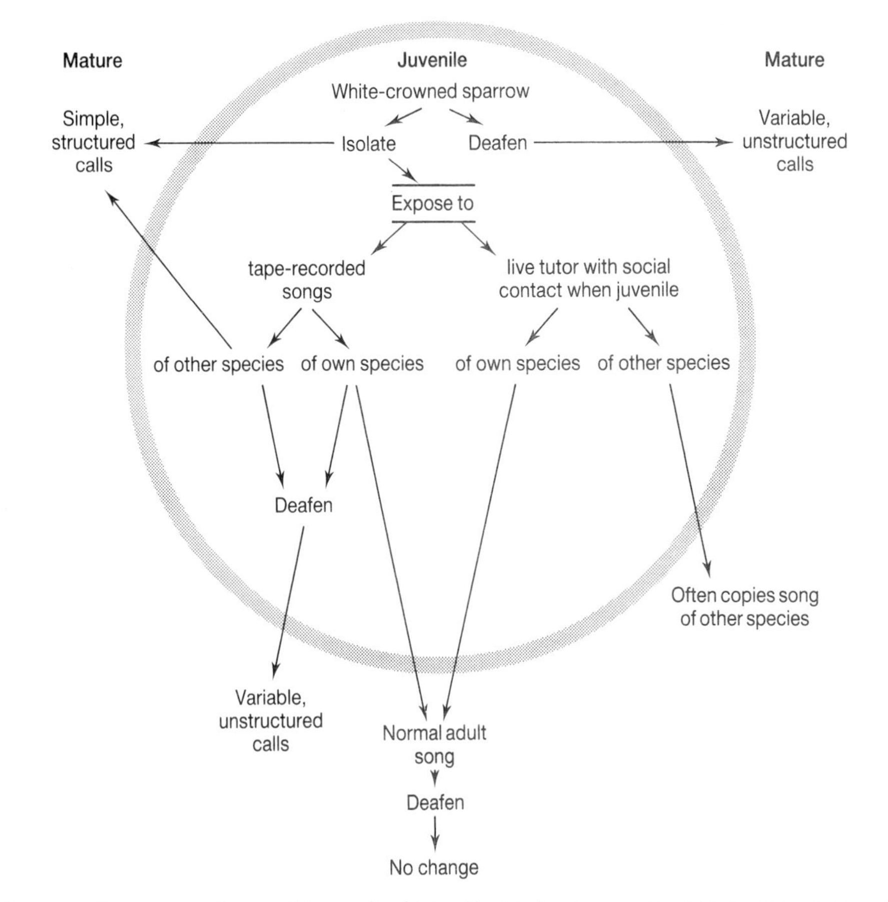

Fig. 2.19 A diagrammatic summary of some of the results obtained by Marler, Tamura, Konishi, Baptista, Petrinovitch and others working on song development in white-crowned sparrows. Events inside the circle represent the juvenile phase before the bird has begun to sing itself. The results when it is mature and is singing are shown outside the circle.

by the song of the one species it will learn to imitate reasonably well from tape recordings. This is the tree-pipit, whose song to the human ear also has a chaffinch-like tone, although it is very different in pattern. The chaffinch may have an inherited tendency to single out this tone from all others and to reproduce the pattern of notes. Marler (1984) and his collaborators have analysed in great detail how, in two closely related species—the swamp sparrow and the song sparrow—young birds hearing tapes for the first time will unerringly pick out the rhythms of their own species' song to copy.

Certainly, as Kroodsma (1982) points out, such selectivity must be important in the song development of many species. A young male will have to concentrate on the song of its own species, often selecting it from a variety of alternatives that can be heard during its sensitive period. Until recently this ability was commonly assumed to be another example of an inherited predisposition to recognize and learn particular things, in this case the characteristics of one's own species song dialects. Recent research suggests that, rather than thinking solely in terms of inherited responsiveness, we should also examine the nature of the learning situation.

Tape recordings offer only a sound stimulus and, as we have just described, they are remarkably effective within certain limits. But provide a young bird with a live song-tutor in the same or an adjacent cage and these limits are easily extended. For instance, isolated white-crowned sparrows have been shown to be capable of learning from live tutors well beyond the 50–55 day sensitive period for tape-recorded tutoring and, more strikingly, they will acquire the songs of other species of caged birds close to them and in sight of them, e.g. the Lincoln sparrow and the strawberry finch. They will certainly not learn these songs from tapes. Nor are such results confined to aviaries because wild white-crowns have been recorded singing Lincoln sparrow song in areas in California where the two species coexist (Baptista and Petrinovitch 1984, 1986; Baptista and Morton 1988). Baptista and his colleagues suggest that the social interactions between tutor and pupil stimulate learning and that aggressive, stressful interactions, which are often seen in aviaries and in the wild between adults and young birds of the same species, may be the most effective.

Social interaction, as a means of focusing attention and thereby affecting the course of behavioural development, is likely to be an important general concept. We have already mentioned Ten Cate *et al.*'s (1984) work on imprinting between zebra and Bengalese finches, where the amount of attention given to nestlings affected their choice of sexual partner. High arousal directing the attention of young animals will accelerate learning in all kinds of social situations. Bird song development is but one example of what we may reasonably call 'cultural transmission', widespread among mammals, where young animals acquire behaviour from observing and copying adults. We discuss this, and its evolutionary implications, further in Chapter 5.

There, and in Chapter 6, when discussing learning in general, we recognize that young animals do not come to new situations with, as it were, a blank mind. All through this chapter we have been emphasizing the interactions between information provided in the genes and that acquired by diverse pathways from the environment, both physical and social. The title of a stimulating review by Gould and Marler (1987) on this topic is a good note to close on—they call it *Learning by Instinct*.

3 STIMULI AND COMMUNICATION

Animals usually respond appropriately to the world around them: a predator elicits quite different behaviour from a mate or food. The first stage in understanding this ability is to discover how the information about the external world is interpreted and used by an animal to give it a representation of what is going on around it. Not all objects or events in the environment are responded to. Some of them are not even detected; others may be detected but have no significance for the animal. Those that are both detected and responded to are called 'stimuli' and can be thought of as goads that ultimately press the animal into action. An example would be the sight of a hawk flying overhead being the stimulus for a small bird to give an alarm call.

Action, however, need not be immediate. Some stimuli have a delayed action and their effects seem to accumulate gradually, changing the responsiveness of the animal over a period of time. The courtship of male doves leading to induction of hormone secretion in the female, which we discuss in more detail in Chapter 4, is a good example of this. The sight of a male dove courting leads to hormonal changes in the female that make her more ready to take part in nest-building, but the stimulus of the male's courtship is only effective if it is repeated many times. One bout of courtship has little effect. It requires several hours of courtship behaviour by the male, usually spread over a number of days, to bring the female into reproductive condition.

Yet another effect that stimuli have on animals is to orientate their behaviour in particular directions. Animals respond almost all the time to the basic physical qualities of their environment—light, gravity, air, water currents and so on—and position themselves to be in a correct relationship to them. Blowfly maggots crawl directly away from light when they move out of their food source to pupate. Fish rest with their heads facing into a water current. From an early age, young rats, even in total darkness, show a 'righting response', which keeps them upright with respect to gravity. Every animal species provides us with such examples.

Orientation can sometimes involve some very complex mechanisms of perception and control. The indigo bunting (*Passerina cyanea*) is a small bird that breeds throughout the eastern United States. In the autumn, the birds fly up to 2000 miles to winter in the Bahamas and Central America. Emlen (1975) showed that the birds use the patterns of stars in the night sky to tell them where 'north' is and to guide them on their migratory routes. He placed indigo buntings in a planetarium where he could control the pattern of stars they saw. By the ingenious method of making them stand on an ink pad and putting blotting paper up the sides of their cage (Fig. 3.1), he could record the direction in which the birds' movements were orientated by the pattern of inky footprints on the blotting paper. By selectively eliminating different parts of the night sky in the planetarium, he was able to show that, as long as the birds could see the pole star and some of the stars within 35° of it, they orientated as they would under a real sky, that is, northwards (towards it) in spring

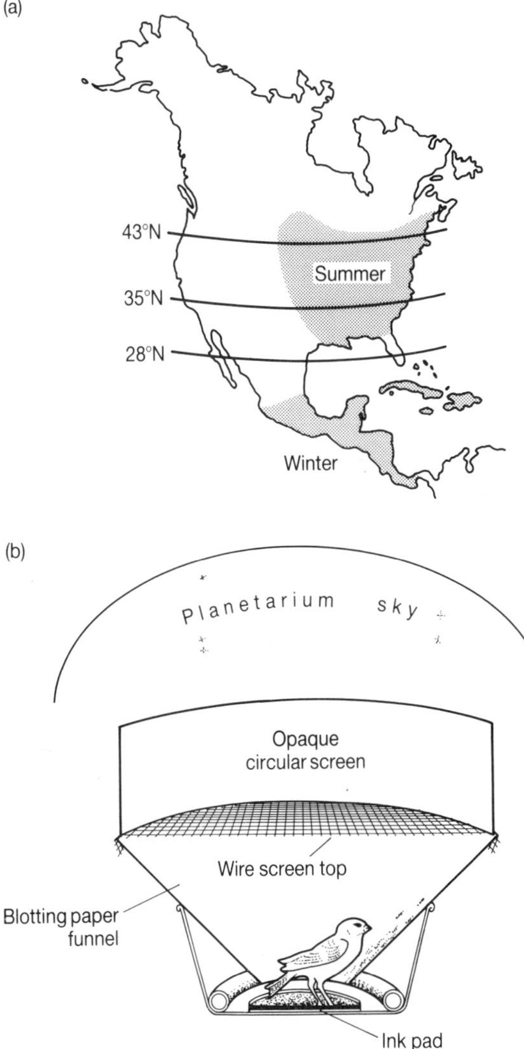

Fig. 3.1 (a) The Indigo bunting's summer and winter areas. (b) The system used to test the buntings' responses to a planetarium sky. The birds stood on an ink pad and left a record of their activity on the blotting paper funnel (from Emlen 1975).

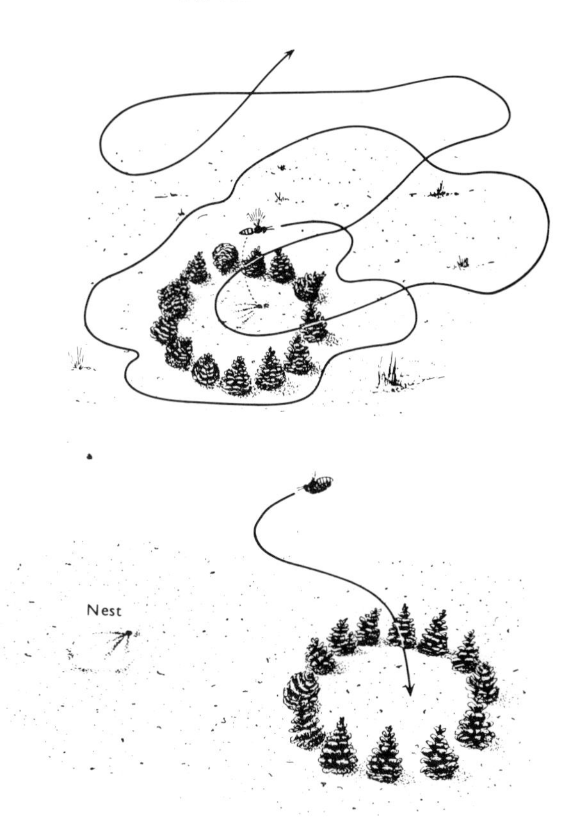

Fig. 3.2 The female digger wasp, *Philanthus triangulum*, builds a nest burrow in sand. Whilst she was in the burrow, Tinbergen placed a circle of pine cones around the entrance. When she emerged, the wasp reacted to the new situation by a wavering orientation flight (upper picture) before flying off. Returning with prey (lower picture), she orientated to the circle, although it had been moved during her absence (from Tinbergen 1951).

and southwards (away from it) in autumn. Whether the birds orientated north or south depended on their own hormonal state, which was in turn dependent on changing daylength.

Orientation can also refer to the way in which stimuli control responses to more transitory features of the environment. Drakes direct their courtship displays according to the position of the ducks on the water. Some displays are given when lateral to the female, others when the male is directly in front of her. The experiment shown in Figure 3.2 shows how a digger wasp, *Philanthus*, uses landmarks such as pine cones to orient and find its burrow. The female wasp digs the burrow and provisions it with insects for her larva. She makes several trips to do this and as she leaves the nest on a foraging trip, she learns some of the landmarks surrounding it. If these are moved while she is away she still uses them to orientate, or rather misorientate, her response on the return trip.

Often it is not really possible to make a sharp distinction between the immediate (trigger), the arousing and the orientating effects of stimuli because they overlap. The stimulus of suckling by the young 'triggers' the final let-down of milk from a female rat, although the prior growth of the mammary glands is accelerated by the cumulative 'arousal' action of the pregnant female grooming and licking her nipples.

Whichever way we classify the modes of action of stimuli, we have to remember that the context in which a stimulus occurs will affect an animal's response to it. A male bird often shows heightened aggressiveness when at home on its territory and may show a quite different response to another bird if it encounters it inside rather than outside its territory. Here context alters motivation. As we shall discuss in this and the next chapter, stimuli and motivation almost always interact to determine the nature and strength of an animal's response to changes in both its internal and external environment.

DIVERSE SENSORY CAPACITIES

Every animal inhabits a world of its own whose character is largely determined by the information it receives from its sense organs. As observers and interpreters of animal behaviour, we would be greatly handicapped if we had to rely solely on the evidence of our own senses. Modern instruments enable us to explore the worlds of animals whose sensory capacities may be very different from our own.

For example, the main visual receptors of insects are their compound eyes, whose construction and properties are very different from those of vertebrates. They provide poor image formation but are excellent for detecting movement and often have a very wide field of view. All insects so far tested prove to have colour vision but their sensitivity to light is shifted towards the short wave end of the spectrum compared to ours, i.e. they can see into the ultraviolet range. On the other hand, with a few exceptions, insects cannot see red as a colour. They confuse it with black or dark grey.

Consequently, bees foraging on flowers may respond to colours and patterns that are invisible to us (and vice versa). Bees fly to the flowers of white bryony (*Bryonia dioica*) because the petals reflect large amounts of ultraviolet, although they appear to us as pale green and provide little contrast with the leaves. Interestingly, not all the flowers that appear plain white to us do so to the bees. Many flowers have 'nectar guides'—radiating lines on the petals that help the insect find the centre of a flower where the nectar is. Some flowers appear, to us, to lack nectar guides altogether, but photographing them through an ultraviolet system (Fig. 3.3) shows that they do have patterns, adapted to the eyes of bees, not humans. Birds, too, are able to see ultraviolet light and the feathers of several species, including gulls and owls, have recently been shown to reflect light in the ultraviolet region of the spectrum (Burkhardt 1989). This raises the intriguing possibility that what we see as a dull or inconspicuous bird may have bright 'colours' to other birds.

Fig. 3.3 Flowers seen in two different lights. The upper picture records what the human eye can see: the flower appears to have no nectar guide patterns. Below is the view that approximates to what a bee sees, photographed through an ultraviolet sensitive system: a striking colour pattern is revealed (photograph by Thomas Eisner).

Another visual faculty outside our own that bees possess is their sensitivity to the plane of polarization of light. Light that reaches us from areas of blue sky is vibrating predominantly in one plane. The angle of this plane changes in a regular fashion with respect to the sun and von Frisch (1967) showed that bees can use this to locate the sun's position, even when it is obscured by clouds or a screen. Birds also appear to be sensitive to the plane of polarization of light.

In the realm of hearing, too, we find some animals possessed of faculties far beyond our own and only detectable with sensitive instruments. Griffin (1958) and others have shown how bats use a 'sonar' system to detect obstacles and flying insects even in complete darkness. They emit a rapid train of very high frequency (50 kHz and upwards) sounds, completely inaudible to human ears, and then listen to the echoes that bounce off objects in the environment. They can tell not only where the objects are, but how big and what sort of things they are too. Dolphins can do the same in water, using even higher frequency sounds (up to 300 kHz) to build their 'sound pictures'. Fish possess a 'lateral line' organ—a row of pressure sensors down each side of the body—that allow them to pick up mechanical vibrations in the water that, in turn, give them quite detailed information about the presence of other fish in the school, predators and so on.

Smell is a faculty that is much more highly developed in some animals than it is in humans. Male silkworm moths, *Bombyx*, are able to pick up the scent of a female when it is diluted to the point of 100 molecules of sex odour substance per cubic centimetre of air. Dogs can smell a variety of substances at concentrations one thousand to one million times lower than humans can. Hurst (1989) describes the complex communication network, based entirely on smell, that exists among house mice. Without necessarily even encountering each other directly, mice can obtain information about the sex and status of other mice from the scent left behind in their urine.

Sixth and even seventh senses are a reality for some animals. The electric fish investigated by Lissmann (1963), Heiligenberg (1977) and others live in very turbid waters where vision is not much use. Instead, the fish set up an electric field around themselves, using especially modified muscle tissue to generate pulses of electrical energy. Distortions in this field tell them that there are objects nearby. Blakemore (1975) showed that the earth's magnetic field can be detected by certain bacteria, which contain pieces of magnetite. Wiltschko and Wiltschko (1988) and others have shown that pigeons and some migratory birds also seem to be using magnetism to orient by, particularly if other cues are unavailable to them.

Information about what stimuli animals are capable of detecting is an essential preliminary to any thorough study of their behaviour. However, even if we know an animals' sensory capabilities in some detail, this alone will not tell us which stimuli it will actually react to. Garcia *et al.* (1966) have shown that rats use taste rather than auditory or visual stimuli in learning that certain food substances make them feel ill, but that they rely on auditory or visual, and not taste, cues when they learn to avoid an immediately unpleasant event such as an electric shock (see Chapter 6).

SIGN STIMULI (KEY FEATURES)

Quite frequently we find animals responding to only one special part of the array of stimuli presented to them. Dragonflies attempting to lay eggs on the shiny metal of a car or the male robin described by Lack (1943), which flew down and attacked a bunch of red feathers on the lawn, are two examples of animals resonding to 'sign stimuli': the animal responds to restricted aspects of its environment, apparently ignoring others. There are some dramatic examples of this. Turkey hens breeding for the first time will accept as chicks any object that makes the typical cheeping call. On the other hand, they ignore visual stimuli in this situation and deaf turkey hens kill most of their chicks because they never receive the auditory sign-stimulus for parental behaviour (Schleidt *et al.* 1960). Roth (1948) describes how male mosquitoes respond with high selectivity to the sound of their females' wings, which beat at a characteristic frequency, different from their own. The males can, in fact, be attracted to a tuning fork vibrating at the correct frequency. It is the delicate hairs on the male's antennae that respond to the sound. Roth suggests that their physical structure is such that they vibrate most readily at the female's wing beat frequency, just as a tuning fork will vibrate if its own frequency is sounded near by.

The calls of male green tree frogs have two peaks of sound energy—a low one at 900 Hz and a high one at about 3000 Hz (Gerhardt 1974). The ears of the females are tuned to pick up these two frequencies in particular: in the female's auditory system, the so-called 'amphibian papilla' is most sensitive to frequencies between 200 and 1200 Hz while the 'basillar papilla' responds best to sounds of about 3000 Hz (Capranica and Moffat 1975). Narins and Capranica (1976) showed a remarkable sex difference in the hearing of a related species of tree frog, which accounts for the fact that males and females respond to different sign stimuli in the call of the male—a double 'co-qui'. By playing back tape recordings of the males' call to the frogs, Narins and Capranica were able to show that males (who attack other males that approach them) respond only to the 'co' note, and females only to the 'qui'. Neurophysiological recordings made it clear that each sex heard only the note of relevance to itself because the neurons of the inner ear were tuned differently for males and females.

Minnows have an extraordinary sensitivity to chemicals from their own species. If a minnow is scratched or wounded in any way, so that some of its blood gets into the water, other minnows show panic flight. They show far less fear when the blood of other types of fish is shed. We have already mentioned the extreme sensitivity of the male silkworm moth to odour produced by the female, but the male moth is not particularly sensitive to odours in general. The scent receptors on his antennae are highly specific to the pheromones released by the female and show little or no response to other chemicals (Schneider 1966).

Ewert and Traud (1979) provide a particularly good example of how selective responsiveness to key stimuli at a behavioural level can be related to underlying neuroethological mechanisms. In response to real snakes, toads fill their lymphatic sacs with air, making their bodies appear much larger than they really are and assume a stiff-legged posture, presenting their flank to the snake (Fig. 3.4(a)). Ewert and Traud found that the toads would perform this stiff-legged snake posture to quite crude dummy snakes made of flexible pieces of cable (Fig. 3.4(b)). Even a model consisting of just a horizontal piece, with a little head attached produced almost as much (93 per cent) response as a real snake.

Ewert and his colleagues have analysed in some detail the sign stimuli that toads use for recognizing both prey and enemies. In both cases, movement and contrast with the background are important, but whereas a toad will turn towards a small, horizontal worm-like object moving in a horizontal plane (prey stimulus), it will not turn towards the same object orientated vertically with the longitudinal axis perpendicular to the horizontal direction of movement (Fig. 3.5). In the toads' view, worms do not walk on their heads! Large, expanding (looming) objects are sign stimuli for escape and a toad will turn away from or freeze to a very crude cardboard model coming towards it.

It is possible to link the toad's responsiveness to prey and enemy to cells in its retina and even to specific parts of its brain. A plan of the amphibian retina is shown in Figure 3.6 and, although it is possible to record the electrical activity of all types of retinal neurons, the most revealing have been made by placing electrodes in the ganglion cells, because these turn out to be of several distinct sorts, distinguished by how they respond to different visual stimuli. In a pioneering paper called *What the Frog's Eye Tells the Frog's Brain*, Lettvin *et al.* (1959) showed

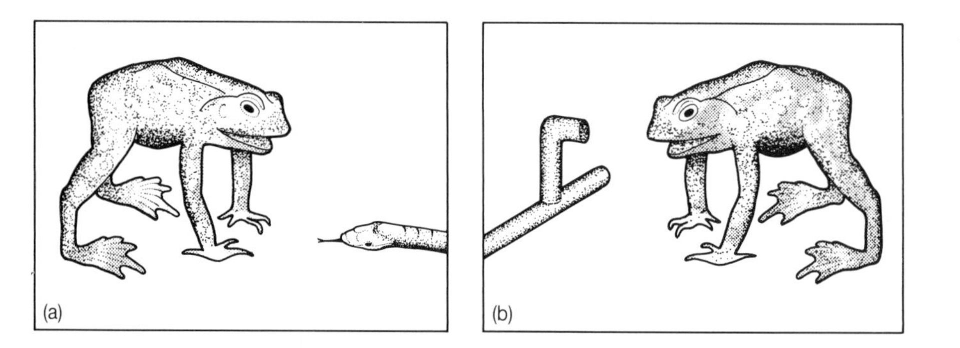

Fig. 3.4 Stiff-legged posture of a toad (a) to a real moving snake and (b) to a moving model made from electric cable (from Ewert and Traud 1979).

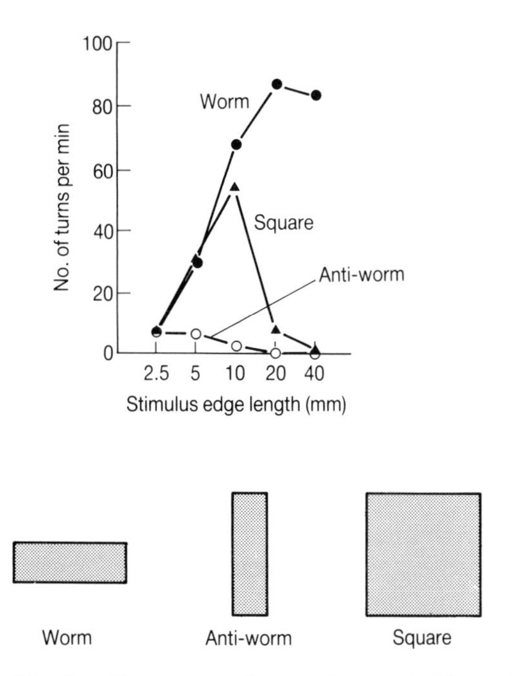

Fig. 3.5 A toad's response to three moving models. The toad turns to follow a model and its response is measured by the number of times it turns to follow the model in 1 min (from Ewert 1980).

that there were six different sorts of ganglion cells, with four being most common, and this has subsequently been confirmed for toads. Class 1 (rare in toads) and Class 2 retinal ganglion cells show a vigorous response whenever the animal is looking at a small moving object. The retinal ganglion cells have a curious property that explains how they can 'recognize' when an object is small and moving. Each retinal ganglion cell responds to a particular section of the total visual field. Thus, any given cell will not respond every time the toad looks at an object but only if the image of that object falls on a particular part of the retina, known as the 'receptive field' of that cell. The receptive field of each retinal ganglion cell is divided into two parts (Fig. 3.7): an inner excitatory centre and an outer inhibitory ring. This means that whenever an object is picked up by the inner part of the receptive field, the retinal ganglion cell is excited and fires, but if the object falls on the outer part of the receptive field, the retinal ganglion cell's activity is **inhibited** (Fig. 3.7). The different classes of retinal ganglion cells have different sizes of excitatory receptive fields, with the Class 1 and Class 2 cells having smaller fields (2°–5°) than Class 3 (6°–8°) or Class 4 cells, which have the largest fields of all (10°–15°) and do not respond at all to a small object unless its visual angle exceeds about 5°.

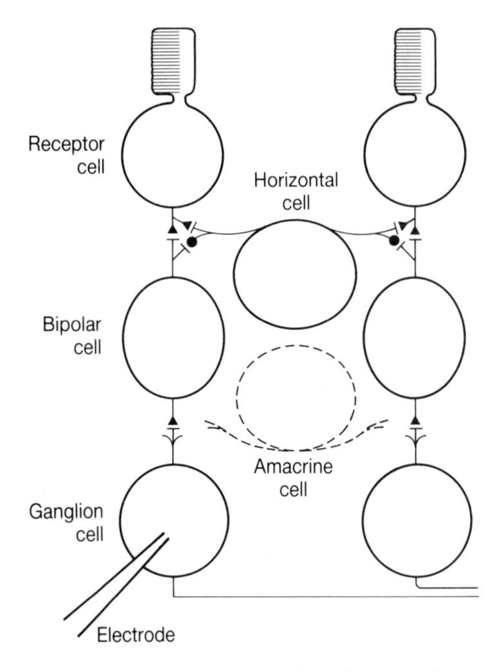

Fig. 3.6 Diagrammatic representation of an amphibian retina showing the connections between the various sorts of cells. Excitatory synapses are shown by —◄, inhibitory synapses by —●.

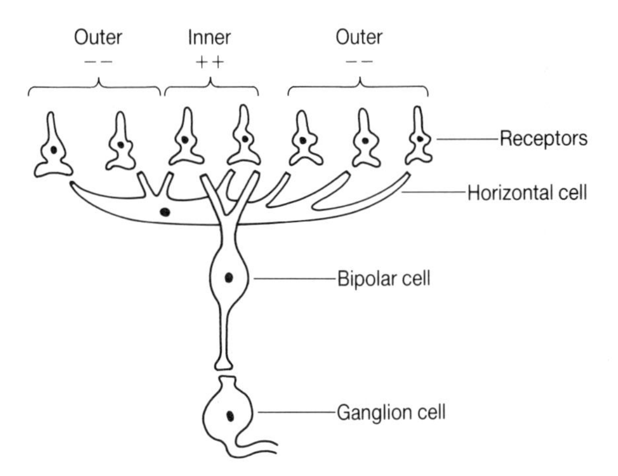

Fig. 3.7 Diagrammatic representation of a retinal ganglion cell's on-centre receptive field. The ganglion cell receives input from a number of different receptors through the horizontal and bipolar cells. Light falling on receptors at the centre of the field excites the ganglion cells and evokes action potentials. If light also falls on receptors on the surrounding (outer) area, this decreases the activity of the retinal ganglion cell and evokes fewer action potentials.

If a square shape is moved in front of a toad's eye, the impulse frequency in the different classes of retinal ganglion cells varies with the size of the square used. In Class 2 cells, with their small excitatory centres, only quite small shapes give maximum response because larger squares cover both the excitatory and the inhibitory areas of the receptive field and this has the effect of reducing the total activity of the cell. Class 4 cells, with their large excitatory fields, are active when the toad is viewing a much larger square, because the inhibitory surround is not stimulated. What this all means is that Class 2 and Class 3 cells are activated mainly by prey stimuli and Class 3 and 4 cells mainly by 'enemy' stimuli. However, it would not be correct to say that there are specific prey or enemy detectors in the toad's retina and that we had accounted for the toad's response to 'sign stimuli' in terms of the physiology of its eye. The function of the retinal ganglion cells is to extract basic information about the size, angular velocity and contrast of the object the eye is looking at, leaving it to the brain to distinguish between food and enemies. Ewert (1980) describes in some detail how the toad's brain performs this task. In fact, the analysis of sign stimuli and their neurophysiological basis has been one of the most successful areas of 'neuroethology' and excellent accounts are also given by Camhi (1984) and Young (1989).

'Supernormal' stimuli

As we have seen, a common method of investigating sign stimuli is to present animals with dummies or models of a natural situation and then to change these in certain ways to see which are most important to the animals. A very curious phenomenon has emerged from a number of these studies: it is often possible to produce a model that produces a greater response from an animal than the natural object does. Examples of such 'supernormal' stimuli are found in all sorts of animals, but some of the oddest have come from the incubation behaviour of birds. Tests have been made with the herring gull, the greylag goose and the oystercatcher and in all three, the larger an egg is, within broad limits, the more it stimulates incubation. Figure 3.8 shows an oystercatcher trying to incubate a giant egg in preference to its own.

Male sticklebacks preferentially court large females with distended abdomens, which indicates that they have a lot of eggs. Rowland (1989) showed

Fig. 3.8 An oystercatcher attempting to brood a giant egg in preference to its own egg (foreground) or a herring gull's egg (left foreground). The bird's original nest was equidistant between the three 'test' eggs (from Tinbergen 1951).

that, when presented with two dummy females, a male would first direct his courtship to the larger and fatter one, even when the 'fat' female had an abdomen distended far beyond the normal range of female sticklebacks. Herring gull chicks will peck at models of the parents bill from which they are normally fed with regurgitated fish (Tinbergen and Perdeck 1950). However, a long thin red knitting needle is more effective at eliciting the pecking response from a newly hatched chick than a more realistic model with a head, yellow beak and red spot on the lower mandible that a real adult herring gull has. A further example of a supernormal stimulus was revealed by the work of Magnus (1958) with the silver-washed fritillary butterfly. Males are attracted to females by the flashing orange wing pattern as they fly by. Magnus could attract males to a revolving drum that 'flashed' a wing pattern at any required speed. The normal wing beat frequency of a female fritillary is about 8 per second, but males show a stronger response the faster the wings of models are made to beat, up to as high as 75 per second.

At first sight, it would appear that by responding to supernormal stimuli, animals are behaving 'stupidly', and against their own interests. But usually, in the natural situation simple 'rules of thumb', such as responding to the biggest or brightest object around, will serve the animal perfectly adequately and be the easiest to evolve given the structure of its sense organs. Sitting on the largest egg around will normally result in a bird's incubating its own egg not a stone. Courting the largest fattest female around will normally (except when she is heavily infested with worms) result in a male stickleback having the greatest number of eggs laid in his nest. And the eyes of insects are constructed in such a way that flickering lights are always attractive and within limits the faster the flicker, the more attracted they are. Supernormal female fritillaries simply stimulate the eyes of males more than normal females. Nor is it only our animal relations that respond in such ways. It is obvious that lipstick makes human lips into a supernormal stimulus (Fig. 3.9).

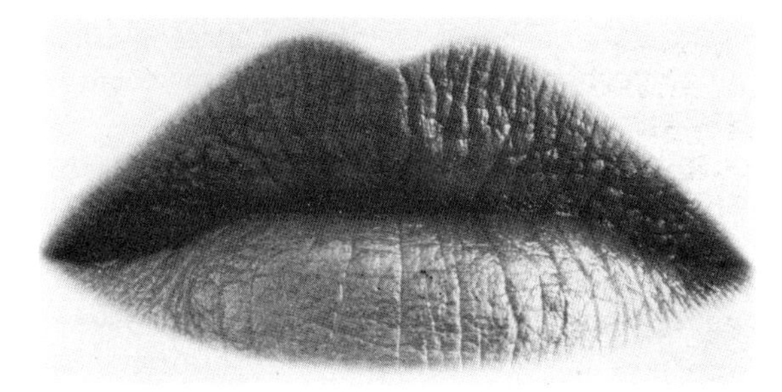

Fig. 3.9 Human lips are made strikingly supernormal by the addition of lipstick.

Conversely, it is also easy to understand why the natural stimuli have not themselves evolved towards the supernormal condition. As we discuss further in Chapter 5, natural selection can rarely lead to perfection; animals settle for the best compromise. Female fritillaries would attract more males if they beat their wings faster, but their wings are also used for flying and there are severe mechanical limitations on the speed with which they can be moved. Similarly, an adult herring gull's bill would probably be highly inefficient in all but attracting the pecks of its chicks if it were as long and narrow as the supernormal model. There will also be a limit on how many eggs a female stickleback can produce.

More complex situations

Sign-stimuli and rules of thumb will operate well when the natural situation and the decision to be made are relatively simple, but this is not always the case and we must not assume that animals always respond to simple 'sign-stimuli'. In fact, one of the striking features of the way animals respond to the world around them is the complexity of their response and the many different features of their environment that seem to be important to them.

For example, Hunter (1980) showed that territorial male great tits are faced with complex choices of where to perform their different activities within their territories. They choose to sing in places that are about 9 m from the ground and very difficult to locate visually. They have a striking preference for hawthorn trees, which have dense crowns and produce leaves early in the year, making them ideal as inconspicuous singing perches. Most of the great tit's other activities, such as foraging, are carried out much lower than this, between 2 and 4 m from the ground. Here too, there is a complex response to changing or uncertain supplies of food. The birds assess such things as the amount of food likely to be gained from one area of a tree relative to another area of the same or different trees. They are certainly not responding just to the simple sign stimulus of 'tree'. As we will see in Chapter 4, sophisticated theories of foraging have been developed to account for the extremely complex discriminations, often involving learning and the assessment of certainty of supply, that many animals show when hunting for their food.

Even insects may make extremely complex responses to their environment. Seely (1985) describes the highly sophisticated assessment made by honey-bee scouts looking for new living quarters. Several hundred individuals simultaneously search the environment near the parent colony in a coordinated hunt. Each scout bee investigates dark places and little holes and may spend up to an hour examining one closely. She inspects a prospective site both inside and out. Inside the hole, she scrambles over the inside walls and Seely showed, by use of a treadmill, that she judges the volume of a cavity by how much walking she has to do to get round it. At least six properties of the site are used by the bee to assess its suitability as the new colony's home:

1. A cavity with a volume between 15 and 80 litres.
2. An entrance facing south.
3. Entrance smaller than 75 cm^2.
4. Entrance low in the cavity.
5. Nest-site several metres above ground.
6. Between 100 and 400 m from the parent nest.

A site is also more attractive if there are old beeswax combs left by a previous colony.

If we look at the natural situation in which the animal's behaviour has evolved, we can often see why its response has to be complex. Black-headed gulls remove broken eggshells from their nests (Chapter 1) but clearly they must have some way of distinguishing eggshells from unhatched eggs. A simple response to an egg-shaped object could be disastrous. Tinbergen and his co-workers (1962a) showed that the gulls are well able to distinguish the two, largely by the presence of serrated white edges on the broken shells.

In species where there is a danger of brood parasitism, we find that egg recognition is even more complex. There is a great advantage in the parents being able to discriminate between their own eggs and those of other birds and their response to different egg characters has become very detailed indeed (Figs 3.10 and 3.11).

Davies and Brooke (1989a,b) investigated the responses of 24 species of songbirds to eggs of the European cuckoo by placing model cuckoo eggs in their nests and recording whether they were accepted or thrown out. They found that most of the species that are currently parasitized by cuckoos, such as reed warblers, meadow pipits and pied wagtails, showed some degree of discrimination in that they would accept model eggs if they were similar to their own but not if they mimicked those of other species. Interestingly, however, there were other species that are not currently parasitized by cuckoos, such as spotted flycatchers and reed buntings, that rejected all model eggs, irrespective of type. These species would seem from their diet and nest locations to be highly suitable as cuckoo hosts, which raises the intriguing possibility that they were once parasitized by cuckoos until their egg discrimination became so good that cuckoo eggs could no longer survive in their nests.

By contrast, other species, such as greenfinches, linnets, great tits and swallows, accepted all the model eggs, whatever they looked like, but these species, either because of diet or nest site, have probably never been suitable hosts for cuckoos and so have never been selected for sophisticated egg discrimination. Meadow pipits in Iceland, where they are not parasitized by cuckoos, were much less discriminating about model eggs in their nest than

Fig. 3.10 Reed warbler feeding a young cuckoo (photograph by J. Horsfall). When the young cuckoo hatches, it ejects the eggs or young of the reed warbler by pushing them out of the nest with its back, so that the parents lose their entire brood and feed only the cuckoo.

Fig. 3.11 Top row (left to right): real cuckoo eggs from a reed warbler nest, meadow pipit nest and pied wagtail nest. Different cuckoos specialize in different hosts and lay eggs that match the host's own. 2nd row (l to r): model cuckoo eggs used in the experiment: reed warbler, meadow pipit, pied wagtail, redstart. 3rd row (l to r): current favourite British hosts of cuckoos: reed warbler, meadow pipit, pied wagtail, dunnock, robin, sedge warbler, wren. 4th row (l to r): suitable but rare hosts: redstart, spotted flycatcher, reed bunting, chaffinch, blackbird, song-thrush. 5th row (l to r): unsuitable hosts: linnet, greenfinch, bullfinch, great tit, blue tit, pied flycatcher, wheatear, starling, swallow (from Davies and Brooke 1989a; photograph by M. de L. Brooke).

the same species in Britain, where the cuckoo is a threat. This illustrates how ability to discriminate is itself under selection pressure.

What all this means is that the neurobiologists' hope that relatively simple sign-stimuli can be identified for each behaviour of each species is balanced by the ethologists' growing realization that many animals, not just birds and mammals but invertebrates too, make multidimensional and very far from simple assessments of their environments. It makes the behaviour of the animals ever more intriguing to study but pushes the hope of a full understanding of underlying neural mechanisms even further into the future.

COMMUNICATION

Many of the most important external stimuli to which animals respond are provided by other animals—stimuli from predators or from prey, stimuli from the parents to which young respond, courtship stimuli from one sexual partner to another and so on. Many animals have evolved 'signals' —special conspicuous behaviour patterns often combined with special structures whose chief function is to send stimuli to another animal. The single large claw of the male fiddler crab (Fig. 3.12) is an example of such a 'signal'. One claw is enlarged way beyond its need for any other function. It is often

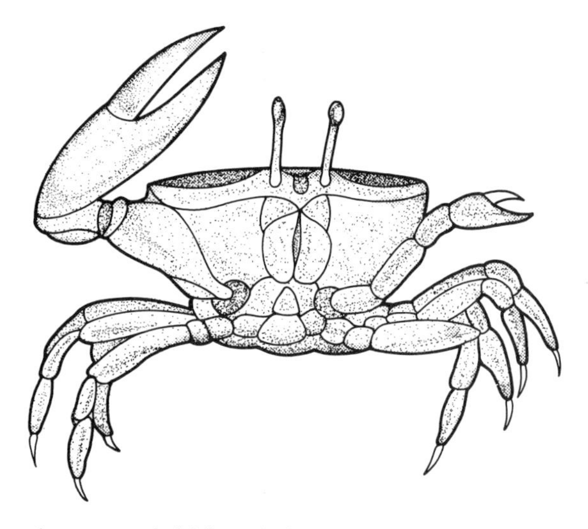

Fig. 3.12 Male fiddler crabs have one claw greatly enlarged and use it to signal to other males, and to females.

brightly coloured and is waved rhythmically in the air during courtship displays to attract females and in defence of the burrow against other males.

'Communication' refers to the fact that one animal responds to the stimuli sent out by another, but it is not a completely straightforward task to give a watertight and universally accepted definition. Like many words used both technically and in ordinary speech, it has gathered several different meanings through constant use. Communication always involves the passage of information and one of the widest definitions would be to accept this and define it as 'any transfer of information' but questions still remain. Should we restrict the definition to transfer of information within a species? Certainly, a moth that suddenly displays the vivid eye-spots on its hind wings in response to a light touch would seem to be signalling to a potential predator. So would gazelles 'stotting', i.e. leaping in the air as they run from a cheetah (Fitzgibbon and Fanshawe 1988), but what about the wildebeest that simply turns and flees when a lion begins its charge? Burghardt (1970), in a most useful essay on this subject of definition, uses an even more revealing example. Foraging ants commonly lay trails of scent that nest mates pick up and follow out to sources of food. A species of small snake (*Leptotyphlops*) also detects the scent trails and follows them back to the ants' nest where it devours the brood. Most of us would probably refer to the scent as a signal, evolved by the ants for communication with each other, but we would not consider that the ants communicated with the snake.

We are approaching a restriction to our definition, which involves the idea that signaller and receiver are mutually adapted to one another. Communication occurs through signals that have evolved to the mutual benefit of two animals. In the example we have just described, ants have evolved to communicate with each other by their scent trails because it is of mutual advantage to recruit other workers to a food source. One of the penalties of this signalling system is that a predatory snake can use the same information, but the advantages the ants derive will certainly outweigh the occasional disadvantages.

Using 'communication' in this way to imply mutual advantage between sender and receiver means that we certainly cannot restrict it to information transfer between members of the same species. The nectar guide patterns on flowers (p. 48) communicate with bees to the mutual benefit of both partners. The skunk communicates to potential predators by presenting its hind quarters and raising its tail. The rattlesnakes' rattling serves a similar function.

Not all workers in animal communication have chosen to use the definition we have used. For example, both Altmann (1962) and Hinde and Rowell (1962) studied social communication in rhesus monkeys and attempted to describe their visual signals, but the former study lists about 50, the latter only 22. The reason for this discrepancy lies in their different uses of the term communication. Altmann's definition of social communication is a 'process by which the behaviour of an individual affects the behaviour of others'. All kinds of movements and postures may do this—the sight of a monkey feeding, for example. Hinde and Rowell, on the other hand, use a definition similar to that given above and they classified as visual signals only those that were likely to have evolved for this purpose. Although we will concentrate on specially evolved social signals in this brief discussion of communication, it is important to recognize that animals do receive a great deal of information in the way Altmann describes. Members of a rhesus monkey troop will respond to a range of diverse stimuli from the dominant male. His general body posture and manner of walking will convey information quite apart from any signals, such as threat movements, he may make. Almost every feature of his body when moving or at rest is in strong contrast to that of a subordinate male (Fig. 3.13). In such cases it may be

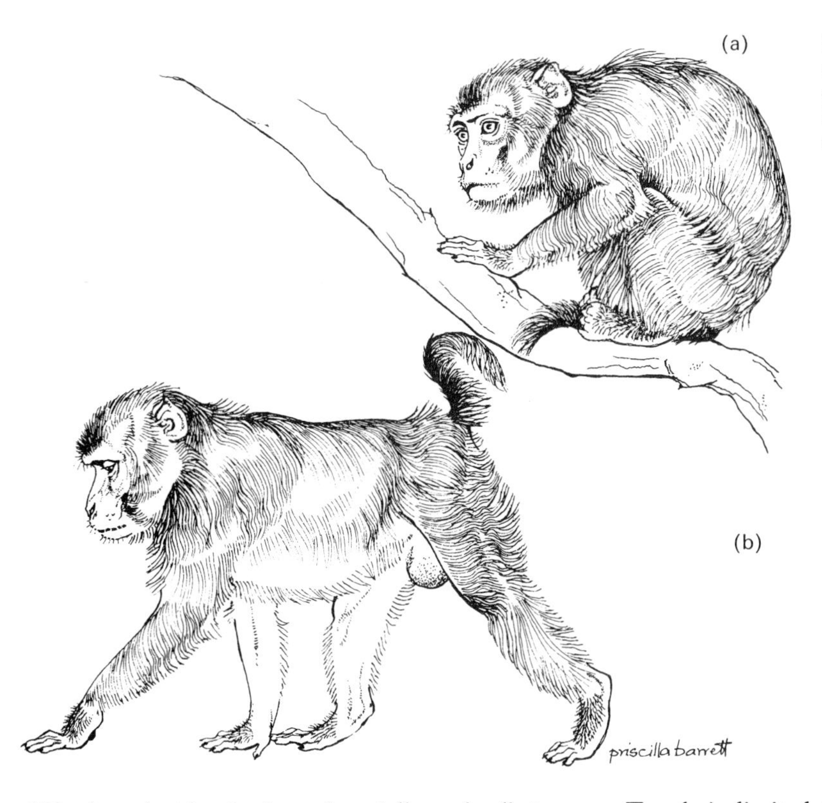

Fig. 3.13 Typical body postures assumed by (a) subordinate and (b) dominant rhesus monkeys (from Hinde 1974. Used with permission of McGraw-Hill Book Company).

difficult to decide whether a 'specially evolved' signal is involved or not.

The diversity of ways in which animals communicate with one another—songs, visual displays, scent trails or slight changes of posture, for example, prompts us to ask why different animals use such different signals. One obvious reason will be the different environments in which they live. Air and water need different sorts of signals, sound signals will be particularly useful if two animals are travelling through dense vegetation and are likely to lose sight of one another and so on. Another reason will be the variety of 'messages' conveyed by signals. Defence of a territory will often involve a loud conspicuous signal, whereas a soft 'contact' call between close neighbours may be enough to keep members of a flock or herd together. A third very important reason, however, will be the sensory capabilities of the animal on the receiving end of the message. Looking at the animal kingdom as a whole, it is clear that different groups rely on different sorts of signals depending on the development of their various sensory modalities. Animals that rely on vision for finding their way about and detecting prey will also tend to communicate through visual signals, and so on.

Touch is limited in its scope for transmitting information but it is in many ways the most basic of the channels of communication. Tactile communication certainly dominates the social interactions of many invertebrates, for example, the blind workers of some termite colonies, which never leave their subterranean tunnels and earthworms, which emerge from their burrow at night to mate. Amongst invertebrates, touch is closely associated with the chemical senses because the specialized tactile organs like the antennae or palps of insects often carry chemoreceptors as well. The social insects transfer a great deal of information through their colonies by a combination of tactile and chemical signals (see Chapter 7). Tactile communication is also important in many vertebrates, particularly in mammals, where some of the more social species spend a good deal of time in physical contact with one another (Fig. 3.14). It is probably not appropriate to think of specially evolved signals in most of these cases. Rather, contact and tactile stimulation, e.g. when one member of a monkey group grooms another, conveys more general information on the social relationship between the two animals.

Tactile communication can, by its very nature, operate only at very close range. The long antennae

Fig. 3.14 Tail-twining—tactile communication by adult dusky titi monkeys (from Moynihan 1965).

of cockroaches and lobsters act as 'feelers' that enable them to explore the world over one body's length ahead, but this is about the limit for touch. Some animals extend the effective range of 'touch' by having mechanoreceptors sensitive to mechanical disturbances in the air or water around them. Male pond-skaters ascertain the sex of other adults by the surface waves they produce. Both males and females produce surface ripples by vertical oscillations of their legs, but whereas females produce only low frequency (3–10 Hz) waves, males produce both low and high frequency (80–90 Hz) waves, to which other males respond (Wilcox 1979). Fish, even when blind, can school, using their lateral line organs to pick up the mechanical disturbances in the water made by other fish in the school (Partridge and Pitcher 1979). Sound waves are also, of course, nothing more than mechanical disturbance that propagates rapidly through air and water. Sound, like scent, has the advantage that it can travel over considerable distances, past and to some extent through, natural obstacles such as dense vegetation. Long distance signals are therefore usually calls or specially produced scents.

Sound, of all frequencies, attenuates as it travels, but high frequencies attenuate and become scattered by obstacles much more than low frequencies in all types of habitat. Low frequencies are therefore generally better for communication over long distances. As mentioned on page 50, Gerhardt (1974) showed that the male green tree frog uses a call with two different frequencies. A female at a distance will pick up the lower frequency and head towards a male. As she gets closer, and the sound gets louder, she responds to both the low frequencies and also the higher ones that she can now hear.

Animals can also signal over longer distances by elevating themselves above ground. Crickets that sing from trees or shrubs can spread their signal over 14 times the area of those that sing from the ground and consequently attract more females (Paul and Walker 1979). The territorial songs of birds are usually delivered from a raised song post, which also increases their effective range (see reviews by Morton 1975; Wiley and Richards 1978).

Sound travels even further in water because there is less attenuation than in air and so aquatic animals use sound extensively for communication. It is ironic to recall that Jacques Cousteau's pioneering book on free-swimming scuba diving was entitled *The Silent World*. The development of the underwater microphone has now allowed us to discover the amazing range of sounds produced by fish and whales. Payne and McVay's remarkable study (1971) of humpback whales suggests that their 'song' could be picked up by other whales several hundred miles away! This is certainly a long distance record for animal communication.

Chemical signals are particularly well developed in insects and mammals. (There are good reviews by Payne *et al.* (1986) for insects, Ralls (1971) and Brown and Macdonald (1985) for mammals and Duvall *et al.* (1986) for vertebrates generally.) One of the drawbacks of chemical communication is the difficulty of changing the signal quickly—it is not normally possible to produce a pattern of scent, whereas patterning is readily achieved with sound and visual signals. Consequently, chemical signals —known as pheromones—are often used to pass a single, relatively stable message, such as the breeding condition of a female or the ownership of a territory. Such a message can persist in the absence of the signaller.

Wilson (1965) and others have discussed the way in which the chemical structure of scents can be adapted more specifically to their function. Sex pheromones and territory markers need to be persistent and therefore their constituents must have a fairly high molecular weight, although it cannot be too high or it will be difficult to secrete and may not disperse well. Moth pheromones strike a compromise between persistence and good dispersal. In favourable wind conditions males can detect them from 4 to 5 km downwind. On the other hand, the chemicals used by some ants to signal alarm must not persist—if they did then the precise localization of a new source of danger would be impossible. These chemicals are volatile and disperse well over a short range of 3–5 m, although they have usually faded below detection levels within a minute, or even less.

Visual signals usually operate only at relatively short range but some alarm signals such as the white tails of rabbits and some deer (Fig. 3.15) are effective over long distances. Fireflies have developed a remarkable form of visual communication, which can also operate over long distances because they manufacture their own light and signal only at night. The electric fish, *Eigenmannia virescens*, has an unusual short range signal: the male discharges its electric organ in a series of 'chirps', which Hagedorn and Heiligenberg (1985), having transposed them into sound for our ears, describe as 'short and abrupt during aggressive encounters' and with a 'softer and more raspy quality during courtship'.

In a variety of ways, then, animals communicate with others both of their own and different species. We can organize our discussion of what is actually going on when they do so around three interrelated questions:

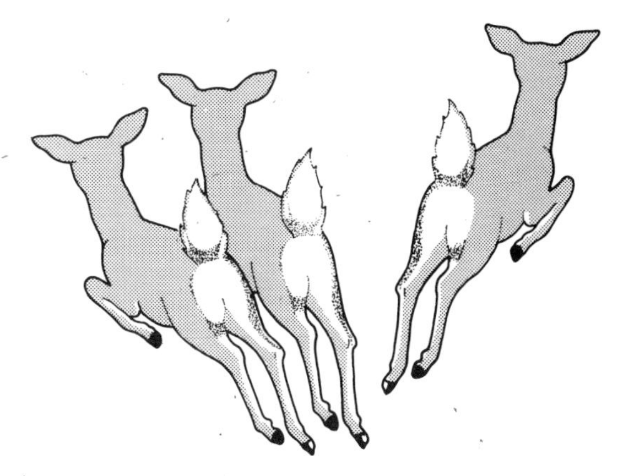

Fig. 3.15 The alarm signal of white-tailed deer. As the deer leap away, the tail is elevated, revealing its brilliant white underside and the white hair of the rump.

1. What is communicated?
2. How do we know when communication occurs?
3. Is communication always 'honest'?

What is communicated?

We have already mentioned some of the types of information that animals communicate: the presence of danger and sexual receptivity are two obvious examples. But even here, the information conveyed by animals may be far from simple. Vervet monkeys (Fig. 3.16) communicate not just the presence of danger to other members of their troop but also what sort of predator is threatening them. One alarm call consists of a series of relatively long tonal units and is given in response to mammalian predators, particularly leopards. Other monkeys hearing

Fig. 3.16 Vervet monkeys live in small groups and communicate the presence of danger to each other.

this 'leopard' alarm call will run up into a tree. A second alarm call, the 'eagle' alarm consists of short tonal units grouped together. Seyfarth *et al.* (1980) showed that if this call is given when a martial eagle flies overhead, the response of other monkeys to the call is to look up into the sky or hide in bushes. A third alarm call, the 'snake alarm' consists of a series of short, widely spaced sounds and on hearing it monkeys look down on to the ground. Tape-recorded playbacks of these three alarm calls give rise to the same responses by other monkeys, as do calls given when real danger threatens, so the monkeys are genuinely using information in the call, not responding to the predator directly. We have, then, the beginnings of a real language of dangers. This is not unique to vervets because other animals, including chickens (Collias and Joos 1953) and ground squirrels (Leger and Owings 1978) have also been shown to convey information about the type of predator they have seen.

Chickens (Marler *et al.* 1986) and honey-bees (see p. 66, later in this chapter) can communicate information about the quality of a food source. Cockerels have a special call that they give when they have found food, and that causes hens to come and feed. The food calls are produced at a higher rate with highly preferred food.

A very interesting category of signals arises when animals come into conflict, for example over a food source or access to mates. Under such circumstances they may give 'threat' signals to one another, either before they come to physical combat or even instead of fighting at all. Sometimes what the animals are communicating here is their own prowess in battle. A particularly clear example of this is shown by red deer stags. During the autumn rutting season the stags frequently challenge one another for ownership of groups of females. These challenges sometimes result in damaging fights, in which one or both stags are injured, but they almost always begin with a 'roaring match', in which the stags signal to each other with roars, giving gradually greater and greater numbers of roars per minute (Clutton-Brock and Albon 1979). Roaring is an exhausting signal to give and the ability to roar at a high rate means that a stag will probably be able to fight long and hard because the same muscles are involved in both activities. If a stag is 'out-roared' by another, he may well decide to abandon his challenge altogether. He uses the information contained in the signal to tell him that the other stag is probably a better fighter than he is and that fighting would be damaging and costly. If, on the other hand, both stags can roar at the same rate, they then proceed to the next stage of the conflict, the 'parallel walk', in which both contestants walk up and down, eyeing each other and apparently trying to judge the other's fighting ability in more detail. The result is that real fights occur only between stags that are so closely matched in fighting ability that each has estimated that it may be able to win. The information contained in the roars (specifically, in the ability to roar at a high rate) is a clear message about fighting prowess because the ability to fight and the ability to roar are so closely correlated (Fig. 3.17). Clutton-Brock and Albon (1979) found that even dominant stags would retreat if a tape-recorder were used to 'roar' at a supernormal rate!

Much communication takes place during courtship. It is most usual to find males rather than females giving the loudest, most conspicuous and flamboyant signals, but it is often not entirely clear what is being communicated here. Certainly, one of the things that animals convey to each other during courtship is what species and sex they are (see Chapter 5). Lizards have species-specific patterns of head-bobbing, different firefly species pulse light in different patterns in time, fiddler crabs wave their claws in ways that differ from species to species and frogs and crickets operate a sort of Morse code—the males sending out pulses of sound in species-characteristic series. Pheromones are another way some animals have of identifying the right species. In all these cases, mates of the right species find each other with signals, but although it is undoubtedly of great importance for them to do so (because hybrids between different species are often inviable or infertile), the signals themselves seem far more elaborate than is required simply to convey information about species identity, or even readiness to mate. The tail of the male blue peacock, for instance, is an extraordinarily elaborate and cumbersome structure, complete with 'eyes' that are waved and shimmered in front of the female. If all the male was doing was communicating information about his sex, his species and that he was a mature male ready to mate, he would surely not need to evolve such a tail—other species convey similar information in much less flamboyant ways.

Darwin (1871) proposed that elaborate male ornaments have evolved to attract females, even though they may be detrimental to the males themselves. Competition for mates—sexual selection—

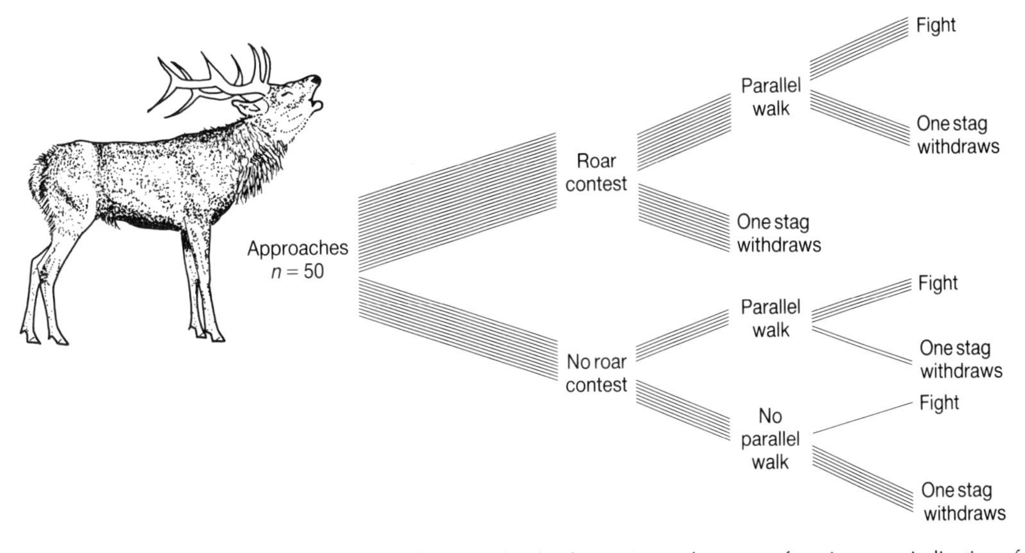

Fig. 3.17 Contests of red deer stags. The males challenge each other by roaring and use rate of roaring as an indication of fighting ability. The figure shows the number of approaches to within 100 m involving two mature stags that lead to roaring contests, parallel walks and fights. Each parallel line represents one encounter (from Clutton-Brock and Albon, 1979).

would lead to the evolution of traits that either helped a male to fight off other males or make him particularly attractive to females, or both. Darwin believed that sexual selection could be such a powerful force that such traits could evolve even though a male possessing them might not survive all that long. As long as he survived long enough to mate and leave a large number of progeny, his characteristics would be successfully passed on into the next generation. Darwin thus saw sexual selection as a special case of natural selection, with the emphasis on mating success.

For a long time, Darwin's theory of sexual selection, particularly its emphasis on female choice, was treated with scepticism. Then, in the 1930s, it was given new impetus by R. A. Fisher. Fisher argued that exaggerated sexual ornaments, such as long tails, could indeed arise from female choice, with the males being caught in a sort of evolutionary trap. He argued that over evolutionary time, tails would become longer and longer: the more females chose long-tailed males, the more male offspring there would be with long tails like their fathers and the more female offspring there would be with a preference for long-tailed males like their mothers.

More recently, a number of alternative theories have been proposed to explain the evolution of sexual ornaments and displays. Hamilton and Zuk (1984) proposed that they enable healthy males to advertise the fact that they are free of diseases and parasites. Since disease is a major source of juvenile mortality, Hamilton and Zuk argue that females should preferentially choose males that have genes for resistance to prevailing parasites, which would then be passed on to their offspring. The female's problem is, of course, how to choose a male that has such genes using his 'signals' as her guide. They argue that females can pick out healthy males by their displays because displays are costly and exhausting to perform and only males that are genuinely healthy will be able to perform them. Long tails, too, give an indication of a male's health because a male is demonstrating his physical strength by showing that he can survive despite the 'handicap' of a long tail (Zahavi 1975; Andersson 1982a; Grafen 1990).

As we will see in the next section, it is often possible to show that females are attracted to exaggerated ornaments in males, such as long tails. What we still do not fully understand is why (in an adaptive sense) they are and what, if any, information is contained in the males' signals. Do the tail displays contain information about a male's health and vigour or are they, as Fisher suggested, simply ornaments for ornaments' sake? And if they do convey information, who are the intended receivers of the information? We have assumed that male 'courtship' displays are directed primarily at females but as Darwin (1871) pointed out, other males may also be the targets. It is possible that some so-called 'courtship' displays are really male–male assessment displays of the type we have already discussed, in which males display to rivals and females simply choose on the basis of their prowess in battle or dominance over other males. Female chickens (Graves *et al.* 1985) and black-billed magpies (Komers and Dhindsa 1989) and ducks (Brodsky *et al.* 1988) all have preferences for dominant males. A dominant male shows his health and vigour by his ability to out-compete other males and such females could therefore choose good quality mates through using signals that were evolved for male–male assessment.

As a final example of the range of animal communication, it is worth mentioning a rather curious type of signal whose function seems not to communicate information in itself, but to qualify other signals that follow it. This phenomenon has been called **metacommunication** and is best known from play situations in carnivores and monkeys. Figure 3.18 shows an adult male lion inviting a cub to play. His posture with forequarters lowered is not seen in any other context and its message is that all aggressive movements that follow are play. Dogs use almost exactly the same posture and may also wag their tails during play fights (as Darwin noted in *The Expression of Emotions in Man and Animals*). Monkeys adopt a 'play face' in similar situations. Playful aggression is a conspicuous part of behavioural development in carnivores and it has obviously been necessary to evolve some convention that allows stalking and attacking to take place without the damage of real fights.

How can we investigate whether communication occurs?

So far in all our discussions of communication we have assumed that social signals are effective in evoking appropriate responses. We see what appears to be a 'signal' given by one animal being 'responded to' by another, but to be sure that communication is taking place, we have to use systematic observation and experiment. The task is not always straightforward, however, for the very reason we discussed at the beginning of this section: the definition of communication itself is not an easy one. Large, conspicuous signals that lead to clear-cut immediate responses in the animals receiving them pose less of a problem than the smaller, more subtle changes of posture and gait that we have seen also form an important part of the social communication of many species.

We have already seen that one powerful method of investigating communication is to use models or dummies. These can be used to reveal the key features of a signal such as the bill spot of the herring gull. In an exactly analogous way, early experiments that showed that female crickets would walk right up

Fig. 3.18 Metacommunication in lions. The male's posture with lowered forequarters is only seen as a preliminary to play. He invites the cub to play and cuffs it gently as the cub joins him (from Schaller 1972).

to the cover of a loud-speaker emitting the male song, first proved the efficacy of sound communication in insects. Cheney and Seyfarth's (1982) playback experiments with vervet monkey alarm calls are a more recent example of the same thing.

Sometimes it is less easy to 'fool' animals with dummies and it has proved more fruitful to make some alterations to the appearance or behaviour of free-living animals. Noble (1936) in a classic experiment, captured the female from a mated pair of yellow-shafted flickers, a type of American woodpecker. The males and females of this species look identical, except that the male has a black moustache. Noble stuck a moustache of black feathers on to the female and the male promptly attacked her and drove her out of the territory. When the moustache was removed she was once more accepted by her mate.

Andersson (1982b) showed that female widow-birds are attracted to the long sweeping tails of the males. He captured wild male widow-birds that had already set up territories and cut the tail feathers. Then he glued tail feathers back on, making the tail longer than before, shorter than before or the same length (control for the effects of tail manipulation). He found that males with artificially elongated tails attracted more females than they had before the experiment while those with artificially shortened tails had fewer. Møller (1988) did a somewhat similar experiment with swallows and again showed that long tails are an important signal to the females (Fig. 3.19). Males with artificially elongated tails attracted females earlier, had more second broods and produced more fledglings in the course of a season.

Sometimes the stimuli used by animals to communicate with each other are so subtle that sophisticated instrumentation is needed to pick them up. For many years, people studying vervet monkeys had heard them give little grunts as they moved around but had classified all grunts as a single type of signal. By recording these grunts and playing them through a sound spectrograph, which analyses frequency composition of sounds over time, Cheney and Seyfarth (1982) were able to show that there were several different sorts of grunts, indistinguishable even to the experienced human ear, which were given in different situations and that were responded to in quite different ways by the vervets. One grunt, for example, was given when the monkeys came to a wide open space, another to a strange vervet monkey and so on.

Identifying the stimuli used in communication is a first step, but getting a measure of their effectiveness is not always easy. If a receiver does not respond it may mean that what the experimenter thought was a signal turned out not to be after all, or it may mean that it is a signal but that the receiver, although it perceived the signallers' message, simply failed to respond for some reason. This is a particular problem with signals that act tonically or over long periods rather than as immediate triggers. Another problem is that animals that know each other very well, such as the members of a monkey troop or a small flock of hens or birds that have defended neighbouring territories for a long time, may show little response to each other. Many animals attack and threaten strangers of their own species (the signals and overt attack will be very obvious) but the lack of response to familiar animals is clearly important too.

For instance, Bossema and Burger (1980) and van Rhijn (1980) studied communication between members of a small group of captive jays. The younger birds tended to give way to the more powerful older members of the group and, once this 'rank order' was established, it greatly affected the way the

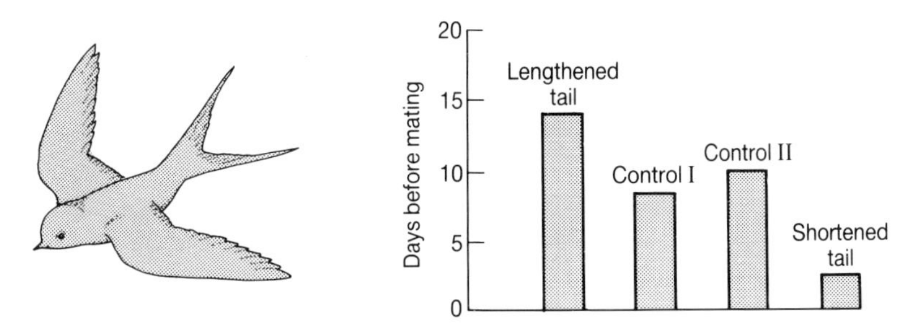

Fig. 3.19 In the monogamous swallow, a male normally takes about 8 days after arriving in the breeding area to find a mate (Control I and Control II). With artificially shortened tails it takes 12–13 days to find a mate but with an elongated tail only about 3 days (from Møller 1987).

birds communicated with each other. In general they used the minimum signal that would suffice. Thus, older birds could sometimes displace younger ones from, say, a piece of food simply by looking at them directly. Only if this glance was ineffective would they 'escalate' to perform a threat display. Similarly, most people who have studied primate groups over a long period can recount episodes where a severe fight between two individuals has repercussions that last for months or years. The loser's responsiveness to the winner may remain perhaps permanently changed and it will flee or show submission to gestures or movements that elicit little or no response from others with different past experiences.

One consequence of this entirely understandable situation is that a new observer, making a spot check on the role of certain signals in this or that social group of a species, may get some perplexing data. Sometimes signal A gets response B and thus appears to communicate C, at other times it gets no response or a totally different response. We shall never be able to interpret a communication system completely unless we know the social relationships of animals and their individual histories as well. There is no substitute for careful and repeated observation of known groups. This is not just a defeatist view because, although it takes time, the picture that will emerge may be far richer and more subtle than we at first anticipate. As we mentioned above, it seems likely that we have underestimated the communicatory capacities of many animals.

Is communication always 'honest'?

Many signals, such as the contact calls that keep the members of a flock together, are clearly of benefit to both the sender and receiver of the signal. Other signals, however, seem better described as 'manipulation', that is, the sender appears to influence the behaviour of the receiver for its own ends. Dawkins and Krebs (1978) and Krebs and Dawkins (1984) cite examples (such as the small fish that snap at the worm-like lure of an angler fish and promptly get eaten or the female hedge-sparrow that feeds a baby cuckoo) as showing that signals 'deceive' receivers into actions that are not to their own benefit. There are also some species of beetle that live parasitically in the nests of ants (myrmecophiles), which trick the ants into carrying them into the nest by mimicking the signals given by an ant larva (Hölldobler 1971). Some ground-nesting birds, such as plovers, lure predators away from their nest by feigning a broken wing. Female fireflies (*Photuris*) can flash in two ways. They can either flash in their own specific pattern to attract males or mimic the flash pattern of another smaller species of the genus *Photinus*. When the *Photinus* males approach in response, they kill and eat them! Lloyd (1965, 1975), who discovered this extraordinary piece of deception, aptly called the *Photuris* females, 'firefly femmes fatales'!

An alternative to this idea of manipulation and deception in animal communication was put forward by Zahavi (1975, 1987) and Grafen (1990), who believe that animal signals are essentially 'honest'. Zahavi argues that signals such as loud calls or energetic displays are costly for the signaller to produce and so the signaller can essentially demonstrate its quality (strength or fighting ability, say) by its ability to survive despite the 'handicap' of the signal. The roaring of the red deer stags, which we discussed earlier in this chapter, is a good example of this. Roaring is exhausting and can only be done at a high rate by stags that are genuinely good at fighting. Roaring rate therefore 'honestly' demonstrates fighting ability.

How can we reconcile these various views? Are signals 'honest' or deceitful? Are signals always mutually beneficial or does one animal manipulate another to its own ends? Maynard Smith and Harper (1988) suggest that it is useful to divide signals into two categories: **assessment** signals, which are always honest and **conventional** signals that sometimes allow cheating.

By assessment signals they mean loud conspicuous signals that are so costly or exhausting to produce that they can only be produced by genuinely fit, strong animals. Cheating here is impossible because weaklings physically could not produce the signal. Roaring in deer, which we discussed, and loud singing in crickets, would be examples of signals that can be used to assess other animals in an honest way. Deceit is impossible.

By conventional signals, they mean signals that usually correlate with the dominance or other feature of quality of the signaller, but do not always do so. The size of the black 'bib' in house-sparrows correlates with male dominance (Møller 1987; Fig. 3.20) but growing a few more black feathers would be possible even for a subordinate bird. It is not that it physically could not have a larger bib in the way that a weak deer stag physically could not roar at a high rate. Cheating is therefore possible but is

Fig. 3.20 In house sparrows, a male's dominance correlates with the size of the black 'bib' (photograph by Mike Amphlett).

probably not widespread because 'cheats' suffer penalties. Møller found that birds with large dominant-style black bibs suffered more attacks from other dominants than birds with subordinate status badges. A bird that 'cheated' and grew a dominant black bib when its own ability to fight was poor, would therefore suffer the consequences of repeated attacks from birds that were stronger than it.

It does seem, then, that both 'honest' signals and manipulation by 'cheats' occurs in animal communication. Natural selection has favoured animals that attempt to manipulate other animals to their advantage, for example to mate with them, to chase them away from a valuable resource or even to eat them. However, selection will also be acting on the receiver not to be manipulated unless it is to its own advantage. Thus a male peacock might be seen as attempting to manipulate a female, but unless it is to the peahen's advantage to respond to his courtship, we would expect selection to act against females that did so.

This coevolutionary arms race between signaller and receiver, as Dawkins and Krebs (1978) called it, has a variety of end results. Sometimes, as in the stags, it results in 'honest' signals. It is to the mutual advantage of both stags to signal 'honestly' what their true fighting ability is because both benefit from avoiding a damaging fight they are unlikely to win. But honest signals are (by definition) costly and in other instances less than honest, and therefore potentially cheatable, signals have evolved (Dawkins and Guilford (1991)). The widespread existence of mimicry forces us to the conclusion that sometimes evolution has led to 'cheats', which are apparently successfully 'manipulating' the receiver. Even here, however, selection will act on the receiver not to be manipulated. A predator that avoids a brightly coloured but highly palatable mimic because it resembles an inedible model is, in that instance, being deceived or manipulated by the potential prey. But overall, a predator that avoids all prey of that colour and pattern will benefit because most of them are poisonous or nasty-tasting. As we saw in the case of cuckoos, which 'deceive' their hosts into caring for them, such manipulation may also lead to a further development in the evolutionary arms race: the receiver becomes ever more discriminating over time, forcing the signaller to become an even better mimic, and so on.

None of the examples of animal deception we have discussed so far necessarily imply that the animals are consciously setting out to deceive other animals. In fact all the cases we have discussed so far are much more plausibly seen as evolutionary deceit, i.e. natural selection favoured animals that behaved in certain ways and elicited particular responses from other animals. In other cases, we cannot be so sure. Munn (1986) described two species of insect-eating birds that appeared to 'cry wolf'. By giving alarm calls when there are no predators present, they induced other birds in the flock to take off and so gained unrestricted access to food themselves. Among primates, baboons have been reported to behave in ways that strongly suggest they are consciously deceiving others. Byrne and Whiten (1988) describe a number of such incidents, including one where a subordinate baboon, being attacked by a dominant animal, gave all appearance of having spotted a predator in the distance. When the dominant animal looked away in alarm, the subordinate ran off. Such incidents must, inevitably, be anecdotal but Mitchell and Thompson (1985) and Byrne and Whiten (1988) both argue that there are now enough comparable examples to support the idea that some animals at least can be deliberately dishonest in their communication with others.

The honey-bee dance

There is no better way to conclude a discussion of communication than by some account of the honey-bee dance. Not only is it one of the most remarkable of all systems of animal communication but it also presents intriguing problems of measurement and interpretation of the type that we have been discussing.

The fact that honey-bees must communicate about flower crops has been known for centuries, but it will now always be associated with the name of Karl von Frisch, who was the first to unravel the nature of this communication and whose book, *The Dance Language and Orientation of Bees* (1967), provides a full survey of the whole dance system. Like others before him von Frisch had noted that if a source of sugar solution was put out to attract bees it was often many hours before the first one found it, alighted and drank. However, once a single bee had located the source it was usually only a matter of minutes before many other foragers arrived—somehow the information had been passed on. It took von Frisch some 20 years of painstaking observation and experiment to work out the bee's communication system to his own satisfaction. The conclusions that he came to were so extraordinary and unparalleled that he himself declared that no good scientist should accept them without confirmation! Following World War II, other zoologists did confirm von Frisch's results and even worked with him on some final experiments. There is now no doubt about the nature of the bee dance itself.

Von Frisch marked foragers as they drank at a dish of sugar syrup and then watched their behaviour when they returned to the hive, using glass-sided observation hives for this purpose. The forager usually contacts a number of other bees on the vertical surface of the comb and gives up her cropful of sugar solution to them. She then begins to dance, and we first consider the case where the food dish she has just visited is close to the hive, within 50 m. Her dance then is rapid in tempo and forms a roughly circular path just over her body's length in diameter. The bee moves in circles alternately to the left and to the right. She stays approximately in the same place on the comb and may dance for up to 30 s before moving on. Other foragers face the dancer, often with their antennae in contact with her body and follow her movements closely, being themselves carried through her circular path (Fig. 3.21). The 'round dance', as this is called, stimulates other workers to leave the hive and search nearby. It appears to convey the information, 'search within 50 m'. It may also convey some olfactory cues because if the food source is scented the dancer will carry this scent on her body and perhaps in the sugar solution itself. If the sugar dish is not scented the forager may 'mark' it to some degree by opening the Nasanoff scent gland on her abdomen as she drinks.

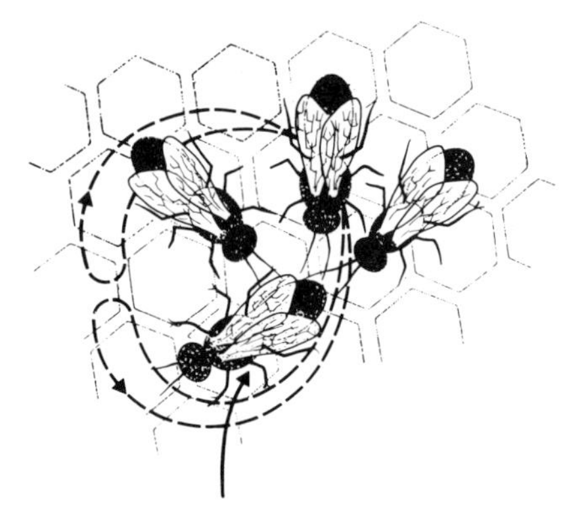

Fig. 3.21 The 'round dance' of the honey-bee worker on the vertical face of the comb: her path is indicated by dotted lines. Note how she is closely attended by other workers.

Thus far the bee dance is not very exceptional—many ants and termites have similar 'alerting' displays and pheromones that help to organize foraging activity when a new food source has been found near the nest (Wilson 1965). The extent of the honey-bees' communication system is not revealed until the food source discovered by the forager is further from the hive—beyond 100 m.

Von Frisch observed that as his food dishes were moved beyond 50 m the forager's round dances gradually changed in form. A short straight run became incorporated between the turns and on this run the dancer wagged its abdomen rapidly from side to side. At about 100 m distant the dance had become the typical 'waggle dance', illustrated in Figure 3.22, and this form remained the same as the dish was moved further, to 5 km or even beyond. More recent work has revealed that during the waggle run the bee produces bursts of high pitched sound (Esch *et al.* 1965). It is this waggle dance that von Frisch claims transmits so much more information and is 'read back' by the dance followers as they follow every move the dancer makes.

The waggle dance certainly contains information about both the distance and the direction of the food source. Distance is correlated with several features of the dance. Von Frisch concentrated on measuring its tempo, which falls off with distance, steeply at first and then more gradually. Thus there are 9–10 complete cycles per 15 s with the food at 100 m, but only 2 when the food is 6 km away. The number of waggles, the duration of the waggle run and the duration of the sound pulses also correlate with distance, all three features increasing with it. Von Frisch had no evidence to enable him to identify which of these distance cues was important for the dance followers. Bees fly at a very constant speed in still air and they interpret distance information in terms of flight time and effort. Von Frisch found that dancers indicated a greater distance for food sources upwind than for those downwind. (It is always the outward flight path that is indicated.)

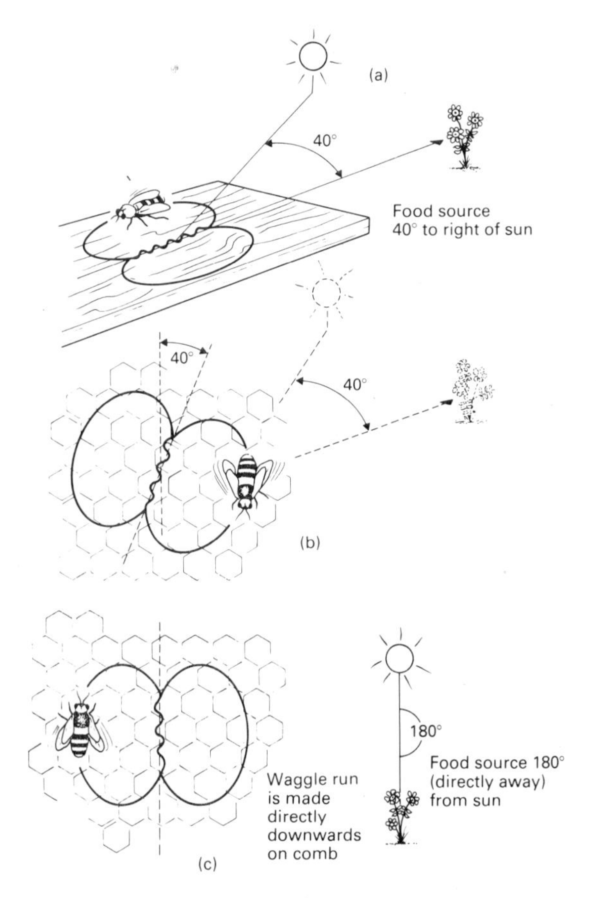

Fig. 3.22 The 'waggle dance' of the honey-bee: further details are given in the text. (a) The dance is occasionally performed on the horizontal entrance board to the hive. The waggle run 'points' directly towards the food source. (b) On the vertical comb the angle of the waggle run to the vertical is equal to the angle the sun makes with the food source. (c) Shows the honey-bee's convention that directly downwards on the comb represents directly away from the sun. In the same way, upwards represents towards the sun.

It is its relation to the direction of the food source that is perhaps the most remarkable feature of the waggle dance. The little south Asian bee *Apis florea* is a close relative of the honey-bee. It builds a single vertical comb on a tree branch in the open with a flattish platform on top where the waggle dances are performed. The waggle run is made towards the direction of the food source, i.e. it operates like a pointer (Lindauer 1961; Gould *et al.* 1985). Now very occasionally honey-bees will dance on a flat platform by the hive entrance. If they do so, then, like *Apis florea*, their dances also point directly to the food source (3.22(a)). However, this observation did not help von Frisch because he watched the dances at their normal site on the vertical face of the combs inside the hive. There he noted that the direction of the waggle run was consistent within a dance and that it was the same for all the foragers who danced after feeding at the same dish. Given that other bees foraging at different dishes made waggle runs at other angles even when the distances were comparable, this was strong circumstantial evidence that the angle related to direction in some way. Now came the crucial observation. Von Frisch recorded dance after dance throughout the day as foragers returned from the same food source and he found that the direction of the waggle run gradually changed. Its mean direction shifted by about 15° per hour and this could mean only one thing; that it relates to the apparent movement of the sun. Figure 13.22(b) and (c) show the way it does so. The foraging bee, like many other insects, uses the sun as a compass and records the position of the food source with respect to it. To get to the food it steers, say, 40° to the right of the sun. When dancing on the vertical comb the sun is not visible but the bee transposes the angle to the sun into the same angle with respect to gravity. The honey-bees' 'convention' takes vertically upwards to represent directly towards the sun. Thus, the forager dances with her waggle run 40° to the right of vertical. She will change this angle to match the sun's apparent movement through the sky. Von Frisch had at last understood the honey-bees' dance language and the world was forced to accept that another animal apart from ourselves—and a humble insect at that—could convey information in a symbolic fashion.

Von Frisch and his co-workers had no doubt that the dance did communicate. Other foragers picked up the dance's rhythm and orientation as they followed through the dancing bee's movements on the comb and they then transcribed back from an angle with respect to gravity to an angle with respect to the sun. This assertion was based on numerous experiments in which an array of food dishes was offered, so arranged as to test the accuracy with which foragers recruited by the dance interpreted its information on distance and direction.

Figure 3.23 shows the two most common types of configuration: a line of dishes at different distances on the same bearing from the hive tested distance communication, a fan pattern of dishes in an arc equidistant from the hive tested for direction. The figures also indicate the results of typical experiments showing that the number of recruits is highest at the dish closest to that at which the original forager fed and recruiting falls off to each side. (Further details of the experiments are given in the captions.)

This type of evidence was very generally accepted as convincing proof that the dance was the effective communication system, although it was clearly not perfect and there is some spread of recruits around the direction and distance indicated by the dancer. This 'noise' is not surprising in view of the crowded, jostling bees on the comb. Commonly, dancers do not have adequate space to keep a consistent line or rhythm. Nor, with natural food sources, will minor inaccuracies matter at all. For the most part the dancers have been foraging at large sources of food —a grove of lime trees in blossom or a field of clover. Provided the recruits get some idea of distance and direction this will suffice, the more so because the foragers are very responsive to scent that the dancers carry into the hive on their body. As mentioned earlier, they may also mark the food source with scent themselves.

Indeed, it was doubts over the importance of scent that led to the later controversy over the interpretation of von Frisch's results. Scent and wind direction tended to be ignored in some of the original tests. It could just be possible that the dance merely alerted and random search did the rest.

This would imply that the information contained

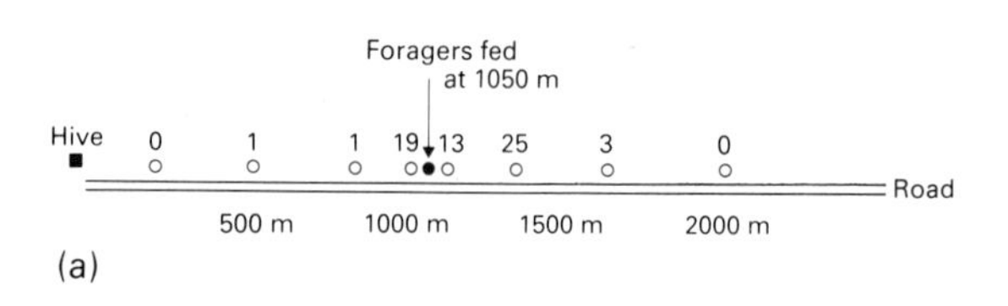

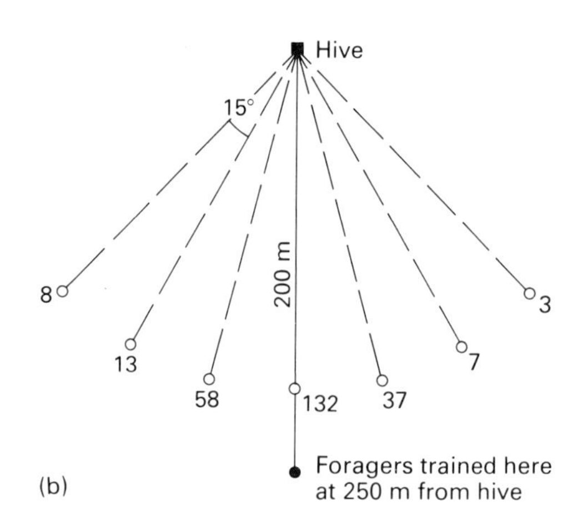

Fig. 3.23 Experiments carried out by von Frisch and his co-workers to test the communication of distance and direction by the honey-bee waggle dance. (a) Distance test. Foragers were trained to a dish of dilute scented food 1050 m east from the hive. Then a series of scented plates without food were put out at distances varying in the same direction from 100 m to 2000 m. Dancing was induced by suddenly increasing the sugar concentration at the feeding dish. Recruits were counted, but not captured, as they approached the different scent plates. The numbers above them record the number of visits made to each scent plate during the test. Most recruits appeared at plates close to the feeding station. (b) Direction test. The procedure is much as for the distance test, but the scent plates are put in an array at the same distance but different directions from the hive. The majority of visits were made to dishes close to the bearing of the training station (from von Frisch 1967).

in the waggle dance relating to distance and direction was there but not communicated. One obvious riposte is to ask why then has such a remarkable relationship evolved? However intuitively reasonable this reply seems, it does not really supply a secure argument because it assumes that every biological phenomenon must be functional. We have other examples of behaviour that certainly contains information but that just as certainly is not used by conspecifics. Dethier (1957) has described the searching movements made by flies after they have located and then exhausted a small source of food, e.g. a drop of sugar solution. Their 'dance' can convey information to a human observer about the 'shape' and concentration of the source, but other flies do not respond. Certainly other bees do respond to the waggle dance but then the 'olfactory' hypothesis requires that they are aroused to search thereby, just as ant foragers are. Von Frisch has already shown that recruits pick up olfactory cues about the food source from the dancer—its scent will adhere to her body.

Gould (1976) summarizes the evidence and the controversy very attractively. He himself did some elegant and better controlled experiments that effectively proved that the information in the dance was communicated as Von Frisch originally described. However, we still lacked the most perfect demonstration, i.e. an artificial bee model to which recruits would respond. There had been a number

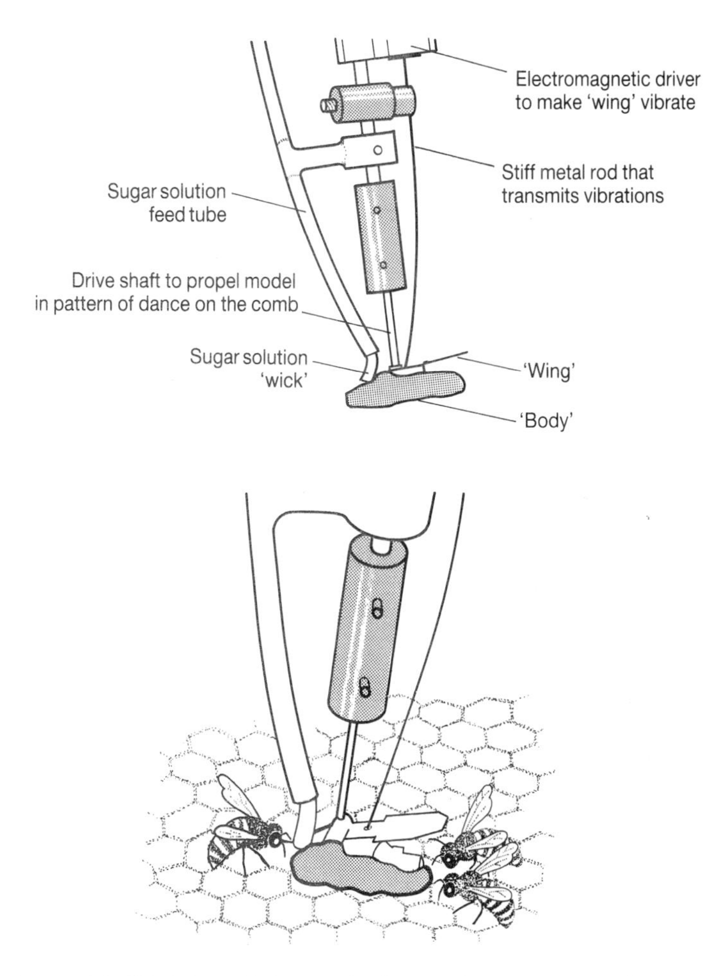

Fig. 3.24 The dancing bee model of Michelsen, Lindauer and colleagues. Its construction is shown above. The body can be made to move on the comb in the form of a dance, 'waggling' during the straight part of the run, whilst the wing (a single piece of razor blade) is vibrated to produce sounds mimicking those of the dancing bee. Below, the model is shown on the comb. Workers fan out at its rear and follow its dance. One is accepting sugar solution which the model offers at its head end (from photographs in Michelsen 1989).

of attempts to make a model bee but none were successful until recently. Now Michelsen, Lindauer and colleagues (Michelsen 1989; Michelsen *et al.* 1989) have succeeded in constructing a relatively crude model of brass covered with beeswax and left in the hive for some hours to acquire the colony odour (Fig. 3.24). The model can be moved through a waggle dance path on the comb and, most importantly, it has an artificial 'wing' attached to it that can be vibrated electrically so as to produce an acoustic field around the model similar to that of a real dancer.

The model works! It recruits foragers to visit food dishes that have not been visited previously and the proportion of bees turning up at different directions and distances follows that of the dance pattern. Sound turned out to be crucial: without it, no bees were recruited and this may be the reason for earlier failures with silent models. There are many pitfalls and teething problems to be overcome, but not only does the fact that the model works at all provide final conclusive proof of true communication in the bee dance, it also gives us the opportunity to investigate the communication process in detail. Already it has given us valuable clues as to the role of sound. Do the sounds themselves convey the information on distance or do they simply ensure the 'attention' of recruits who then acquire this information from tempo or some other aspect of the waggle dance? We should get the answer to this soon, and to many more fascinating questions concerning this most remarkable bee communication system.

4 MOTIVATION AND DECISION-MAKING

For a wild animal life is a constant battle against different sources of danger—not getting enough to eat, falling victim to a predator, being overheated, dehydrated, frozen and so on. As we have already stressed, most behaviour is beautifully adapted to overcome these hazards and to help animals to survive and reproduce successfully despite them. But obviously it is not sufficient for an animal to produce behaviour in random order. Much depends on it doing the right behaviour at the right time and in the right place. This choosing of behaviour in time and place is known as **decision-making**, and the various internal factors that contribute to an animal behaving in a particular way are often called its **motivation**. Although both these terms might be taken to imply a conscious, deliberate working out of future courses of action, they are used in ethology without any necessary implication of this. A robot that had a fixed amount of time before its batteries ran out could, in the same sense, be said to make 'decisions' about whether to start a new task or to go and plug itself into the mains, with no implication that it consciously thought about what it was doing.

Whether conscious or not, we often find that the decision-making processes of animals are extremely complex and subtly dependent on the exact situation the animal finds itself in. A house sparrow that discovers some food will often give a 'chirrup' call from a nearby perch. This call has the effect of recruiting other sparrows to come and join it (Elgar 1986a; Fig. 4.1(a)). All the birds then fly down to the food and feed together, each gaining the advantage of being surrounded by other individuals that are all watching out for danger. Often the bird that has spotted the food in the first place will not fly down to the food at all unless it is joined by several other birds. This is particularly true if the food is near a potential source of danger, such as a human being. The birds are evidently sensitive to the conflict between feeding and being caught.

Other factors apart from the presence of danger also affect how much 'chirruping' a sparrow does when it has discovered food, e.g. temperature. During cold weather, sparrows chirrup at a much higher rate than when it is warmer (Elgar 1986b; Fig. 4.1(b)). The adaptive significance of this is probably that food requirements are higher in cold weather and so the advantages of feeding in a flock with its many pairs of eyes are increased: an individual bird can concentrate on feeding without having to be so vigilant. Another factor that affects chirruping is whether or not the food source is divisible. If the food is divided up into portions, so that several different birds can feed without interfering with each other, a sparrow will chirrup vigorously, stimulating others to come. But if the same amount of food is in one, indivisible lump, it will chirrup much less and be more likely to feed alone. Under these circumstances the advantages of being in a flock are apparently outweighed by the interference resulting from feeding with other sparrows (Elgar 1986a).

From this example we can see something of the complexity of animal decision-making processes. A sparrow evidently responds to the sight of food but

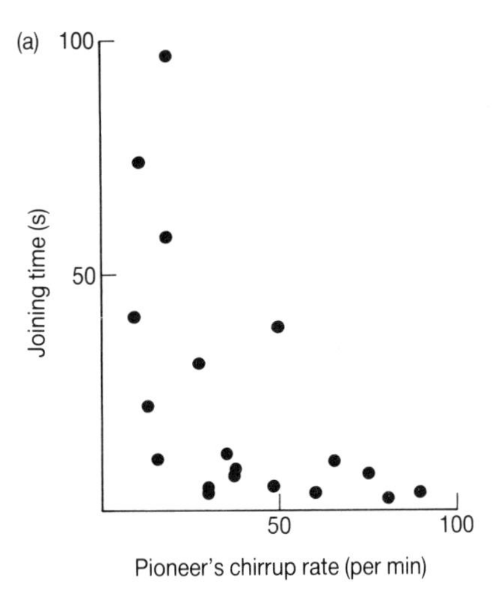

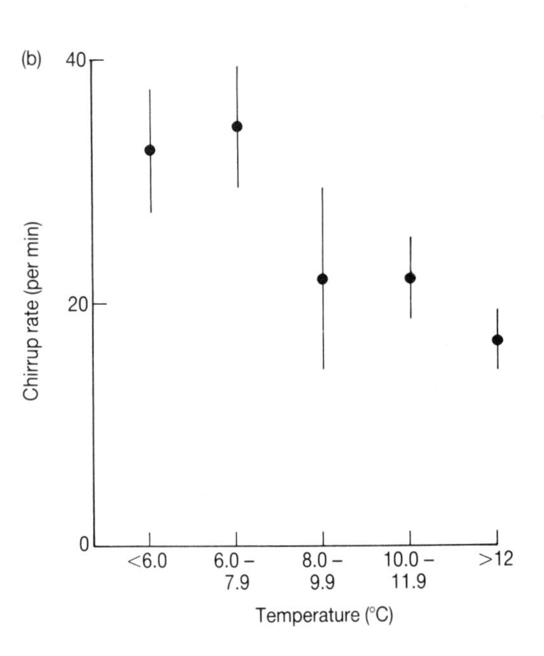

Fig. 4.1 (a) The time for a first-arriving (pioneer) house sparrow to be joined by another sparrow is inversely proportional to the pioneer's 'chirrup' rate (from Elgar 1986b). (b) The rate at which the 'chirrup' call is given is higher when the ambient temperature is lower (from Elgar 1986b).

whether or not it 'decides' to feed depends on many different stimuli from the environment—food location, food type, temperature, presence of predators and so on—as well as stimuli from within the animal's body, such as degree of food deprivation. We are at present a long way from understanding exactly how this decision-making process works and there are, as we shall see in this chapter, many different opinions as to how it is best to go about trying to find out. Some people believe it can be achieved only through the techniques of physiology—by looking inside the animal and recording nerve impulses, muscle movements and hormone levels. Others believe that such a 'complete' explanation is still so far off that we would do best to invoke intervening variables such as 'motivation', at least as a temporary measure, in our search for causal explanations. We will look at these different ways of explaining behaviour, what 'motivation' means and how controversies have arisen. We will do this in the way we stress throughout this book—by looking at different levels of explanation as well as constantly keeping our eyes on the interplay between mechanism and function in the explanation of behaviour. We start by looking at decision-making on a behavioural level and then go on to discuss what is known of its physiological basis.

DECISION-MAKING ON DIFFERENT TIME-SCALES

Time budgets and daily routines

Almost all animals show different behaviour patterns depending on time of day, and many show great regularity in their habits. Male red junglefowl crow just before dawn, then descend from the trees where they have spent the night to forage on the ground in the company of hens. They rest in the heat of the day, forage again in the late afternoon and then go to roost at dusk. In general, the daily routines of wild animals and the amount of time they spend on different activities (their time budgets) can be seen as strategies for coping with changes in the environment such as variations in temperature, activities of predators and availability of food.

To some extent, these daily changes in behaviour depend on animals having an 'internal clock', which times their behaviour more or less independently of events in the external world. Figure 4.2 records the activity of a flying squirrel living in a rotating cage.

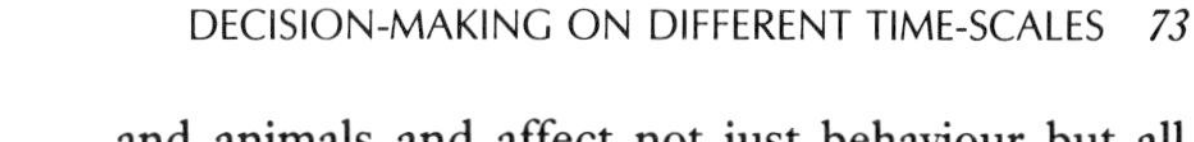

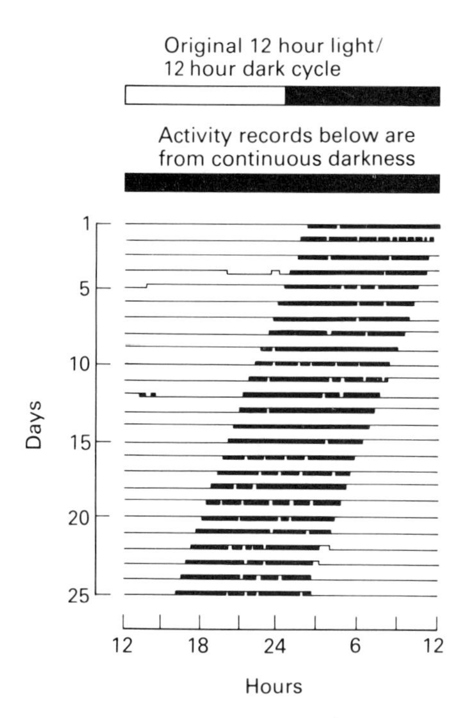

Fig. 4.2 The spontaneous wheel running activity of a flying squirrel in total darkness. Each line is the trace from an event recorder over 24 h. Active periods appear as broad dark lines as the pen of the event recorder moves frequently up and down on the paper. Before these records were taken the squirrel had been kept on a regular cycle of 12 h light/12 h dark—indicated at the top. In total darkness it retains its rhythm and is active only in the period that had previously been dark. In the absence of external cues, however, its natural circadian rhythm, which is slightly less than 24 h, asserts itself and the period of activity begins a little earlier each night (from DeCoursey 1960).

Each row represents 24 hours and the dark bands show when the animal was actively moving around. The squirrel is a nocturnal animal and it was normally active from just after 'dusk' until the lights came on again at 'dawn'. At the beginning of the records shown in the figure the lights were switched off permanently. Now, even though it was living in complete darkness, the squirrel nevertheless maintained fixed periods of activity. These lasted almost exactly the same time as the original dark periods and occurred at extremely regular intervals. Interestingly enough, these intervals are not 24 hours, but just over 23 hours, so that in real time, the squirrel begins its activity a little earlier in each 24 hour period. A rhythm of activity of this type is called **circadian**, from the Latin *circa diem* 'about a day'. Circadian rhythms are widespread amongst plants and animals and affect not just behaviour but all aspects of metabolism (Aschoff 1981, 1989).

Wild animals spend a surprisingly high proportion of the 'active' part of their days resting and apparently doing nothing (Herbers 1981) but this too can be seen as adaptive. Periods of inactivity may be essential for digestion (Diamond *et al.* 1986), energy conservation or simply keeping out of the way of predators. Indeed, spending certain parts of each day sleeping in a hidden place may be a way that animals have of removing themselves from potentially dangerous situations (Meddis 1983). Figure 7.6 (p. 156) illustrates what a high proportion of rest time there is in the life of a honey-bee.

Decision-making from minute to minute

Most studies of animals have involved studying their activities over periods rather shorter than that of a whole day. For example, if we watch a bird foraging through the trees or an animal building a nest, the behaviour may change from one minute to the next. If we look closely, we can see that the animal is not just 'feeding', but that its feeding is divided into searching, capturing food, preparing the food, eating it, flying to another tree and so on. The animal is constantly making decisions about the next stage in its sequence of behaviour—when to stop doing one thing and go on to the next.

An important idea in the study of these short term sequences has been that of the 'goal', which can be defined objectively as that situation that brings a whole sequence to an end. The female great golden digger wasp digs burrows that it then provisions with katydids (large grasshoppers) as food for its offspring. The wasp first digs a downward-sloping main burrow and then a side tunnel, at the end of which is the nesting chamber (Fig. 4.3). Something must switch the wasp from digging the main tunnel to beginning to widen it out and starting to tunnel sideways. The wasp appears to have a 'goal' of a main tunnel of a particular length. Brockmann (1980) altered the depths of main tunnels artificially and showed that the wasps altered their behaviour accordingly. If a wasp was confronted with an artificially lengthened main tunnel she switched to making the side tunnel much sooner, evidently satisfied with the construction, although she had not made it herself.

However, it is not always easy to classify behaviour into 'goals'. When a blackbird builds a nest it

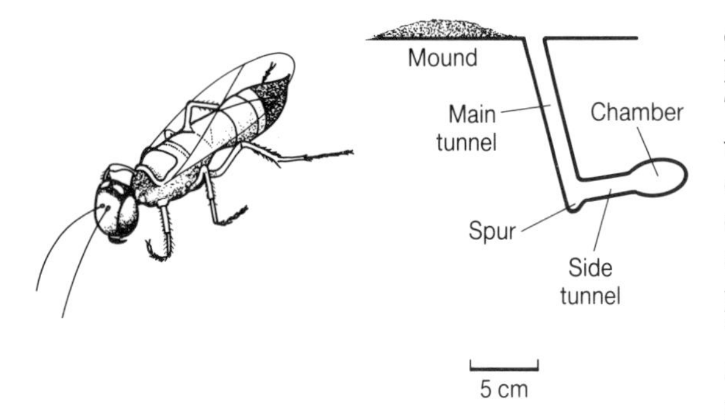

Fig. 4.3 A female great golden digger wasp digs a burrow consisting of a main tunnel sloping downwards with a side tunnel ending in a nest chamber (from Brockmann 1980).

begins by searching for large twigs to form a foundation. Then the sides are made from finer material. When this is complete, mud is collected and shaped to form the cup, and finally this is lined with fine grass and hair. The bird could be said to have one ultimate goal—the completed nest, but also a series of subgoals—nest foundation, sides, cup, etc. along the way. In some cases, the definition of a goal might be extended to imply sort of mental image of a desired situation. Does a rat running a maze, for example, have a mental image of its goal? Do homing pigeons have a mental image of their loft? We will return to such elusive questions in Chapter 6.

Leaving aside the question of whether animals have conscious goals or are just behaving 'as if' they did, we often find it useful to describe complex behaviour in terms of the goal that brings a sequence to an end, particularly when the behaviour itself is very variable and the end-point of apparent goal relatively constant. A dog shown his bowl of food on the other side of a fence will go through all sorts of different behaviour (running, jumping, scratching the fence, etc.) in order to get to it but, in each case, the behaviour would cease once the dog had achieved the 'goal' of getting to its food. It would be difficult to give a concise description of the dog's behaviour without calling it goal-directed.

The searching or striving phase of a sequence has sometimes been called **appetitive behaviour**, with the achievement of the goal referred to as the **consummatory act**. However, these terms derive from a particular model of animal motivation—Lorenz's 'psychohydraulic' model, which, as we will see later, has its critics and the terms are less commonly used today. One reason is that it is often difficult to fit an animal's behaviour neatly into the categories of 'appetitive' and 'consummatory'. A horse grazing in a field literally has its food at its feet. The appetitive phase would have to be described as very short or non-existent and the consummatory act of feeding as continuing without a break for an hour or so until the animal is full. The situation is totally different for the feeding behaviour of a small bird picking minute insects off leaves. Here each brief 'consummatory act' (eating) is followed not by quiescence as the goal is achieved, but by a further phase of appetitive searching. Eventually, but only after some hundreds of such sequences, the appetitive behaviour ceases.

Sequences of behaviour are now much more commonly described not as having just two (consummatory and appetitive) phases, but as involving a whole sequence of decisions. The small foraging bird is thus seen as deciding to fly to an area likely to contain food and then deciding which prey items to select from those available within the patch, and so on. Here, no mention is made of goals. Rather, an attempt is made to specify the circumstances that lead an animal to 'decide' on one behavioural option rather than another.

One widespread explanation of such behavioural decision-making is optimal foraging theory (OFT), which in its original form was based on the idea that when an animal is feeding it makes its decisions in such a way that it will maximize its net rate of energy intake. So, as an animal eats up the food available in a given patch, it will become less and less worth its while continuing to search for food there and more and more worth its while moving on to another patch, which has not been depleted. Its decision to move on to the next part of its sequence of foraging behaviour is, according to OFT, dependent on the balance of energy considerations between staying and flying away. If the next patch is a long way away and demands a long energy-consuming flight, a foraging bird may actually do more for its energy reserves by staying and feeding in the depleted patch than moving on. As it consumes food, there will come a time when there is so little food left in the first patch that it is now worth the bird's while to fly to the next one. Clearly, more distant patches, needing longer flights, will become 'worth it' only when the present patch is severely depleted (Fig. 4.4; see also Krebs and Davies 1987). OFT predicts that an animal should stay in one area, even though it is gradually depleting the food there, until the net rate of energy intake drops to the average for the environ-

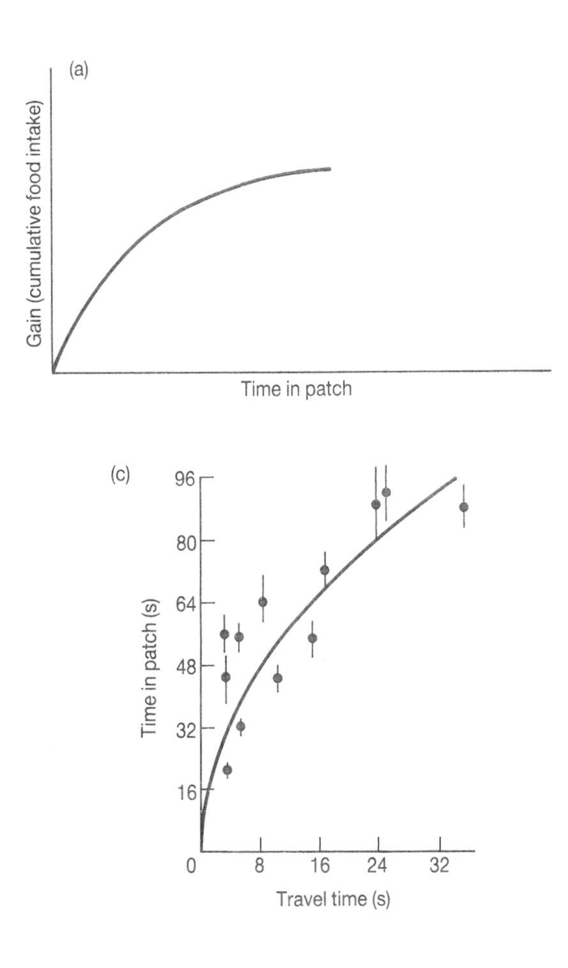

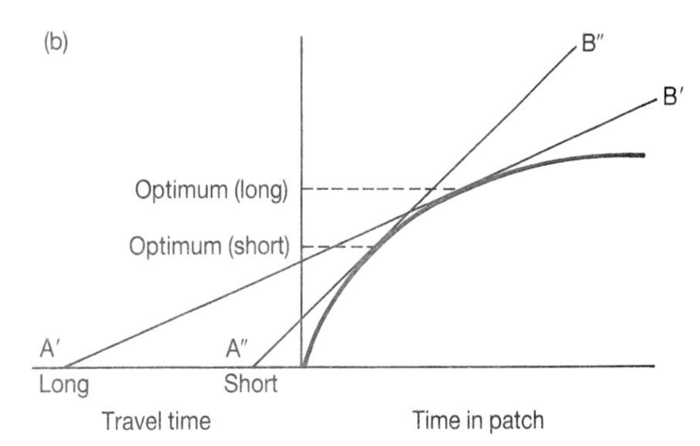

Fig. 4.4 Optimal foraging theory (OFT) applied to the behaviour of a great tit feeding in a patch of food. (a) The curve represents the average net energy gain as a function of time in a patch. The longer it feeds in one patch, the more depleted that patch becomes and so the gain (food intake) begins to go down. Food intake is plotted cumulatively over time, so this results in a 'levelling off' of the gain curve. (b) Calculation of the optimum time the bird should stay in a patch that is being depleted depending on whether there is a long or a short travel time to another (undepleted) patch. The optimal predator should stay in the first patch just long enough to maximize the slope of the line AB (representing the average food intake for the habitat as a whole). So, the optimum time to remain in the patch is given by the tangent of the line AB to the gain curve. (c) Cowrie's (1977) test of OFT using great tits searching for food hidden in pots. The solid line is the result predicted taking into account the cost of travelling and searching in a patch. The twelve dots are the means and standard error of six birds tested with both long and short 'travel times'. Each bird had 6 trials in each environment, and so was able to learn the optimal time allocations in each one.

ment. At this point, the animal should decide to move on to seek out something different.

It will be obvious that this model assumes that the animal knows what conditions are like elsewhere in the environment and that it has some way of comparing the net rate of energy gain it is experiencing now with what it would experience if it flew to another patch and started eating there. The fact that, at least in some cases, OFT does describe behaviour remarkably well, suggests that even if animals do not consciously know these things, they are behaving 'as if' they do. Cowie (1977) studied great tits foraging for pieces of meal worm hidden in 'patches' consisting of sawdust-filled pots in different parts of an aviary. He made moving on to the next patch more or less difficult by putting lids on the pots that were easy or hard to remove. So a difficult lid, by taking more time and effort to remove, was taken to be the equivalent of a long distance between patches. The fit between the time the birds spent in a patch depending on 'distance' to the next one and the time predicted by OFT was remarkably good (Fig. 4.4(c)). The birds were behaving as though they knew how much food was in each pot and how difficult each sort of lid was to prise off. The sequence of their decisions—to fly to an area where food is likely to be found, to prise off a lid, to push away sawdust, to eat, to continue searching in the same spot or to fly off to another one—certainly suggested that they had this knowledge, otherwise the real decisions of the real birds and the predicted decisions of the model would not have coincided so well.

But do birds, or any animals, 'know' about their environments? Perhaps all that is happening is that a bird moves to another patch whenever it has failed to find a food item for a certain length of time, that length of time varying with its past experience of hunting in that environment. It may be following such a relatively simple 'rule of thumb' and not even unconsciously working out 'travel times' or 'energy gains' at all (and certainly not understanding the equations of OFT!). We do not consciously calculate the trajectories of balls thrown into the air, but we often catch them all the same. An automatic process takes over and we behave 'as if' we knew all about the physics of flight and the aerodynamics of moving objects. Some possible 'rules of thumb' that foraging birds follow might be 'leave after catching *n* prey', 'leave after *x* seconds', 'leave after *y* seconds of unsuccessful search', etc. depending on the exact conditions the bird finds itself in (Stephens and Krebs 1986).

Not all cases of animal decision-making fit the predictions of OFT as well as Cowie's great tits. A possible reason for this discrepancy arises from one of the basic assumptions of the model we encountered earlier, namely that the animal is doing all it can to maximize its net rate of energy gain: suppose this is not always the case. In winter, when food is scarce and the survival of a small bird may depend critically on its obtaining enough energy and not expending too much in the process, it might be reasonable to assume that energy gains are paramount. Energy and survival become practically synonymous. But when food is plentiful or when an even greater danger is posed by dehydration or being eaten by a predator, why should any animal go all out for maximizing its net rate of energy gain? Houston and McNamara (1988) point out that there are other threats to survival and reproduction and we might therefore expect natural selection to have favoured animals with wider horizons and more flexible 'rules of thumb'. In other words, we might expect that their decisions about what to do next would be influenced by factors other than just those to do with feeding. We have already come across the example of Elgar's house sparrows, making different decisions about feeding depending in the nearness of danger, the divisibility of the food source, etc. Milinski and Heller (1978) showed a similar effect with sticklebacks feeding on water fleas offered to them in two swarms, one high density and other low density. When the fish were very hungry they preferred to feed at the high density swarm, not unnaturally, as they could feed at a higher rate here. However, when a model kingfisher was flown over a tank containing hungry fish, the sticklebacks preferred to feed on the low density swarm (Fig. 4.5). A plausible explanation for this effect is that kingfishers are predators of sticklebacks and if a stickleback feeds on a high

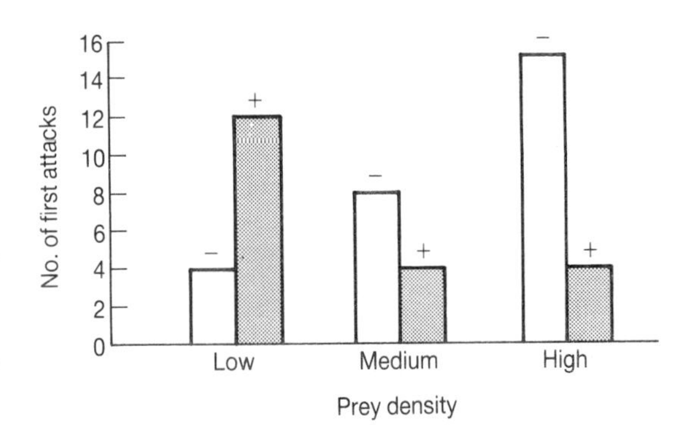

Fig. 4.5 Number of first attacks by sticklebacks when presented with prey (water fleas) at different densities either with (+) or without (−) a model kingfisher overhead.

density swarm, although it may get a lot of food it finds the many, jerking water fleas so distracting that it is less able to keep a look out for predators. The low density swarm of fleas, while giving less food, allows a stickleback to keep an eye out for danger so the decision to feed on the swarm of lower density paves the way, in turn, for a series of even shorter term decisions, to snap at a water flea or to look upwards to see if a predator is about to strike (Milinski 1984). Doing the right behaviour at the right time and place can obviously be considered on many different time-scales.

Decision-making from second to second

We have so far looked at decision-making at the level of gross categories of behaviour such as 'feeding' or 'vigilance' or 'flying to a patch'. But it will be obvious that if we are to make the connection between these categories and the output of muscles, i.e. if we are to have any chance of seeing how behaviour and neurophysiology are really connected, we will have to be much more specific than this. When an animal is 'feeding', what does it actually do? Does it hammer open a mussel like an oystercatcher, graze like a sheep or stalk its prey like a lion? And what does 'hammering open a mussel' involve? What movements does the animal make? If we are ever to understand how the action of muscle systems leads to behaviour, we need to look in more detail at these movement themselves—the movement of limbs and heads and bodies. We should not underestimate the complexity of the machinery underlying animal movement. Even an apparently simple behaviour such as a water snail opening its mouth and scraping its radula along the surface of a plant—an action that takes only 1–2 s—involves the contraction, in the correct sequence, of 25 different pairs of muscles controlling the mouth (Kater and Rowell 1973). There are, of course, many other muscles controlling the head and the snail's foot as it moves along. There is a veritable 'symphony beneath the skin' involving hundreds of muscles, each coming into action at a particular time. If this symphony is complex for water snails, it is even more so for a male mallard duck courting a female, the oystercatcher, the sheep and the lion, all feeding in their idiosyncratic ways. We should remember that no man-made robot comes anywhere near the complexity, the finely tuned adjustment and the flexibility of the behaviour of an animal.

In attempting to investigate the immediate causation—to find how the machinery gives rise to the symphony—it is often useful to focus on easily recognized and distinct sequences of behaviour lasting only a few seconds or so, for then we have some chance of seeing the results of the action of a manageable, if still large, number of muscles. These short sequences are the behaviour patterns such as the courtship strut display of the sage grouse that we discussed in Chapter 1. They have descriptive names like 'pecking', 'drinking', 'wing flapping' and together they constitute the **ethogram** or behavioural repertoire of a species.

Early ethologists laid great stress on these short sequences of behaviour, which they called fixed action patterns, implying that they are immutable and always performed in exactly the same way. Certainly, the courtship patterns of some ducks studied by Lorenz (1941) do have this rigid 'clockwork' quality (Dane *et al.* 1959). However, not all behaviour patterns are so stereotyped. There is enough similarity between different instances of 'drinking' in chicks or tail-wagging in dogs, that we can recognize these behaviour patterns on different occasions, but they are not exactly the same from individual to individual or even from occasion to occasion in the same individual. For this reason, the term 'fixed action pattern' gives an inaccurate impression and a more neutral description is preferable. Most 'behaviour patterns' are recognized intuitively, but film and videotape analysis allows us to be a bit more objective. Dawkins and Dawkins (1974) used slowed down videotape to show that the behaviour pattern of 'drinking' in chicks was very similar from one drink to the next (Fig. 4.6). Drinks were much the same whether or not the chick was thirsty and even whether it was drinking water or a substance that turned out to be distasteful. There was much more variation in the intervals between drinks than in the drinks themselves, suggesting that a decision about whether or not to continue to drink was made between drinks, not during the drink itself.

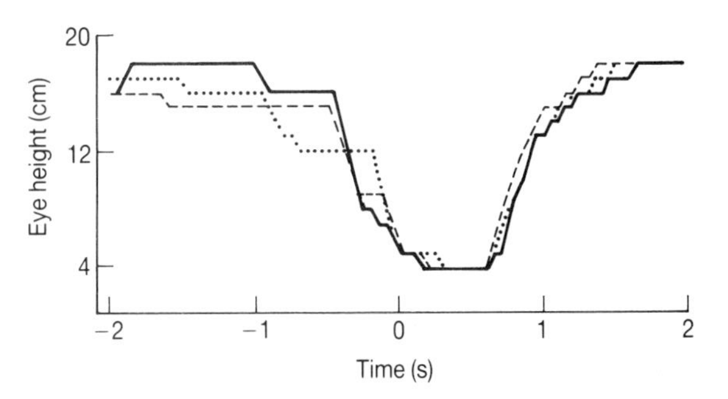

Fig. 4.6 Superimposed graphs of the height above the ground of a chick's eye during three successive drinks, lined up on the moment when the bird's bill strikes the water (time 0) (from Dawkins and Dawkins 1973).

Sometimes behaviour patterns that themselves last only a few seconds can be strung together into sequences that have a high degree of predictability that lasts much longer. The 'song' of the humpback whale (mentioned in Chapter 3) is a long series of sounds that is then repeated in precisely the same way. Each repetition lasts about 40 min—about the same length of time as Beethoven's 5th Symphony!

MECHANISMS OF DECISION-MAKING

So far, we have seen that behaviour occurs in sequences in time-scales ranging from days right down to a few seconds. These sequences are the result of decision-making processes going on inside the animal but it is clear that these decisions are not simply whether or not to perform a particular behaviour. Animals have to weigh up all sorts of conflicting demands, such as whether to flee when a predator appears or continue to feed. As Hinde (1970) pointed out, animals are probably in some such conflict for most of their lives and this means that decision-making is an almost constant feature of animal behaviour. We will now look at how decision-making works, paying particular attention to the parallels in mechanisms at the neuronal and behavioural levels. Following Hinde (1970), we can list the outcomes of behavioural decision-making as follows:

Inhibition of all but one response

A small bird is feeding when a predator appears overhead. It immediately freezes into immobility. In other words, its decision is to inhibit all other activities except the antipredator response, and only when the danger has passed will other behaviour be resumed. As we saw in Chapter 1, inhibition operates at all levels of the nervous system. At the level of neurons, one nerve cell is actually suppressed because of the action of another. At a behavioural level, inhibition implies that behaviour that would otherwise have occurred is prevented from occurring.

In the giant sea-slug, *Pleurobranchaea*, we see a dramatic example of the power of inhibition, mediated through a hormone. As in many other animals, escape behaviour, when it has been activated, inhibits all other behaviour, but in this species, egg-laying also has powerful inhibitory effects. When the sea-slug is laying its eggs, feeding behaviour is inhibited. It is a carnivore and normally eats whatever animal matter it can find, including other sea-slugs, and inhibition of feeding is necessary to prevent it from eating its own eggs. Davis and his colleagues (1977) have shown that the inhibition is brought about by a hormone released when the animal is laying eggs. The hormone affects the working of the buccal ganglia and inhibits the movements of the mouth (Fig. 4.7).

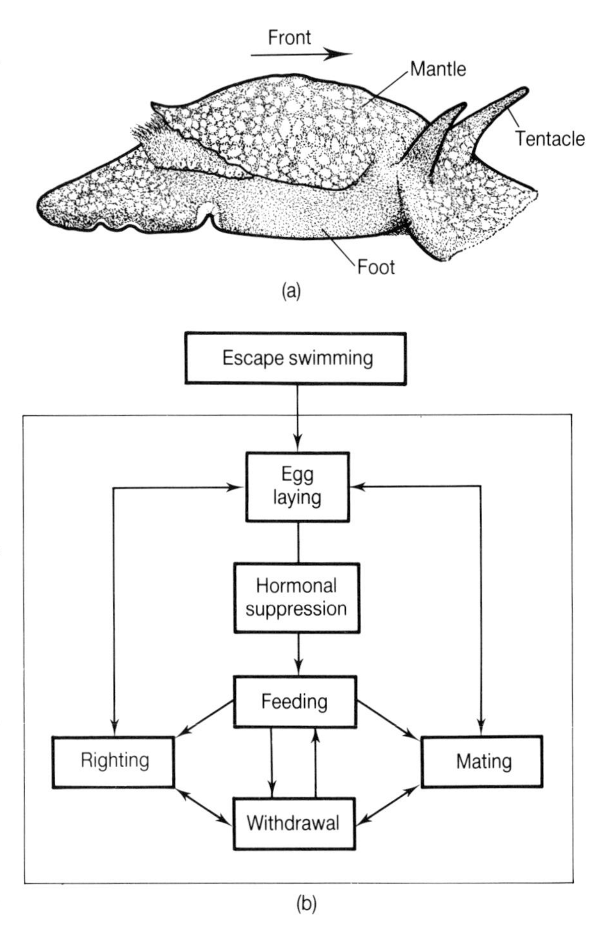

Fig. 4.7 (a) The carnivorous sea-slug *Pleurobranchaea*. (b) The behavioural hierarchy of *Pleurobranchaea*.

This example illustrates inhibition acting in a unidirectional way. *Pleurobranchaea*'s decision-making is straightforward and fixed: if the top priority behaviour is occurring, others are inhibited and do not occur until the inhibition is removed, when they are 'allowed' to occur (by disinhibition). In this case, the priorities are clear and understandable: escape is given top priority because of the serious and immediate consequences of not getting away from imminent danger and it is not advantageous for a sea-slug to eat its own eggs.

However, not all behavioural decisions are as rigidly fixed as this and we often find that inhibition can be two-way, with first one behaviour and then another assuming priority, according to the needs of the animal (reciprocal inhibition). It would be an ill-adapted animal that never drank unless its requirement for food had been totally met, or that did not get enough to eat through constantly fleeing from potential danger. Instead of a rigid hierarchy of feeding always inhibiting drinking, say, we would expect to find a priority system flexible enough to shift according to the needs of the animal. We might expect that even fleeing from danger should not always be given the very highest priority unless the animal is in imminent danger of being eaten. It is not surprising to find that animal decision-making sometimes involves reversals of motivational priority.

Reciprocal inhibition: shifting motivational priorities

When the oral veil of a sea-slug is touched, the animal withdraws it defensively. However, this withdrawal response is inhibited when the animal is feeding by neurons in the buccal ganglia that are active during feeding and inhibit the motor output to the veil (Davis *et al.* 1977). While this looks like a straightforward case of inhibition, the inhibitory interaction can be two-way. If the animal is satiated with food the inhibition of withdrawal normally brought about by the presence of food does not happen. When it is touched, the animal withdraws the front end of the body, this time giving priority to defence over feeding. The inhibition between withdrawal and feeding is clearly reciprocal.

Another example of changing decisions according to circumstance is shown by the courtship of the smooth newt. Newts breathe air but perform their courtship under water. This poses a particular problem for the male newt, because his courtship consists of vigorous side-to-side movements of the tail, which demand a lot of oxygen, while the female's role is much more passive. During courtship, the male rapidly finds himself in a conflict: he is stimulated both to go up to the surface to breathe and to continue with a sequence of behaviour that may lead him to a successful fertilization. As the female newt may lose interest or pick up the spermatophore of another male if he takes time off to breathe, he has to make a decision between these two important activities. Halliday and Sweatman (1976) altered the behaviour of a female newt by encasing her in a 'strait-jacket' (Fig. 4.8). This immobilized female provided enough stimulation for the male to go through the initial parts of his courtship display, but since the female could not move towards him, he did not receive the stimuli necessary to move on to the next stages. He was thus trapped, repeatedly displaying to a female who did not respond. Under these circumstances, the male held his breath for much longer than he would normally do under non-sexual (i.e. no female present) circumstances. However, he would immediately go up for air if the strait-jacketed female was removed. This suggested that sexual behaviour was inhibiting breathing and that breathing occurred by disinhibition when the sexual stimulus was removed.

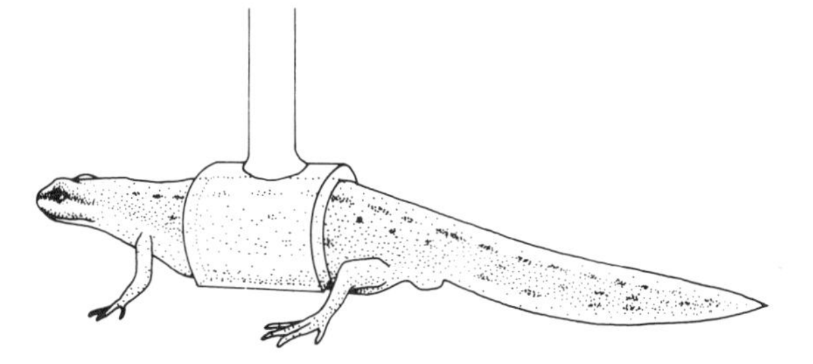

Fig. 4.8 Female smooth newt in a 'strait jacket'. By controlling the behaviour of the female, the male's courtship can be prolonged (drawing by Tim Halliday).

Clearly, the need to take in air imposes an upper limit on the amount of time that a male can devote to sexual behaviour and eventually the inhibition of breathing by sexual behaviour breaks down and he goes up to breathe, leaving the female behind. The interaction between sexual and breathing motivation is, therefore, not fixed, but swiftly adjustable to the immediate needs and opportunities of the animal.

Inhibition is of very great importance in animal behaviour, at all levels. In Chapter 1 we saw how the most basic movements of limbs depend on inhibition: flexor muscles must inhibit extensors and vice versa or they would cancel each other out and no movement would occur. At a behavioural level, too, effective behavioural decisions depend critically on inhibition to resolve conflicts adaptively. An animal's food and water will often be located in different places with a distance between the two, so that inevitably there will be a cost (in energy, time, conspicuousness to predators, etc.) to changing from feeding and drinking. The worst possible solution would be for the animal to 'dither' between the two—taking a drink and then immediately running to feed and then going back to the water and so on. Rapid alternation would carry a high cost of constant changes. There has to be a degree of 'authority' with inhibition of competing responses, at least for short periods of time before the next behaviour occurs. The nature of this authority (that is whether the inhibition is one-way, reciprocal or subtly shifting from moment to moment), is clearly very variable and is considered in more detail by McFarland and Houston (1981). The decision-making shown by most vertebrates is far less rigid and inflexible than that described above for *Pleurobranchaea.* In fact, because many vertebrates have highly complex decision-making rules, they may not make their decisions immediately. Instead of a rapid resolution of conflict, the state of indecision may persist for considerable periods of time.

Prolonged conflict: unsettled motivational priorities

There are two reasons why an animal's decision-rule might not result in immediate action. The first is that it may not be able to carry out the behaviour that it has 'decided' to do. For example, it may be highly stimulated to drink and all set to give top priority to drinking behaviour over anything else, only to find that its water is covered with an immovable glass lid. Such an animal is said to be 'thwarted'. The second reason for prolonged conflict is that the animal may need time to collect information about its various options and cannot make an instantaneous decision. This is particularly likely to happen in encounters with other animals where an animal may be in a conflict between attacking an opponent and fleeing from it. It cannot decide what to do because initially it does not have sufficient information about its opponent's fighting ability or willingness to fight.

'Displacement activities', frustration and stress

Tinbergen (1951) and Kortlandt (1940) described a very curious category of behaviour, which they called **displacement activities**, the most striking common characteristic of which was their apparent irrelevance to the situation in which they occur.

For example, in the middle of a fight, two cockerels may stop and display, each turning aside briefly and pecking the ground, sometimes picking up stones or grains, which they allow to fall again. A male stickleback that has been courting an unreceptive female will suddenly swim to his nest and perform the characteristic parental 'fanning' movements, which ventilate the nest with fresh water, even though there are no eggs present. A tern that is incubating eggs in its nest makes preening movements just before it takes off at the approach of an intruder. The thirsty dove, which is prevented from getting to its water bowl by a sheet of glass, pecks at the ground near by, or preens itself.

In all these examples, there is good reason to suppose that the animal is either thwarted or in a conflict between two opposing tendencies. The appearance of, say, feeding in the middle of a fight is surprising because it seems irrelevant to the conflict (presumably between attack and escape) that the cockerels appear to be in. Similarly, when a tern is faced with the decision of whether to escape from a predator or stay and incubate its eggs, preening itself seems 'irrelevant'.

Rowell (1961) provides some evidence that displacement activities occur when an animal is in a state of prolonged conflict and unable to decide what to do. He worked with chaffinches in aviaries and used two methods to produce a conflict between approach and avoidance. In one, a stuffed owl placed just outside the aviary provoked the chaffinches to mob it. The chaffinches were in a conflict between attack and escape and could 'decide' to settle on any one of a row of perches at different distances from the owl. The second situation involved hungry birds that had previously been trained to use a food dish at one side of the aviary. During these tests, when they reached the dish, a bright light was flashed on inside it and this then provoked escape. Here, the conflict was between hunger and escape. In both cases, the birds tended to make short flights between perches, resting briefly and gradually getting closer to the owl or food dish and then retreating. Rowell found that, for both conflict situations, the birds paused more often and for longer at intermediate perches neither too near nor too far. He also found that chaffinches did most preening and bill-wiping on intermediate perches. While this suggests that displacement activities are most likely to occur at a balance point between two conflicting tendencies, it is not conclusive. One would get the same result if preening simply occurred at a fairly constant rate, wherever a bird made a pause and was doing nothing else—the longer the pause, the more preening.

In control observations of birds under no conflict, Rowell found that this was indeed the case, but still the amount of preening per length of pause was significantly greater when the birds were in a conflict, which justifies labelling this extra amount 'displacement preening'. Roper (1984) gives a more recent discussion of displacement activities and addresses the vexed question of whether displacement activities really constitute a special category of behaviour at all.

If conflict or thwarting are prolonged for days or weeks, with the animal allowed no chance to escape, then stereotypies such as bar-biting in confined sows (Fig. 4.9) and head swinging movements in caged elephants or polar bears may develop. Broom (1986) discusses these stereotypies in the context of animal welfare, and we return to this important topic on page 91. Other 'abnormal' behaviour has been produced in the laboratory. Masserman (1950)

Fig. 4.9 An example of a stereotypy: bar-biting by sows confined in stalls (photograph by Mike Appleby).

trained cats to open a box for a food reward when a signal light flashed. Later, when the food box was opened, the cats sometimes received a strong blast of air. Under these conditions, the animal's behaviour often became severely disturbed. Some of the cats became hyperexcitable, others moped in corners for days on end. They nearly all showed signs of acute stress, with raised blood pressure, hair erection and gastric disorders.

Many bodily changes are likely to occur in conflict and thwarting situations. We can lump them under the term 'stress' because the body's response to a wide range of 'stressors' (thwarted escape, overcrowding, extreme cold and burns, for example) is very similar. Barnett (1964), Archer (1979) and Fraser and Broom (1990) describe the physiological changes involved and point out that most of them are attempts to restore the delicate balance of the body's metabolism when it has been upset.

In moderate stress, we can detect increased activity in the autonomic nervous system, which supplies the viscera and smooth muscles. It also supplies the adrenal glands, endocrine organs close to the kidneys with a double structure—an internal medulla (supplied by autonomic nerve fibres) and an external cortex. When stimulated via its autonomic nerve supply, the adrenal medulla releases the hormone adrenalin into the body. This causes changes in numerous parts of the body; the sweat glands of the skin begin to secrete, hair becomes erected, the heart beats faster, breathing becomes more rapid and deeper and blood gets diverted to the muscles from the alimentary canal. These changes also accompany the strong arousal tendencies such as attack, escape and sex—not just conflict. They prepare the body for violent action of any type required.

In brief conflicts, there will perhaps be a rapid flush of adrenalin through the animal, which then subsides, but if the stressful situation persists then a further reaction begins. This involves the other part of the adrenal glands, the adrenal cortex, which is stimulated to release its hormones, not directly by nerves as is the medulla, but by another hormone, adrenocorticotrophic hormone (ACTH) produced by the pituitary gland. Here, as with the release of adrenalin, the nervous system initiates the response. Stress activates cells in the hypothalamus, which itself then stimulates the pituitary to release ACTH. It is not fully clear how the adrenal cortex hormones help the animal to adapt to stress. Some of them are concerned with glucose metabolism and may serve to mobilize the body's long term food reserves. Whatever their action, the release of adrenocortical hormones is most dramatic. The cortical cells become drained of their contents and, if stress persists, the adrenals enlarge, sometimes by 25 per cent.

Animals under chronic stress become really ill and may die. Barnett (1964) has shown how a wild rat, unable to escape from the territory of a dominant resident male, may die after a few hours of intermittent attack, even though it has no wounds.

We know that the stress resulting from a prolonged conflict is often severe enough to produce physical damage. Animals may develop gastric ulcers or tumours of the pituitary gland. In such cases, it is impossible to regard the response as being in any way adaptive. The unfortunate animal finds itself in a situation to which there is no solution. It is not surprising that no mechanism exists to deal with chronic conflicts. It could provide little selective advantage in the natural world where escape from conflict situations is almost always possible as a last resort.

Conflict and display

Many of the conflicts that persist for long periods of time arise in social situations—particularly aggressive encounters—where an animal is in a conflict between attack and escape. In the roaring duals of red deer, for example (discussed in Chapter 3) two males may remain in a state of unresolved conflict for half an hour or more as they assess each other's ability to fight. The roaring and parallel walks that the males use to assess each other both take time.

The conflict between attack and escape that the challenger stag experiences will not be resolved (that is, neither attack nor escape will predominate) until he has received information about the likely fighting ability of his opponent. Hearing the harem holder deliver a high number of roars per minute—something only a physically fit and strong stag can do—will probably tip the balance in favour of retreat. He turns away, no longer in a conflict.

Many other displays can equally be seen as arising from unresolved motivational conflicts, unresolved, that is, until the contestants have assessed each other for their relative fighting abilities and other factors we discussed in Chapter 3. In classical ethological theory, these displays were called 'threats' but although present day workers emphasize the information about the sender that a signal conveys (status, fighting ability, etc.) rather than its threatening or intimidating qualities, the idea is essentially the same. Tinbergen (1952) saw many signals resulting from dual motivation, conflict between simultaneously aroused tendencies to attack and to escape, when neither can find separate expression. Several lines of evidence pointed to this conclusion, long before modern ideas about game theory or assessment were conceived of.

Firstly, it was noticed that 'threat' often occurred at the boundaries between territories, where there is good reason to believe that both tendencies to fight and to flee are shown simultaneously. This might be called evidence from situation or context.

Linked with this conclusion, there is some evidence on the mechanism of threat that comes from independent manipulation of the levels of attack and escape tendencies. Blurton-Jones (1959) had a group of completely tame Canada geese, which ignored him if he was dressed in familiar old clothes. If he wore a white coat, the geese attacked him uninhibitedly, whilst if he appeared carrying a broom (used to drive the geese into their house for the night) they would flee. The familiar threat postures of the goose (lowered head on outstretched neck, hissing, etc.) only appeared when Blurton-Jones combined wearing a white coat with carrying a broom!

Finally, there was also evidence gained from a close examination of the forms of threat postures that can sometimes be analysed into elements belonging both to attack and to escape behaviour and for this reason such postures are called 'ambivalent'. For example, Tinbergen (1959) gave a detailed analysis of the threat postures of the lesser black-backed gull (Fig. 4.10(a)). The bird moves towards its rival with the neck stretched upwards and slightly forwards and the head and bill turned down. The wrist joints of the wings are lifted well clear of the body and the plumage is slightly raised. Now, gulls normally launch an attack by beating with the wings and attempting to peck down at their opponent. The raising of the bill and the down-pointing bill look like actual attack. Elements of escape can also be seen, particularly as the two rivals come very close. Now the head moves increasingly back, the bill becomes lifted and the plumage sleeker; the bird may turn sideways to its opponent and more parallel, not towards it. This turning aside looks like an element of escape. Gulls draw back their heads and sleek the plumage preparatory to taking off so these may be escape elements also. It is interesting to note that the so-called 'appeasement' display adopted by submissive birds (Fig. 4.10(b)) is almost an exact antithesis of the threat posture. Compare these gull displays with the functionally equivalent postures of the dog (Fig. 4.10(c) and (d)), with which there are remarkable parallels.

Not all threat postures can be interpreted as mixtures of attack and escape and indeed, as we have seen in Chapter 3, animals may use quite different displays as the only 'honest' way of signalling their fighting ability, rank, etc. The conflict between attack and escape may still be there but not always manifest in ambivalent 'patchworks' of display. Baerends (1975) gives a stimulating review of the interpretation of displays as a result of a balance between two conflicting tendencies.

Maynard Smith and Riechert (1984) have also used this idea to develop a model of the behaviour of the desert spider, *Aegelopsis*. This spider is very aggressive and fights over access to web sites. In Maynard Smith and Riechert's model there are two variables, fear and aggression, which fluctuate during the course of an encounter between two spiders according to such factors as who owns the disputed web site, relative body weight of the two, territory quality, etc. There are also some genetic factors involved, because it is known that some spider populations are more aggressive than others, even when the external situation is kept constant. In their model, if fear greatly exceeds aggression, the animal simply withdraws. If fear is equal to or less than aggression, then the contest continues, but what behaviour is performed (locate–signal–threat–

Fig. 4.10 Threat and appeasement postures. (a) Shows the 'upright threat' of the lesser black-backed gull and (b) the hunched appeasement posture of the same species. This latter represents an almost perfect opposite to threat. The head is held low on a shortened neck and the bill points upwards, whilst the wings are pressed close into the flanks so that the wrist joints—so conspicuous in the threat display—are completely hidden. Now compare the gull postures with those of the domestic dog—illustrations taken from Charles Darwin's *The Expression of the Emotions in Man and Animals* of 1872. He entitled them (c) 'Dog approaching another dog with hostile intentions' and (d) 'The same in a humble and effectionate frame of mind.' The parallels with the gull postures are remarkable and both animals exhibit clearly what Darwin called the principle of antithesis in their communicatory behaviour. (a) and (b) by Niko Tinbergen; (c) and (d) by permission of the Syndics of the Cambridge University Library.

contact) depends on the absolute level of the two variables.

The model accounts well for many observed facts about spider fighting behaviour, such as the fact that owners of web sites tend to win, larger spiders tend to win, etc. Maynard Smith and Riechert argue that a model with two variables is the simplest one that will account for all the observed data. It does appear, then, that the traditional idea of agonistic encounters (including threat) as extended conflict has stood the test of time, although there has perhaps been a shift of emphasis. Animals may display when they are in a state of conflict, not to signal that they are in a conflict but, by displaying and seeing the displays of others, to get themselves out of the conflict. They make their decision on whether to attack or to flee on the basis of new information received during the cause of the encounter. Their decision to do one or other behaviour is then made in an adaptive way that would not have been possible without the period of extended conflict and the exchange of information it made possible.

'DRIVE', 'MOTIVATION' AND PHYSIOLOGY

As we mentioned at the very start of the book, the physiological approach to understanding the mechanisms of behaviour involves making direct measurements of nerves, brains, muscles and hormones. We are making some progress in our understanding of the physiological mechanisms of animal decision-making, as the previous section showed. In *Pleurobranchaea*, for example, the decision to feed and not to escape is mediated through particular known neurons, and the decision to inhibit feeding when the animal is laying eggs is mediated via a specific hormone. But with many other animals, the decision-making processes are so complex that we are still a long way from any adequate physiological explanation. The foraging great tit, deciding whether or not to continue to search in any area where it has found no food for 3 min or to fly to another tree, at present defies physiological analysis and there is not much point in pursuing the endeavour—it remains in the realm of behavioural analysis. The same could be said of the red deer stags deciding whether to fight or the tern parent deciding whether a predator constitutes a sufficient threat that it should fly away from its nest.

The evident complexity of animal behaviour and the awesome nature of a full physiological explanation even if we had it (a snail has 10^6 neurons; a cat 10^9) has led many workers to study the mechanisms of behaviour without even attempting to confront all the physiological details. Instead, they invoke intervening variables such as 'motivation' to explain the, as yet, unknown causal factors that are at work inside the animal. We have already seen some examples of this approach in action, even though the word 'motivation' has not yet been used. In Maynard Smith and Riechert's account of the behaviour of the desert spider, two unspecified variables, 'fear' and 'aggression' were invoked. Despite the fact that no-one yet knows the physiological basis of these variables, their model has considerable explanatory power in predicting the outcome of fights, so that it seems reasonable to say that two 'somethings' inside the animals are varying independently. Those 'somethings' are what is meant by fear motivation and aggressive motivation. Sometime in the future we may have learnt so much about physiology that we will have no need to invoke such motivational variables. But for the moment, they enable us to proceed with behavioural analyses without waiting for all the physiological details to be filled in.

The same approach can be used for a wide range of different situations, for example, if an animal stops feeding and starts drinking. If the external environment has remained constant and the behaviour has changed, something inside the animal must have changed and this change can be studied and fruitful deductions made about its nature. We will first look at how 'motivation' and 'drive' have been used as explanatory concepts in animal behaviour and then look at how far they are understood in physiological terms.

Changes in 'motivation' are deduced when we can eliminate other factors that may cause changes in behaviour such as 'fatigue' (p. 7), 'maturation' and 'learning' (Chapter 6). Motivational changes occur over a short term and are usually reversible. The animal shows a changed threshold of response to particular types of stimuli: sometimes a very slight stimulus is adequate to evoke a powerful response, at other times far stronger stimuli are ineffective. Which motivational state the animal is in may change over minutes or hours and then change back again.

Another characteristic of motivational changes is that it is often not just the animal's response to one specific stimulus that fluctuates, but a whole range of responses that are functionally related to one another. Thus an animal's threshold of response to all stimuli connected with food and feeding behaviour will rise and fall together, so will those connected with sexual behaviour and so on. It is because of this association of effects with different sets of stimuli that ethologists and other workers have talked about 'specific motivational states'—a set of internal causal factors that affect not just one behaviour but a whole functional group. Instead of invoking a different motivational state for every single element of its behavioural repertoire, such as a 'chasing motivation' and a 'swallowing motivation' the animal is said to have 'feeding motivation', implying that most or all of the behaviour functionally related to obtaining food (stalking, chasing, biting, etc.) are affected by many of the same causal factors. Feeding motivation thus refers to factors that, whatever they are, affect the whole group of feeding-related behaviour patterns. Evidence for the operation of common causal factors comes from work such as that of Baerends and his collaborators (1976), who studied the changes in various behaviour patterns that occurred when herring gulls were disturbed at the nest while incubating an egg. In the 90-min period following the interruption, three behaviours (building, preening and resettling) all decreased at first and then increased at very much the same time, suggesting that all three shared at least one common causal factor, the level of which was falling and then rising again. Baerends (1976) also used an analysis of what behaviour precedes or follows another one to identify which ones might have common causal factors. He found that there were two sets of behaviour patterns. One set involved resettling, pecking, picking up material and sideways building. The other set was composed of turning, head-shaking, scratching, shaking, yawning and looking around or looking down. If a gull had just performed one of the behaviour patterns in the first set, its next action was most likely to be one from the same set, not one from the second set. If it had just performed a second set behaviour, the converse was true. This suggested that, within one set, the various behaviour pattern shared a common causal factor and that this was different (or at least different in effect) from the other set. Possibly the two causal factors acted antagonistically.

Here, then, we can begin to say something about what appears to be going on inside an animal, using the temporal correlations between its different behaviour patterns, although the exact nature of the various causal factors has not been determined. In the next section, we will look at some more detailed motivational models and the sort of understanding they can give us about animal behaviour, but first a word of caution, and a further clarification.

It would be misleading to think of animals as though they had fluctuating sets of causal factors for 'feeding', 'drinking', 'sex', etc. all operating independently of one another. Many stimuli produce non-specific 'arousing' effects, which render animals more responsive to a wide range of stimuli and this could be described as a rise in 'general motivation' or, as it is called by some psychologists, 'general drive'. Although it might seem a straightforward question, it is in fact very difficult to collect really conclusive evidence as to whether motivation is general or specific; there is a good discussion of the problem in Chapter 9 of Hinde (1970) (see also Grossman 1967; Toates 1986; Colgan 1989). In what follows, we shall assume that there is a considerable degree of specificity, although it is certainly wrong to think of motivation as being rigidly compartmentalized. Different motivational systems certainly interact with one another and these interactions have major effects on behaviour.

Even more confusion surrounds the term 'drive' than 'motivation'. 'Drive' was originally introduced by R. Woodworth who saw drive as 'energizing' behaviour. He distinguished between the energizing (drive) aspects of motivation and the directing aspects of motivation. This idea of drive as an urge to perform behaviour was taken up by Lorenz (1950) and various schools of animal psychology in the United States. Ever since it was introduced, it has provoked objections and has been criticized, sometimes because it is too vague and imprecise to be useful, sometimes because animals seem not to be 'driven' at all (is there a sleep drive?), but mostly because the whole idea of anything energizing behaviour seemed to be an error (Hinde 1970; Bolles 1975; Kennedy 1987). The general consensus of opinion now seems to be that 'drive' is not a useful term at all. In the general discussion of motivation that follows, we will touch on it only briefly in describing certain models of motivation that have made use of it.

MODELS OF MOTIVATION

We have now outlined some of the behavioural observations that have led people to postulate 'motivation' or internal factors common to several different behaviour patterns. Obviously these internal factors are not the sole cause of behaviour. Animals **react** to external stimuli (Chapter 3) and the ways in which internal and external factors work together to produce behaviour has been the subject of much speculation. In particular, people have developed models of animal behaviour, the purpose of which is to devise a system of hypothetical components that are seen to be connected together in such a way that their behaviour mimics that of real animals.

Good models can help us to organize our thinking and suggest experiments that can then test how good they are. If we find that one model consistently explains an animal's behaviour under a wide range of conditions, that may tell us a great deal about the **principles** upon which the animal's nervous system is working. It can tell us little or nothing about the **means** of operation. The nervous system uses large numbers of interconnected neurons for its operation and the interpretation of how the principles are put into practice remains a problem for neurophysiologists.

This distinction between principles and means is well illustrated by Lorenz's (1950) famous 'psychohydraulic' model of behaviour, which is illustrated in Figure 4.11. Lorenz drew an analogy between the way a water tank with a spring valve operates (e.g. immediately after flushing it cannot work until the water has had time to build up) and animal behaviour (e.g. immediately after feeding, it may be difficult to elicit more feeding behaviour). This model is now not considered to be satisfactory, not because someone discovered that animals do not have water rushing round inside them (that had never been claimed), but because animals do not behave like water tanks with valves. For example, as long as the tank has not been recently emptied, it should be able to discharge water and should only cease to be able to do so when all its water has been used up. If animals operated like this, they would stop behaving in certain ways only when their motivational equivalent of water (which Lorenz called 'action specific energy') has been used, i.e. through previous performance of behaviour. However, there are a number of cases where behaviour ceases without the full sequence having been performed. Sevenster-Bol (1962) showed that male sticklebacks do not need to perform the act of fertilization to stop showing sexual behaviour. The male stickleback builds a nest and, after a female has laid eggs in it, he enters the nest and fertilizes them. Sevenster-Bol showed that the male stickleback's sexual behaviour is reduced for a while after he has fertilized one clutch of eggs (Lorenz's model and the real stickleback being apparently similar in this respect) but it is not the performance of the act of fertilization that causes this reduction. It is the sight of eggs in the nest. Placing a new clutch of eggs in the nest had exactly the same effect, showing that it is the usual result of sexual behaviour (eggs in the nest) that reduces sexual behaviour. Similarly, animals stop feeding not because they have performed the act of eating but because there are various consequences of eating (such as the taste of food in the mouth and food in the stomach) that normally follow. Janowitz and Grossman (1949) showed that, if food is placed directly into the stomach of a hungry dog, thus by-passing the act of eating, the dog does not eat nearly so much when subsequently offered food. It is

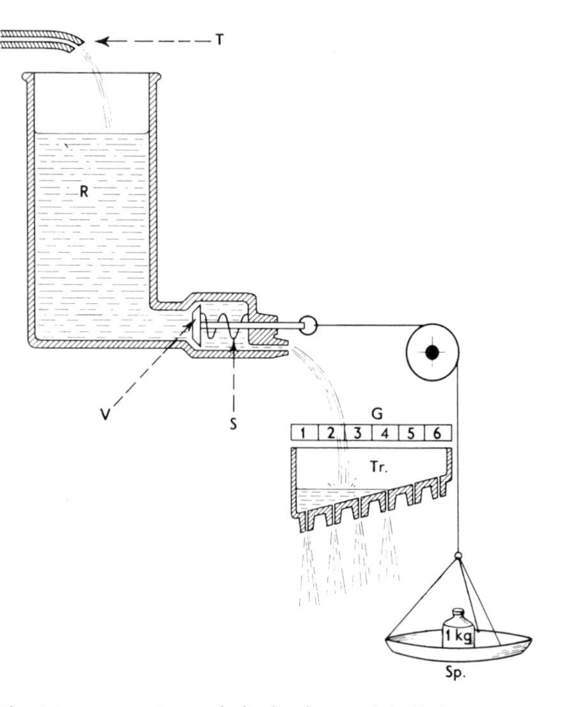

Fig. 4.11 Lorenz's 'psychohydraulic' model of behaviour in which water was envisaged as building up inside the reservoir R, to be released by stimuli (represented by weights on the scale pan) (from Lorenz 1950).

the stimuli from the distended stomach, not the act of eating, that is most important in switching off feeding behaviour.

Lorenz's model fails not because it involves water and plugs and weights but because it does not behave sufficiently like a real animal. Specifically, it lacks a feature that is now known to be important in many aspects of animal behaviour—feedback from the environment as a result of earlier actions (see Chapter 1). Feeding, drinking and nest-building are just some of the behaviours that are drastically affected by what the animal itself has done earlier, through the animal's own action on its environment. Such feedback effects have now been incorporated into a number of motivational models under the general heading of **homeostatic models** (McFarland 1971; Toates 1986).

'Homeostasis' was the term coined by Cannon in his book *The Wisdom of the Body* (1974) to describe the relative stability of the body despite the changes that go on in the outside world. Our internal body temperature does not depart much from 37°C, even when the external temperature may be much hotter or much colder. We constantly lose water by evaporation, urination and so on and yet the body's fluid volume remains roughly constant. Homeostasis implies that the body has some means of correcting deviation, so that if temperature or fluid volume falls, steps are taken to restore the balance. The starting and stopping of drinking behaviour can be seen as part of this homeostatic mechanism. Body fluids decrease, this stimulates the animal to drink and this in turn helps to correct fluid loss, returning the body fluids to some 'ideal' or normal value. Figure 4.12 shows a simple homeostatic model of motivation in which this process is represented diagrammatically.

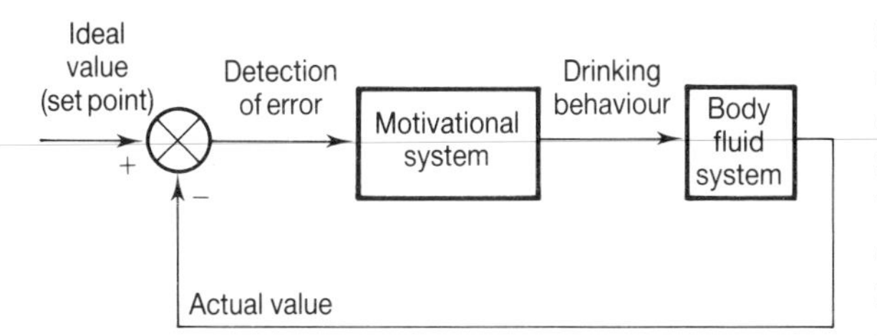

Fig. 4.12 Simple homeostatic model of drinking behaviour. The actual state of the body fluids is compared to an ideal 'set-point' and any difference activates drinking. Water reaches the body fluids via the mouth and gut (part of the 'body fluid' system) and corrects the deficit, switching off drinking.

Homeostatic models assume that there is an ideal state or set point for the animal. If there is a difference between this set point say, for body fluids and the actual state in the body, this so-called 'error' or 'discrepancy' is said to provide the motivation for drinking. You may see such a system described as a 'negative feedback' loop (see Chapter 1) because error is 'fed back' into the system and stimulates the behaviour or physiological response to operate until the error itself is reduced. The system is thus self-correcting.

In many instances, such a negative feedback model of motivation seems to provide a reasonably good analogy to the behaviour of a real animal. Fitzsimmons (1972) showed that rats injected with salt, which dehydrates them, drink just enough water to restore their fluid balance. The effects of placing food directly into the stomach, which Lorenz's model does not explain, are also understandable on a homeostatic model: loading the stomach with food is enough to 'turn off' eating because a full stomach is normally part of the negative feedback loop of the homeostatic feeding system.

It would be quite wrong, however, to think of animals like simple thermostats, switching behaviour on and off whenever they are in particular states. Homeostasis is in practice much more complicated than this (Rolls and Rolls 1982) and for good functional reasons. Consider the problem of an animal having to maintain a constant fluid volume and composition in its body despite losses of water due to urination, perspiration, etc. and despite variations in the external temperature, the composition of its food and the availability of water in its environment. Its first difficulty is the time-lag that exists between behaviour (drinking) and the eventual effect on fluids in the body. When rats and other mammals are deprived of water for a long time the cells of their bodies become dehydrated and their extracellular fluid (such as plasma) also diminishes in volume. One of the main reasons why we become thirsty when we take in salt is that the salt stays outside cells (because it cannot get through the cell membrane) and draws water out of the cells by osmosis. Water loss from cells is then detected by specialized 'osmoreceptors' in the brain, which are spread out over quite a wide area of the lateral hypothalamus (Fig. 4.13). Blass and Epstein (1971) injected small amounts of saline or sucrose solutions into the lateral hypothalamic areas of rats and found

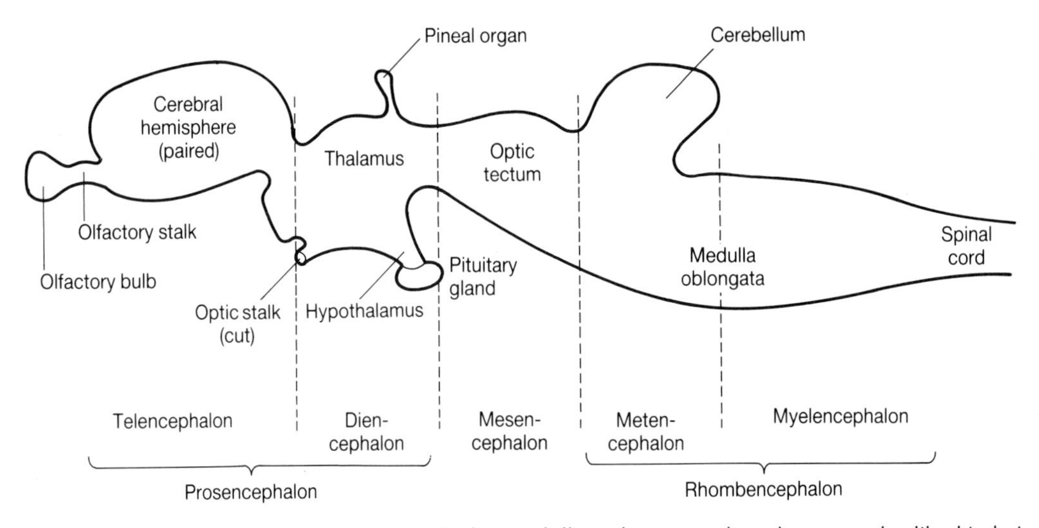

Fig. 4.13 The basic divisions of the vertebrate brain. The brains of all vertebrates pass through a stage rather like this during development, but in mammals and birds in particular, the adult brain is dominated by the enormous growth of the cerebral hemispheres and cerebellum. These come to overlay all the rest and obscure the original layout (from Romer 1962).

that the rats started drinking, but that they did not do so following injection with urea. Since urea, unlike salt or sucrose, is able to cross cell boundaries, it therefore does not draw out water from the cells and the stimulation of drinking therefore seems specifically related to such cellular dehydration.

Now, on the very simplest sort of homeostatic model, drinking would occur when cells become dehydrated and cease when they become rehydrated again. But since most animals drink much more rapidly than the fluid can be restored to their cells, the animal would drink far too much if it drank until its tissues and plasma were fully rehydrated. If human beings are deprived of water for 24 h and then allowed to drink, they will drink almost all that is needed to restore fluid balance within 2.5 min, although changes in plasma dilution cannot be detected for 7.5 min and are not back to normal for about 12.5 min (Rolls and Rolls 1982).

There must, therefore, be means of detecting that water has been taken into the body before the full physiological consequences have made themselves felt. Miller *et al.* (1957) showed that at least some of these water detectors are in the mouth and oesophagus and that activation of these causes termination of drinking even before fluid balance is restored. They allowed three groups of previously water-deprived rats to drink for 18 min. The first group had just had 14 ml of water loaded directly into their stomachs (i.e. bypassing the mouth and oesophagus altogether), the second had been allowed to drink 14 ml in the normal way and the third group of rats had had no water at all. Putting water into the stomach had relatively little effect on the amount drunk immediately—the stomach-loaded rats drank almost as much (average 16 ml) as the completely deprived rats (average 21 ml), whereas the rats that had been allowed to drink for themselves drank only 6.7 ml during the 18 min of the test (Fig. 4.14). Water is thus detected in the

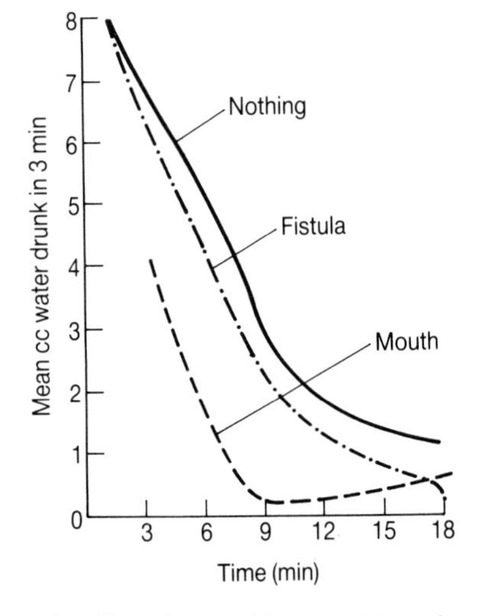

Fig. 4.14 The effect of giving thirsty rats 14 cc of water by mouth or by stomach fistula on the amount of water they subsequently drank in each of 3 min intervals immediately afterwards compared to control animals given no water (from Miller *et al.* 1957).

mouth and throat during the course of normal drinking and reduces the subsequent tendency to drink. Distension of the stomach and stimulation of the intestine (together with possible hepatic-portal factors) may also play a part in terminating drinking, the effects being different in different species. For example, in dogs gastric distension is relatively unimportant for drinking, but it is very important in monkeys. Furthermore, some drinking occurs not because of a fluid deficit at the present moment but because of an anticipated fluid deficit in the future. Fitzsimmons and LeMagnen (1969) showed that, when feeding, rats drink in anticipation of a water deficit (dry food will make them thirsty), a phenomenon that McFarland (1971) called 'feed-forward'. The state of the body fluids is certainly detected but it constitutes only one among many other factors in determining whether an animal will drink at a given moment. But whatever the precise mechanism, it is quite clear that a simple homeostatic model, involving only the detection of fluid imbalance to initiate and terminate drinking, would be a poor predictor of the drinking behaviour of real animals.

Simple homeostatic models similarly fail to explain all aspects of feeding behaviour. Animals fed to satiety on one kind of food will often resume feeding if given a wider range of foods. Wirtschafter and Davis (1977) showed that rats used to a normal laboratory 'rat food' diet and then give a 'supermarket' diet (a variety of sweet and palatable foods) will eat a great deal more than usual and may even become grossly obese. As we have already seen, feeding behaviour is also affected by the availability of food, the presence of predators, whether other conspecifics are present or not and so on.

All models are simplifications. Nevertheless, the attempt to identify certain principles, such as negative feedback, can be an important step in understanding some aspects of animal behaviour, although we may be fully aware that it is not a complete explanation. Some people interested in motivational models have made a distinction between behaviour that is homeostatic and behaviour that is not separating systems, such as feeding and drinking where there is an identifiable set-point and non-homeostatic systems such as sex or aggression where it is difficult to see what the set-point would be. On the other hand, McFarland (1971) argues that there are no fundamental differences between these two categories on the grounds that all behaviour has consequences for the internal state of the animal performing it.

MEASURING MOTIVATION

One of the reasons for the limited success of simple motivational models is that, as we have seen throughout this chapter, one motivational system may be affected by many others. Feeding motivation is affected by drinking, for example, and decisions to perform one kind of behaviour may involve inhibiting another. To understand how these interactions between motivational systems might work, and to attempt to predict which behaviour will occur at any one time, ethologists have therefore tried to measure the relative strengths of an animal's motivation to perform different behaviour. Measuring the strength of motivation has other, more practical applications, too. Many people are concerned that animals in farms or zoos suffer because they are deprived of the opportunity to perform much of the behaviour they would be able to do in the wild. Other people believe that as long as the animals are well fed and healthy, there is no real cause for concern. If we could have some way of measuring the strength of an animal's motivation to perform behaviour that it is 'deprived' of, we might go a long way towards resolving this particular issue. So, both for an understanding of animal decision-making in general, and for dealing with practical problems in animal welfare, measuring how strongly motivated an animal is can be important. It is usually impossible to measure the internal state of an animal directly, so we have to be content with measuring a response of some kind. Unfortunately, different responses do not always give the same answer, but the following are possible ones we could use:

Amount of behaviour performed

Perhaps the most straightforward way of measuring an animal's motivation is to give it the opportunity to perform a response and to see how much or for how long behaviour is performed. In the case of feeding behaviour, we might measure the amount of food eaten—it is usually easier to weigh the amount eaten than to count the number of feeding movements. Drinking motivation can be measured by the amount of fluid drunk and even sexual and aggressive behaviour can be measured with a little ingenuity. Sevenster (1961) put an 'object' stickleback into a glass tube and counted the number of bites directed at the tube by a test fish as a measure of its aggression. Similarly, the number of 'zig-zag' courtship movements directed by a male towards a tube containing a female was used as the measure of sexual motivation. Vestergaard (1980) measured the motivation that hens have to dustbathe by how long the hens dustbathed when they were moved from wire floors (where they could not dustbathe) to a litter floor (where they could). Interestingly, he found an upsurge in the amount of dustbathing the hens showed under these circumstances, suggesting that their motivation to dustbathe rose during the period of deprivation.

In fact, we commonly find that when complex behaviour, such as courtship, has not been elicited for some time, it has a lowered threshold and is performed at a high intensity when it is at last elicited. This phenomenon is sometimes called the 'rebound effect', implying a parallel with the 'reflex rebound' described by Sherrington at the level of the reflex arc. (As we discussed in Chapter 1, when inhibition is removed from a reflex, it often returns at a higher intensity than before—in other words it 'rebounds'). It is possible that the motivation for courtship or dustbathing is inhibited by other activities (p. 8) or the absence of usual stimuli and shows something akin to reflex rebound when the inhibition is removed or the right stimulus is provided.

How aversive a stimulus can be made before it is avoided?

The aim here is to attempt to prevent the animal from performing a behaviour and to see how far it will persist in spite of this. For example, quinine is an intensely bitter substance to us, and other mammals seem to find it equally so. If quinine is put into food pellets or drops of condensed milk, at increasing concentrations, a rat will eventually reject usually attractive food as too bitter. The concentration of quinine it will tolerate can be used as a measure of feeding or drinking motivation.

An alternative version of this method is to place a stimulus, such as food, in full view of an animal and then contrive that, in order to reach it, the animal has to overcome some obstacles or 'run the gauntlet' of something it would normally avoid, such as electric shock or a blast of air. By varying the intensity of the shock and seeing how much an animal will accept in order to reach the food, we have a measure of motivation. Female rats will cross an electrified grid to reach a male. The strength of shock needed to stop them shows cyclical changes corresponding to their oestrous cycle; it is highest during their 'oestrous' or 'heat'. As a variation on this theme, Duncan and Kite (1987) gave cockerels access to hens only if they pushed through doors with weights. The weight of the door a bird was prepared to push was used as a measure of its motivation.

Rate of bar-pressing or key-pecking

The 'Skinner box' is a useful piece of apparatus for studying both learning and motivation (see Chapter 6). An animal is put into the box when hungry and it is taught that it will receive a small food or water reward when it presses a bar that protrudes into the box or, in the case of a bird, when it pecks at a key. The apparatus is so arranged that rewards do not follow every bar-press or key-peck but come at irregular intervals, averaging out at, say, one reward every 30 s. In psychological jargon, this is called a 'variable interval reinforcement schedule' and it means that the animal never knows whether a given bar press will give it food or not. Somewhat surprisingly, perhaps, animals will often press their bar (or peck their key) much more reliably when they get their rewards only irregularly than if a reward comes after every response.

The rate at which the animal presses the bar under a variable interval schedule is so predictable that it can be used as a measure of motivation. The rate at which water-deprived rats will bar-press under these circumstances is reliably related to the length of time for which they have been deprived of water. In some cases, an animal's motivation appears so strong that it will show 'compensation' or 'resilience': thus, if the number of bar-presses required for a given amount of reward is gradually increased, hungry rats may be prepared to work harder and harder to obtain it. Hogan (1967) showed that Siamese fighting fish can be taught, in an analogous way, to swim through a loop to get a food reward and they will similarly show 'compensation' if they have to swim through the loop many times to get the same amount of food. However, aggressive behaviour does not show the same effect. Although fish will readily learn to swim through a loop for the reward of gaining access to a rival male fighting fish, which they then display to, they will not show compensation. They stop swimming through the loop when the number of responses they have to make for each sight of the rival gets too high.

Vacuum activities and stereotypies

When very highly motivated, animals sometimes carry out behaviour even when the appropriate stimuli are not present. Lorenz (quoted in Tinbergen 1951) describes the case of a starling that went through all the movements of catching and eating an insect even though no insect was present—so the behaviour was performed 'in a vacuum'. Hens kept in wire-floored cages sometimes go through all the movements of dustbathing, even though there is nothing for them to dustbathe in. Vestergaard (1980) calls this 'vacuum' dustbathing and argues that when behaviour is performed in the absence of suitable stimuli, or at least with very minimal stimuli, it suggests that the animal is very highly motivated to do the 'frustrated' behaviour.

As mentioned in our discussion of conflict and stress (p. 80) animals confined over a period of time may develop 'stereotypies'—fixed sequences of behaviour, quite unlike that shown in wild animals, that are performed over and over again in the same way and that have no obvious function. For example, sows confined in small stalls may repeatedly bite the bars of their stalls; polar bears even in quite large zoo enclosures may pace around a set route or sway their head and neck from side to side. Stereotypies are often individually distinct and acquired during the lifetime of one animal. One sow may take two steps forward, move her mouth to left and then right on the bar in front of her stall six times, rub her nose on the side of the stall and then take two steps backwards over and over again, whereas another individual pig may have a quite different routine. The presence of stereotypies in some zoo and farm animals has been taken by many people to suggest that the animals doing them are highly motivated to perform some behaviour they are unable to carry out in the confines of a cage.

Rushen (1985) suggested that at least some stereotypies may indicate high motivation to feed when the animals anticipate food: many of the stereotypies of sows are particularly likely to occur when the animals can hear their food trolley coming but cannot yet get at the food. The difficult issue of the interpretation and welfare implications of stereotypies is discussed by Broom (1986) and Fraser and Broom (1990), but it is generally agreed that they do indicate a high level of motivation.

These, then are some of the ways in which levels of motivation have been measured. Superficially, they all appear to be measuring the same thing and we would expect that when, say, an animal is deprived of food, all these measures would increase the same way. However, Miller (1957) describes a number of experiments that show that at least three of the measures of feeding motivation—amount eaten, quinine accepted and rate of bar-pressing—do not all rise together. Over the range from 0 to 54 h of food deprivation, the amount of quinine that rats will accept steadily rises, so does their rate of bar-pressing, but their food intake reaches a maximum after only 30 h and actually declines slightly thereafter. Thus, a rat appears to be becoming 'hungrier' in that it will accept food that is more and more bitter and yet it eats less. There are also some situations in which rats will eat abnormally large amounts of food (hyperphagia) and yet do not show any other signs of hunger, such as being prepared to overcome a barrier or press a bar for food, or eat quinine-treated food. Hyperphagia can be produced by certain brain lesions and it would seem that what we commonly lump together under the term 'hunger' may in fact be a conglomerate of factors; brain damage elevates some of these factors and depresses others.

There is evidence for a similar multiplicity of factors under the heading of 'thirst'. Very much in parallel with Miller's experiments on feeding mentioned above, experiments by Choy (quoted by Miller 1956) show that different measures of drinking motivation do not correspond with each other very well either (Fig. 4.15). Choy had rats with tubes implanted directly into their stomachs so that they could be given liquids without having to drink. He measured drinking in rats, previously satiated with water, after 5 ml of concentrated salt solution was put into their stomachs and he recorded three measures of thirst over time. Up to 15 min after giving salt, bar-pressing does not increase, even though the amount of water drunk does. Yet the amount of water drunk levels off 3 h after the salt is given, even though bar-pressing and tolerance of quinine in the

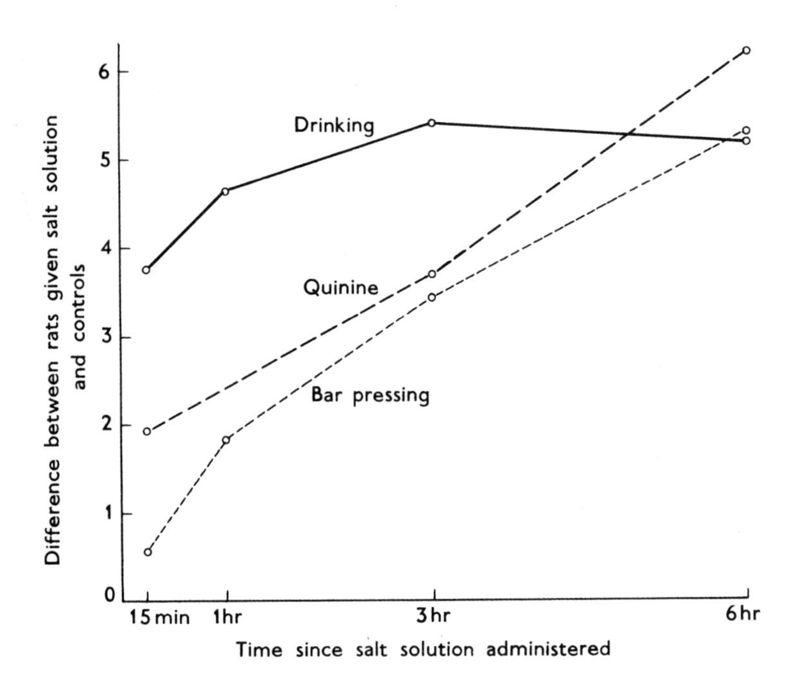

Fig. 4.15 How three different measures of thirst change in the period following the placing of 5 ml of strong salt solution directly into the stomach of a previously water-satiated rat. The units on the vertical axis are arbitrary; they are simply the difference between control and experimental rats on the various measures of thirst (from Miller 1956).

water continue to rise. It is obvious that we need a combination of measures to get a reasonable picture of the effect of salt on drinking motivation, which is clearly complex and not a simple, single entity. 'Motivation' as we have seen, covers a range of causal factors inside the animal (hormones, activation of specific parts of the nervous system, etc.) and so it is hardly surprising that its component parts do not always change exactly in step.

We now turn to the physiological bases of the behavioural systems we have been discussing.

THE PHYSIOLOGICAL BASIS OF MOTIVATION AND DECISION-MAKING

We have seen that, for the purposes of behavioural analysis, the concept of 'motivation' is likely to be with us for some time to come. Nevertheless, considerable progress has been made in moving from behaviour to physiology. This is a large field of study and here we can pick on just three topics. Good reviews of some of the other work that has been done are given by Bolles (1975), Toates (1986) and Colgan (1989) on animal motivation and by Ewert (1980), Camhi (1984) and Young (1989) on neuroethology.

Invertebrate neuroethology

We have already discussed decision-making in invertebrates such as *Pleurobranchaea*. One particularly good example where we can go so far as to locate the cells responsible for decision-making is the escape behaviour of the crayfish (Fig. 4.16). When a crayfish is touched, it has a sudden and dramatic way of escaping: it flexes its abdomen so that the body is drawn into a tight curve. This has the effect of propelling the crayfish rapidly backwards. This behaviour is brought about initially by the activation of receptor cells that are sensitive to high frequency water movements and to touch. These receptor cells are connected to sensory interneurons (SIs), which in turn connect to a lateral giant interneuron (LGI) (Fig. 4.16). There are over 1000 tactile sensory cells and over 20 SIs. The LGI is a very distinctive nerve cell and is unusually large—about 100 mm in diameter. There is one LGI per abdominal segment and each one connects to the LGI in the segment in front, all the way up the nerve cord. In each segment,

hormone from the pituitary, prolactin. Prolactin not only leads to crop growth but also inhibits the secretion of FSH and LH, which in turn shuts off the secretion of sex hormones. As a result, sexual behaviour between the pair dies down as they incubate the eggs and feed the squabs. After 10–12 days, the young birds leave the nest, by which time parental feeding and prolactin secretion have begun to decline. As prolactin levels in the blood fall so FSH and LH are once again secreted and the male begins to court the female again, ready for the next cycle of reproduction. Both visual and auditory stimulation are important for the female's hormonal response (Barfield 1971), but physical contact is not essential. Castrated males are less stimulating for females unless they are given an injection of testosterone and it seems to be the male's display that has the stimulating effect (Erikson 1985).

Some of the experiments involved in working out this sequence are quite dramatic. Thus, the secretion of prolactin normally follows when the doves have been sitting on the eggs for a few days but the male dove's crop develops even if he does not actually incubate—provided he can see his mate incubating! Here, then, a specific visual stimulus leads to the secretion of a specific hormone, but it will only do so if the male is in a particular physiological state. He must previously have participated in nest-building because, if he is separated from his mate earlier in the cycle, his crop does not develop, even if he can see her incubating.

These effects of hormones can be directly demonstrated by implanting specific hormones into different parts of the brain. Controlled amounts of hormone can be placed in particular sites through a hollow needle without interfering with an animal's freedom to behave. Hutchinson (1976) and Komisaruk (1967) showed that castrated male doves would show courtship behaviour to females and aggression to other males if testosterone were implanted in the anterior hypothalamus and the preoptic nucleus (so called because it lies just anterior to where the optic chiasma enters). Progesterone implanted in the same places suppresses these behaviour patterns and causes an increased tendency to incubate—a direct demonstration of antagonism between progesterone and testosterone.

Harris and Michael (1964) have used similar hormone implantation techniques to demonstrate the role of oestrogen in the sexual behaviour of female cats. If oestrogen is injected into certain parts of the hypothalamus, even castrated cats, whose reproductive system remains completely undeveloped, show full oestrus behaviour. Castrated female cats are usually completely unreceptive to males and will lash out viciously if a male gets too close. Implanted castrated females will elevate the rump, deflect the tail to one side and make treading movements with the hind legs. They will assume this posture as soon as a male approaches and will submit to being mounted.

There are some remarkable variations on the basic theme of the intimate relation between hormones and the nervous system. Cuckoos do not build nests but respond to the sight of their hosts' nesting behaviour. A female cuckoo's ovulation is so timed that it can lay an egg in each of several nests just as the host bird completes its own clutch (Chance 1940). Another quite different effect of hormones is the 'Bruce effect', named after Hilda Bruce (Bronson 1979), who found that pregnant female mice abort their litters and reabsorb the embryos if a strange male mouse (not the father of the litter) comes into contact with them. This even happens if pregnant females are put into cages with material soiled by a strange male and do not have direct contact with him at all. The effect is mediated by smell, and females with olfactory bulbs removed remain pregnant even when exposed to alien males. The pregnancy-blocking effect is brought about by an inhibition of prolactin secretion and consequent reduction in progesterone.

AGGRESSION: DECISION-MAKING AT BEHAVIOURAL AND PHYSIOLOGICAL LEVELS

We have now touched upon the motivation of a variety of different behaviour—feeding, drinking, courtship and comfort movements like dustbathing. We have also, in various contexts, mentioned a kind of behaviour that arouses perhaps more interest among scientists and non-scientists alike than any other. That behaviour is aggression. Fighting in animals is often very destructive and ethologists have recently given a great deal of attention to trying to understand the circumstances in which animals fight and when they refrain from fighting. Psychologists and psychiatrists have also become interested in animal aggression because of its possible relevance to human aggressiveness, one of the major problems of modern society (Groebel and Hinde 1989). In this final section of a long chapter on motivation and decision-making, we look at what is known about aggression and use this to exemplify a point that we make throughout the book: the importance of studying both causation and function as separate but interconnected aspects of behaviour.

Much of the controversy about human aggressiveness stems from conflicting responses to Lorenz's views, which were developed in detail in his book *On Aggression* (1966). Lorenz argued that aggressive behaviour follows the pattern illustrated by the psychohydraulic model we have already discussed (p. 86). Just as feeding or drinking motivation build up over the time that has elapsed since their last occurrence, so he saw aggressive behaviour as inevitably rising over time, until the accumulated drive was discharged by the act of fighting. Using this view, then, aggression in both human and non-human animals was inevitable, something genetically built-in and demanding expression. We have already seen that Lorenz's model is inadequate when it comes to explaining behaviour such as feeding and sex, and it seems equally unsatisfactory in the explanation of aggression. If Lorenz's interpretation is correct, there should be a period of quiescence after the performance of aggressive actions, which should last until aggressive motivation (which is supposed to be discharged by performing the behaviour) has built up again. Sevenster (1961) and Wilz (1970) measured levels of aggression in male three-spined sticklebacks by using the number of bites delivered to a test-tube containing another male. They both found that the male's aggressive motivation appeared to be higher at the end of a 10 min test period than at the beginning, i.e. performing aggressive behaviour resulted in an increase in the tendency to fight. Similar increases in aggressiveness following a spell of fighting have been reported for many other species including spiders (Riechert 1984) and lizards (Rand and Rand 1978). Indeed, escalation of fighting in the course of an encounter seems to be a common and highly adaptive strategy, as we will see. In sticklebacks, Wilz (1970) showed that even if the rival male is removed after being attacked, the remaining male continues in such a high state of aggression that he attacks even females, and seems unable to respond sexually to them for some time. Thus, a quiescent phase following the performance of aggressive behaviour is certainly not at all evident.

There are a number of experiments that are often quoted as showing the inevitability with which aggression expresses itself, but they are all open to other interpretations. Animals kept in total isolation are often very aggressive. This has been clearly shown in fish, mice and in junglefowl, where Kruijt (1964) found that, after months of isolation, the birds would attack feathers and have prolonged, circling fights with their own tails. However, the behaviour of isolated animals is profoundly affected in many ways—for example, they may be generally more active and extraordinarily excitable, even by trivial events so that it is not just their aggression that is increased. Heiligenberg and Kramer (1972) tested Lorenz's idea that animals become more aggressive if they cannot find an outlet for their aggressive behaviour. They studied a species of cichlid fish, *Pelmatochromis*, where the males are highly aggressive. They kept males on their own without the opportunity (or stimulation) to fight and found, contrary to the prediction of Lorenz's model, that the tendency to fight decreased over a few days when the fish were not stimulated to fight (Fig. 4.19). This makes evolutionary sense, since if there is no opponent, there is no advantage to be gained from being aggressive. Heiligenberg and Kramer's results lend no support whatever to the idea that aggression is accumulating in the absence of an opportunity to fight.

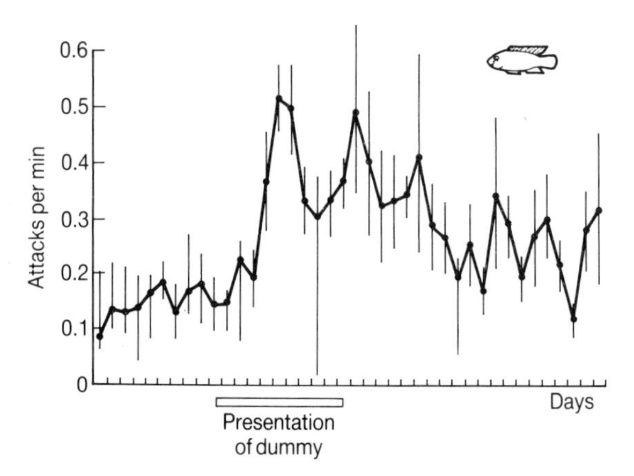

Fig. 4.19 Attack rate by male cichlid *Pelmatochromis* before, during and after presentation of a dummy fish (from Heiligenberg and Kramer 1972).

Lorenz's ideas have also been criticized because he appeared to neglect the profound effect of experience on the development of behaviour. Scott (1958), Brain (1975) and Brain *et al.* (1978) all survey numerous experiments with rodents that show how levels of aggressiveness can be changed in a most dramatic fashion by regulating early experience. It is relatively easy to train one mouse always to attack a strange animal, whilst another of the same strain can be trained to remain completely placid. Both male and female rats reared in isolation are more aggressive, whereas those that have been handled gently by humans when they are young become less aggressive than normal. Namikas and Wehmer (1978) showed that male mice reared in litters with a number of male siblings grow up to be less aggressive as adults than those reared solely with sisters. In primates, too, we can see the effects of early experience on later aggressiveness. Chamove (1980) showed that young rhesus monkeys that experience a high level of conflict with their mothers show higher subsequent aggressiveness. In view of such clear effects of upbringing in other animals, there seems no reason why we should accept the expression of aggressiveness in any species, including humans, as being inevitable. We certainly have to accept that many species, including ourselves, have evolved a potential for aggressiveness. This certainly causes problems for human societies, but at least the long period of childhood development and the powerful influences that both parents and the rest of society can bring to bear on the individual offer us hope of a solution.

The general consensus now seems to be that Lorenz's view of aggression is not only incorrect but seriously misleading. There is nothing inevitable about aggressive behaviour and that whether or not it occurs depends as much upon experience and external factors as on inheritance and the internal state of animals. Montagu (1968) collected together a number of different essays on Lorenz's book, while Huntingford and Turner (1987) and Archer (1988) provide useful surveys of the whole field of aggression. Groebel and Hinde (1989) look at aggression and war in humans.

One of the problems in studying aggression in any species is that it is difficult to arrive at a satisfactory definition that fits all cases. Aggression has the universal effect of displacing other individuals, although it may take a range of forms, from overt fighting to subtle postures that involve no physical contact. Various authors have proposed that, in humans, behaviour as diverse as nail-biting, verbal insults, suicide and war are all manifestations of aggression (Carthy and Ebling 1964). The aggressive behaviour of a street gang may involve actual physical violence, whereas modern war may involve no more than the act of pushing a button, with the button-pusher not having been aggressively aroused in a biological sense. This means that we must be wary of drawing too close a parallel between animal aggression and modern warfare, where technology allows a soldier to burn people alive without even seeing them or having any human contact that might inhibit action (Lorenz 1966; Tinbergen 1968).

A second difficulty with studying aggressive behaviour is that it occurs in many different contexts and it is often not clear whether motivation in these different situations is the same. Animals may fight to win a territory, to win a mate or to obtain food. Some species have a social organization based on a stable hierarchy of dominance (Chapter 7) and they may fight to achieve their status within such a group. Pain arouses aggression: rats given small electric shocks will attack a cage-mate they have previously ignored. Frustration of various types has a similar effect. A rat in a Skinner box will attack another rat tethered nearby if the bar it has learnt to press ceases to yield to it the expected food reward. In man, too, we are familiar with the effects of frustration—bad tempers abound in traffic jams. Some psychologists used to suggest that all aggression is the result of frustration (Miller 1941) but it is unrealistic to explain all aggression in these terms, nor can we regard it as one simple category of behaviour (Brain and Jones 1982).

Aggression towards members of the same species—often called social aggression—is reasonably distinguished from the interspecific fighting shown in catching and killing prey, and also from the defensive behaviour shown by an animal faced by a predator. This is partly because the behaviour patterns used in these latter situations are often different (although both may show some overlap with social aggressive behaviour) and partly because the external and internal factors giving rise to them seem to be different. However, we should not assume that the motivational systems are entirely unrelated. Huntingford (1976) has shown that, in sticklebacks, there must be some common causal factors between the aggression shown to rival males and the aggression—or perhaps we should call it defensive behaviour—shown towards predators such as small pike. The fish that behaved in a 'bolder' fashion towards pike were also those that showed the highest levels of fighting in territorial disputes.

Despite these complications, there are some generalizations that can be made. We have already seen that early experience has a major effect on the expression of aggressive behaviour, but there is also now a great deal of evidence that, whatever their previous experience, many animals have a tendency to respond aggressively when first placed in certain situations or given certain stimuli. Here we should note that aggressive responses are far more common amongst males (the male hormone testosterone tends to increase aggressiveness) and that the stimuli to which they respond are often those provided by a rival male. Thus, Cullen (1960) showed that male sticklebacks, reared in complete isolation, set up territories and attacked rival males in the same way as normally reared fish. Now, for the reasons we discussed in Chapter 2, if all members of a species behave in the same way regardless of their upbringing, it is effectively impossible to analyse the genetic basis of their behaviour. However, we commonly find that individuals and natural populations of some species do differ quantitatively in the level and expression of aggression. Hybridization shows that some of these differences are genetically based. For example, desert spiders (*Aegelopsis*) from different populations differ genetically in how aggressive they are in defending web sites (Riechert and Maynard Smith 1989) and lake char are genetically less aggressive than brook char (Ferguson and Noakes 1982).

Selective breeding, which we discuss further in Chapter 5, is another way of examining the genetic basis of aggressive behaviour. Selection in game fowl and in Siamese fighting fish has been carried out for centuries and has resulted in breeds with abnormally high levels of aggression. In the laboratory, Lagerspetz and Lagerspetz (1971) have successfully bred for both high and low levels of aggression in mice. The two selected lines differ greatly in their readiness to attack others and the intensity of their fighting behaviour. Finally, we may mention the remarkable series of selection experiments carried out by Bakker (1986). He bred sticklebacks for high and low levels of territorial aggression between males. He got a significant reduction in his low line but less in the direction of high aggression, where the selected line did not differ significantly from unselected normal fish. This immediately suggests that natural selection had already 'used up' much of the variability for genes promoting high aggression.

This is an important conclusion for, as Lorenz (1966) himself pointed out, the potential to be aggressive is often very beneficial to animals. Although animals may risk injury or even death in a fight, without accepting this risk, many animals would die or fail to reproduce because they would not obtain a vital piece of food, a territory or a mate. So we should not think of aggression as some sort of abnormality or disease. Rather, aggression evolved because it benefited the animals concerned. At the same time, it may have such costly and dangerous consequences that many animals have evolved sophisticated mechanisms for 'deciding' whether or not to fight. Animals as diverse as spiders, deer and fish are able to assess each other's fighting ability and often 'decide' not to risk a costly fight if their opponent is seen as stronger than they are: 'He who fights and runs away lives to fight another day!'. With aggression, as with all other behaviour, natural selection has been at work and it is now time to look in more detail at how we can study the adaptiveness of behaviour.

5 EVOLUTION

Evolution by natural selection has become the great unifying concept of biology, so much so that most biologists now feel that, without it, none of the phenomena they study really make sense. Dawkins (1986) in his book *The Blind Watchmaker* provides an excellent modern survey of the great explanatory power of what has come to be called neo-Darwinism. Natural selection, operating on random, inherited variation has, over the generations, shaped animals to match the environments in which they live. All through this book we have emphasized the adaptive role of behaviour in an animal's life and so the concepts of 'evolution' and 'adaptation' have been implicit in much of what has already been said. Now we look at them in more detail. We will look at the genetics of behaviour and the various remarkable ways in which natural selection has led animals to behave—to cooperate and care for each other, as well as to fight and to kill. We will also look at the way behaviour is known to change over evolutionary time and the way it, in turn, influences the evolutionary process itself. We begin with one of the most basic —and yet most elusive—evolutionary concepts of all, that of adaptation.

THE ADAPTIVENESS OF BEHAVIOUR

Most people have a general idea that animals are adapted to their way of life. A camouflaged insect sitting quite immobile on a background that it matches perfectly, for instance, would be described as adapted to conceal itself from predators. But we may need more detailed studies to show just how perfect adaptation can be.

Male mole crickets (*Gryllotalpa*) dig burrows underground and then sit in them and sing to attract females flying overhead (Fig. 5.1). As with many other species of crickets and grasshoppers, the song of each species is very distinctive and plays a role in the sexual isolation between species (p. 118). The sound of the song is determined in part by the structure and size of the forewings, which are rubbed rapidly together to produce sound. A scraper or plectrum on the hind margin of the left wing rubs along the toothed undersurface of a vein on the right wing (the 'file'). In two species studied by Bennet-Clark (1970), the fundamental sound frequency within the pulses of the song was about 1600 Hz in *G. gryllotalpa*, a species with small wings, shallow teeth on the file and a quiet song, but 3500 Hz in *G. vinae*, which has large wings, deep teeth on its file and a much louder song.

Here, then, we can see a direct correlation between behaviour (the singing) and morphology, but the mole crickets' adaptiveness goes much further than this. The male of each species excavates its burrow in a different way. Bennet-Clark showed that the burrow forms an exponential horn with a bulb at its base so that when the male sings with his

Fig. 5.1 Side views of males of two species of the mole cricket *Gryllotalpa* sitting in their burrows, head downwards in the singing position. The shape of their burrows approximates to an exponential horn. *G. vineae*, on the right, has a smoother tunnel and a much louder song, but in both cases the shape of the burrow is adapted to their song frequency (oscillograms of which are illustrated below) so as to emit with maximum efficiency (from Bennet-Clark 1970).

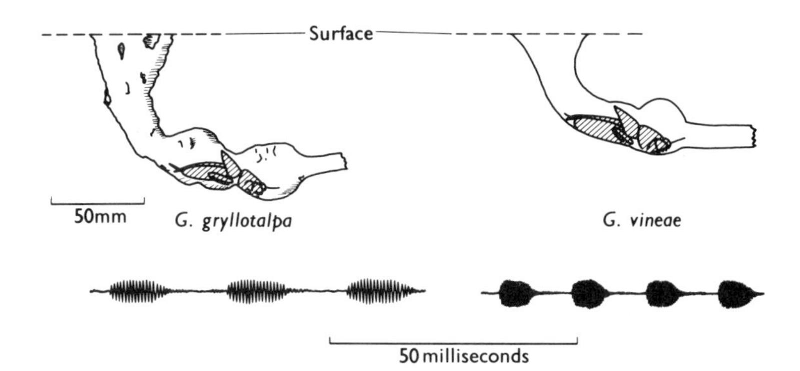

head just at the origin of the horn, his song is amplified, as in the horn of an old-fashioned loud-speaker. The size and shape of the burrow are, however, quite different in the two species. *G. vinae* builds a double-mouthed horn-shaped burrow to sing in. The walls are smooth and there is a bulb that acts as a resistive load for the vibrating wings and concentrates the sound in a disc-shaped patch of sound energy above the burrow. *G. gryllotalpa* also builds a horn, but it is much larger and is very effective at amplifying the much lower sound that this species makes. Thus, the wing structure, singing and burrow-digging behaviour of the mole crickets are all precisely coadapted, each with the others, so as to provide the most efficient sound production, 'aimed' at the females, which are attracted to the males of their own species. Similar coadaptation is shown by the 'eyes' on the hind wings of the eyed hawkmoth, normally hidden beneath the forewings but revealed when the moth is disturbed—the fore-wings are moved away, revealing two big eye-spots. The moth moves up and down so that there is rhythmic movement associated with the eyes. In the butterfly *Caligo*, there are even a few white scales to portray the reflected highlights on the bulging 'eye' of what to a small bird can look like an owl predator (Fig. 5.2).

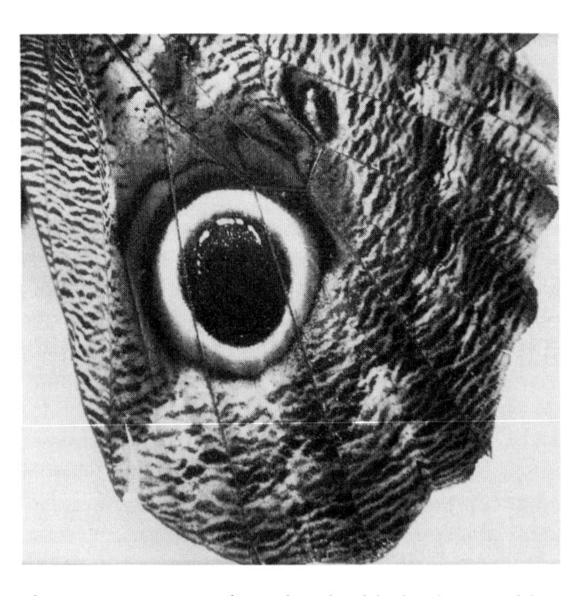

Fig. 5.2 'Eye spot' on the underside of the hind wings of the butterfly *Caligo*. Their resemblance to large vertebrate eyes, such as those of owls, is remarkable.

As a further example of the precision with which behaviour has evolved to meet the demands of an animal's environment, we can look to the habit of black-headed gulls of removing empty eggshells from their nests soon after their eggs hatch. Even gulls breeding for the first time carry away eggshells promptly, although they never pick up eggshells in any other situation. It is an activity that occupies them for about 5 min each year and might, for this reason, seem of trivial importance. Nevertheless, Tinbergen and his colleagues (1962b), in a classic study, clearly showed that eggshell removal is important for keeping the nest camouflaged. By placing eggs out on the sand-dunes, with or without eggshells near them, they showed that eggs with eggshells near them were much more vulnerable to predation by crows and herring-gulls. Both aerial and ground predators discover nests and chicks more easily if empty shells, with their smell and white shell insides are nearby. Natural selection has operated to ensure that parent gulls always perform this brief but important behaviour. Clearly, we must never write off as functionless—however trivial it may seem—any piece of behaviour that we regularly observe in a natural situation.

It is worth pausing for a moment to consider how we can feel completely confident that natural selection has been operating on a trait and that we are not just telling 'just-so' stories as Gould and Lewontin (1979) have accused ethologists of doing sometimes. In the case of the mole crickets, we have two pieces of evidence—the shape of the burrow and the difference in shape of the burrows of the two species. If a human engineer was asked to design simple devices for amplifying sound, he would probably come up with horns (as in ear trumpets) of varying shapes, depending upon the frequency of the sound to be amplified. The fact that the mole crickets dig burrows of precisely the shape to maximize the sound produced by their particular song frequency suggests that sound amplification was the feature that natural selection favoured. Crickets that built burrows of a different shape would not be able to emit such a loud sound (an engineer could tell us that) and, by implication, would be less successful at attracting females. The burrow appears to have been 'designed' to amplify the male's song.

This conclusion is confirmed by the second piece of evidence—the comparison between the two species. If a burrow is 'designed' to amplify a particular frequency with maximum efficiency, then we should expect that, in a species with a different song frequency, the burrow shape should be correspondingly, and predictably, different too. This is exactly what we find. In fact, comparison between species that are all successful, but achieve success in slightly different ways, has become one of the most important methods by which ethologists can document the adaptiveness of behaviour.

Comparisons between just two species (like the two species of mole crickets or between ground-nesting black-headed gulls that do remove eggshells and the cliff-nesting kittiwakes that do not) may certainly be instructive and tell us something about adaptation. Thus, on looking more closely, we find that kittiwakes nest on tiny cliff ledges where no predator can land and, as Cullen (1957) pointed out, this probably explains why their nests are conspicuous and messy and why eggshells are not removed. But we can gain even more information if, instead of just comparing two species, we look at the habits, morphology and environments of many different species (Clutton-Brock and Harvey 1984).

For example, goshawks may copulate up to 100 times before egg-laying and, in trying to understand the adaptive significance of this behaviour, Birkhead *et al.* (1987) looked at the mating habits not just of goshawks, nor even of hawks in general, but of 131 species of birds from many different families. They found that copulation rates were highest in those genera where the female was most likely to be mated by more than one male, for example, in colonial species where there are many opportunities for mating and/or in species where the male leaves his mate alone for long periods of time. In solitary nesters or where the male guards his mate for most of her fertile time, copulation frequencies are lower. This suggests strongly that the adaptive significance of very high copulation rates is to prevent a female's eggs being fertilized by the sperm of another male, as the highest rates are found where there is the greatest risk of such sperm competition. Sperm competition explains Birkhead *et al.*'s data much better than an alternative hypothesis, namely that more copulations are needed to fertilize larger clutches of eggs, because there is no correlation between clutch size and copulation frequency.

To go back once more to the black-headed gulls removing eggshells, notice that Tinbergen *et al.* did not rely primarily on comparisons between species to reveal adaptive significance. They did an experiment with eggs and eggshells, artificially recreating a situation in which they could see what would happen if black-headed gulls did not remove eggshells (the problem is, of course, that all existing gulls do, so well has their behaviour become adapted!). Their experiment shows yet another method for removing the study of adaptiveness from the realm of 'just-so' stories and putting it firmly to empirical test: artificially creating a situation that does not exist natur-

Fig. 5.3 Frog-eating bats locate male frogs by listening to their courtship calls.

ally. In this case, they put painted hens' eggs on to the sand-dunes. The increased predation on the eggs with shells near them demonstrates what would happen (and probably has happened in the past) to nests if shells are not removed.

But, however effective natural selection can be in the shaping of behaviour, it often results not in the best but in the best compromise, because one single aspect of behaviour can rarely evolve in isolation. The elaborate displays or loud songs that many male birds use to attract their mates, for example, are also likely to attract a predator. Most males stop singing as soon as their mate begins to incubate and some lose their breeding plumage at the autumn moult. There is obviously a compromise between reproductive success and danger from predation.

The neotropical Tungara frog demonstrates this compromise in a particularly dramatic way (Ryan *et al.* 1982). Females are attracted to the sounds of males calling at the edge of ponds. But frog-eating bats use the males' calls to home in on the males and eat them (Fig. 5.3). Unfortunately for the frog, those aspects of male vocal behaviour that are most attractive to females (calling more intensely and producing calls with 'chuck' sounds) are also the ones that most increase the predation risk.

EVOLUTIONARILY STABLE STRATEGIES

An important idea in the study of adaptation is that of the **evolutionarily stable strategy** or ESS, which we owe to Maynard Smith (1982). A 'strategy' is simply a specification of what an animal does, such as 'remove eggshell from nest'. An ESS is a strategy that, if adopted by most members of the population, cannot be bettered by another. In the simple case of the gull removing eggshells, the strategy of 'removing' is stable. But suppose that the situation was somewhat more complicated, so that the benefits of what one animal did were dependent on what other animals around it were doing. The golden digger wasp (discussed in Chapter 4 in connection with the goal-directedness of its burrow-digging) is a good example. Females do not always dig their own burrows: sometimes they take over the burrows dug by other females. The two strategies of 'digging' (own burrow) and 'entering' (someone else's) appear to coexist side by side but clearly the success of each is dependent on what other females are doing. If there were very few 'diggers' around, 'enterers' would have a hard time finding burrows and 'diggers' would be at an advantage. But the more 'diggers' there are, the easier it will be for 'enterers' and the more successful enterers will be because they save the time and effort of digging their own nests (Brockmann *et al.* 1979). The success of both strategies is therefore frequency dependent—it depends on the relative frequency of the two strategies at any one time. The idea of ESSs has now been applied to a wide variety of situations and has been particularly fruitful when applied to aggression because here the best strategy (e.g. 'attack' or 'retreat') is very dependent on what strategy an opponent takes up. Dawkins (1980) gives a non-mathematical and Parker (1984) and Maynard Smith (1982) more mathematical treatments of ESSs.

CULTURAL TRANSMISSION OF BEHAVIOUR

Behaviour can evolve only if animals vary in what they do and if the variation between them is transmitted from one generation to the next. Then natural selection can come into operation, favouring some variants, which thus spread, and eliminating others. With most morphological and physiological characters, there is only one way in which such inherited variations can be produced—by genetic changes that will involve new mutations or recombinations of genes already present.

Behavioural evolution has an extra and unique dimension because an animal may learn from its parents or others in its group. Later it may in turn act as a model from which its own offspring will modify their behaviour. Suppose one animal learns a completely new behaviour pattern, which proves more successful for some purpose than the hitherto typical behaviour. This pattern may be copied and thus passed on to succeeding generations and gradually replace the old pattern, without any genetic changes being involved.

We begin our discussion of cultural evolution by

considering some examples of this less familiar, but potentially very important, manner of behaviour transmission in animals. We are familiar with such 'cultural' evolution in human behaviour and many would argue that almost everything of importance in human behaviour is transmitted in this way from one generation of a society to the next. The various human languages are an obvious example of a continuous cultural tradition that maintains different types of behaviour in different populations. Reviews of cultural evolution in non-human animals are given by Galef (1976), Bonner (1980) and Roper (1986).

Clearly, cultural evolution is possible only among animals with considerable ability to modify their behaviour by copying and practice. We might not be surprised to find examples from the other primates, our closest relatives. Intensive, long term observations on the Japanese macaques of Koshima Island have shown that some of the differences between the behaviour of different monkey troops are indeed of cultural origin. The monkeys are attracted to feed on maize, sweet potatoes and other foods artificially put out for them. The potatoes are often caked with earth, which the monkeys used to rub off with their hands. One day observers saw a young female dip her potato into a stream and wash it. She persisted in this habit, which was copied first by one of her infants and subsequently by nearly all the younger members of her troop, which can be clearly distinguished as a washing 'subculture'. We have already mentioned in Chapter 2 that patterns of grooming in chimpanzees can be passed on by copying. Chimpanzees also show similar group-dependent feeding habits. In some areas, chimpanzees 'fish' for termites by poking a twig into the termite mound and letting the insects crawl up the stick when they can be eaten. In other areas chimpanzees use leaf stalks or grass stems for the same purpose (McGrew *et al.* 1979; Fig. 5.4).

Fig. 5.4 Chimpanzees in different areas have different techniques for extracting termites from their nest. Here a chimpanzee inserts a long segment of twig into the termite hole, lets the insects crawl up the twig and then eats them (from Jolly 1988).

There is now increasing evidence for cultural transmission in other groups besides primates (Bonner 1980). It is certainly cultural transmission that maintains the song dialects of many birds such as white-crowned sparrows (Baker and Cunningham 1985). As described on page 42, the young male models the finer details of his song on that of his father and neighbouring males singing close by; he would pick up a different dialect and sing quite differently if he grew up in another area.

One of the best known examples of cultural transmission in birds is the habit of opening milk bottle tops by blue tits (Hinde and Fisher 1952). Starting in the 1920s, it was noticed that milk bottles delivered to doorsteps were being plundered by birds that fed on the cream (Fig. 5.5). By plotting the spread of this habit across Britain, it seemed that it was 'invented' in three separate areas of London and then spread outwards apparently by imitation. However, Sherry and Galef (1984) subsequently showed that such cultural transmission could occur without imitation, i.e., without a bird actually seeing another bird open a bottle. They showed that chickadees (which are closely related to tits) learn to open containers with lids even if all they saw was just the opened container. Naïve individuals could learn what to do simply by seeing the results of behaviour, not the behaviour itself.

Fig. 5.5 A blue tit opening a milk bottle to obtain cream.

Another important area of cultural transmission is the recognition of danger. Young animals often learn what is dangerous by the responses of adults, particularly their alarm calls, in the presence of predators. Curio *et al.* (1978) showed that young European blackbirds gave alarm calls to objects such as a stuffed owl, or even a plastic bucket, if they had heard their parents' alarm calls while looking at them. Curio devised an ingenious arrangement where he could present both adults and young with different objects at the same time, with neither being able to see what the other saw. So, while the adult would be giving an alarm call to a stuffed owl, the young bird saw a plastic bucket and heard the alarm call. Thereafter it associated buckets with danger! The example given in Chapter 2 of the Addo elephants' aversion to people, which is still apparent 70 years after the events that caused it, shows how long-lived cultural traditions in animals can be.

Cultural and genetic evolution are not separate processes—they are bound to interact in many ways. A cultural trait may expose an animal to a new range of selective forces, for example. The Japanese macaques that started washing their sweet potatoes in a stream then began going into the sea to wash their food. Once they were in the sea, some of them started swimming and eating seaweed. Such cultural changes in behaviour can be a major driving force for evolutionary change (Hardy 1965; Plotkin and Odling-Smee 1979; Wyles *et al.* 1983) and social behaviour in particular offers many opportunities for cultural traits to affect selective pressures. Here, however, we are concerned with the evolution of behaviour by natural selection operating on inherited variation. Consequently, before we can profitably discuss what types of changes have occurred during evolution, we have to look first at what is known about the genetics of behaviour.

GENES AND BEHAVIOUR

The evolution of any trait by natural selection implies that there exists—or at least there has once existed in the past—genetic variation between individuals. Evolution is about changes in gene frequency: one genotype becomes more frequent in the population because it makes the body it is in taller, fatter, able to run faster than another, or succeed in some other way. Many people are quite able to accept the idea of 'genes for' eye or coat colour but find the idea of genes for behaviour much more problematical. How could there be genes for eggshell removal or courtship behaviour? The special problems of bringing a genetic analysis to bear on behaviour were discussed in Chapter 2 and, while it is certainly true that the causal links between genes and biochemical traits such as blood groups or skin pigmentation are often closer and clearer than between genes and behaviour, nevertheless, the links are still there. Behaviour is the result of an enormously complex interaction of nerves, muscles, sense organs, hormones and so on. Genes can affect all of these. Sometimes they must act rather specifically on the way the developing nervous system grows and forms connections, sometimes they may have more indirect and general effects by altering, say, the amount of a hormone secreted or the sensitivity of a sense organ.

Behavioural genetics is not an easy subject to deal with succinctly. Our knowledge is extremely patchy because so few animals are suitable for even the simplest of genetic investigations. Human genetics has attracted a great deal of analysis because, although we cannot carry out deliberate breeding programmes, we can often trace parents and other relatives (who will share a proportion of the same genes) and compare their behaviour with those of our subjects. Identical twins are, understandably, much sought after by psychologists interested in behaviour genetics. A huge literature on mental abilities, especially 'intelligence' has accumulated and much of it still attracts bitter controversy over the proportional influence of genetic and environmental factors, and their interaction, in human development. Apart from humans, mice and the fruit-fly, *Drosophila*, tend to dominate the behaviour genetics literature, at least in the number of papers written about them. Fuller and Thompson (1978) remains the most complete source book, whilst Hay (1985) and Vaysse and Médioni (1982) both provide attractive introductions to this difficult field.

Many hundreds of mutants have now been described in *Drosophila*. A gene called *Bar* reduces the number of facets in the compound eyes, while another, *white*, reduces pigmentation in the eyes. *Bar* and *white* both affect vision: male fruit-flies that carry these genes cannot see as well as normal males and have trouble locating females and in receiving visual stimulation from them during courtship. *Vestigial* and *dumpy* genes both alter the form of the wings, making them grossly malformed. An important part of the male fruit-fly's behaviour in courtship is when he vibrates his wings close to the female, something that *vestigial* and *dumpy* males cannot do adequately. So it is quite clear that all of these genes are affecting courtship behaviour through their intermediary effects on the eyes or wings. Looked at in this way, the idea of 'genes for' courtship behaviour becomes much more reasonable, but at the same time, many genes will be involved and will interact in development.

Also in fruit-flies, Benzer and his colleagues (Hotta and Benzer 1976) have made use of some extraordinary genetic manipulations that are possible in this insect to produce 'mosaic' flies—flies that are genetically different in different cells of their bodies (Fig. 5.6). The most interesting mosaics are **gynandromorphs**, individuals in which some cells are male and some female. Some mutant genes are carried on the sex chromosomes, and it is therefore possible to use these gynandromorphs to discover which part of a fly must be male if it is to show masculine behaviour. When male fruit-flies vibrate their wings in front of females, there are, in fact, two sorts of vibrating 'song' that they make: pulse song and sine song. Only if the mesothoracic ganglion has genetically male cells with the right genes will the insect produce pulse song, even if all its other cells are male—unless this particular section of the body is also genetically male, no pulse song is produced. This shows that the presence of genes in particular cells affects how the fly behaves. Such a remarkable result is a reflection of the way insects are organized; each cell determines its own sex, for example. We could not expect such results in vertebrates, where sexual behaviour develops as a result of hormones, secreted by the gonads, that affect the whole body (Chapter 2, p. 27).

Fig. 5.6 Diagrammatic representations of mosaic or gynandromorph fruit flies. Female cuticle is represented dark, male cuticle light. The varying proportions result from the different stages of development at which an X chromosome is lost from one daughter cell following cell division in a female embryo. This cell and its descendants will become male. The top left hand example is exactly divided into male and female halves: the X chromosome must have been lost from one daughter cell at the very first division of the fertilized egg.

The study of single gene effects is one important approach to behaviour genetics, but we also need to study the inheritance of behaviour patterns themselves if we want to understand the mechanisms of behaviour evolution. We need to look not just at the often bizarre mutants produced in the laboratory but also at naturally occurring genetic variation in the world outside, because it is this that has been the basis for natural selection. Ideally, we would use the methods of classical genetics—setting up crosses between animals whose behaviour differs and counting numbers of animals of each type in the F1 hybrids and in the F2 hybrids, and so on. But the methods that work so well for blood groups and pea plants of different heights often run into difficulties when applied to behaviour.

One immediate difficulty is the choice of suitable units of behaviour for genetic analysis, a problem that has always faced geneticists. Mendel's stroke of genius was to choose characters in his pea plants that were clear-cut and could be counted. In this case, each 'unit'—tallness, seed colour, texture of seed coat, etc.—corresponded to a single gene, but there is little probability of getting such a simple correlation with behaviour.

One of the few examples of naturally occurring behaviour showing a classical Mendelian pattern of inheritance is Gwadz's (1970) work with mosquitoes. The females of different populations of these mosquitoes become sexually receptive at different times after emergence. The females of one strain, called GP (because it was originally collected at Gunpowder Falls, Maryland) are receptive to insemination quite soon—a mean of 38 h after emergence. Females of another strain, called TEX (because it was originally collected at Austin, Texas) takes much longer to accept males—with a mean of 120 h (Fig. 5.7). Hybrids between the two strains (either GP/TEX or TEX/GP) have an intermediate time to insemination, with a mean of about 54 h, but

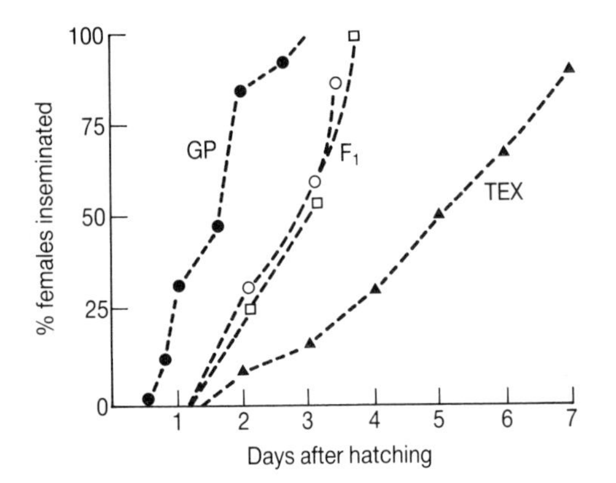

Fig. 5.7 Onset of receptivity to insemination for two parental (GP and TEX) strains of mosquito, and the F1 hybrids. (Open circles show GP/TEX and crosses show TEX/GP hybrids.)

with a slope more similar to the GP strain. When the F1 hybrids were backcrossed to the GP and TEX parental strains, the results were compatible with the idea that early receptivity was due to a single, autosomal, semidominant gene. Gwadz also showed that the late-emerging females of the TEX strain could be made receptive much earlier by application of the juvenile hormone, normally produced by the corpora allata. With the hormone, they behaved like GP females. The genetic difference in behaviour between the two strains seems therefore to be due to a difference in the rate of production of a hormone.

But such clear-cut Mendelian results are rare in behaviour genetics. As we mentioned before, we must expect most behavioural traits to be affected by more than one gene. Large numbers of genes may be involved in one trait and, furthermore, they will interact in ways that are not always straightforward. This means that the variation between individuals in a population tends to be continuous, and not discontinuous with regular ratios 3:1 or 1:1. Thus, mice cannot be classified as 'aggressive' or 'non-aggressive'; rather, they show widely varying degrees of aggressiveness. Continuous variation of this type requires the analytical methods of quantitative genetics and even then will give us only a rather general overview of how behavioural potential is organized genetically.

Nevertheless, some interesting information on the genetic control of instinctive behaviour has come from hybridizing populations, or sometimes species, that differ in some easily measured way. For example, Berthold and Querner (1981) have investigated the phenomenon of migratory restlessness in different populations of the blackcap. Blackcaps fly to Africa in the winter but birds from populations that spend the summer in the Canary Islands obviously have less far to migrate than birds that breed in Finland. Berthold and Querner found that birds from these different populations differ genetically in how far they are prepared to fly. They took nestling blackcaps from four different populations (S. Finland, S. Germany, S. France and the Canary Islands) and hand-reared them. The young, inexperienced birds were kept in aviaries away from adults. When the time came when they would, in the wild, be undertaking their migration to Africa, the birds showed migratory restlessness, fluttering and jumping in the aviaries. The interesting thing was that there was a correlation between how far the birds would have had to travel had they been left in their natural populations and the amount of migratory restlessness (its intensity and how many days it persisted). Birds from the Canary Islands (which are relatively close to the wintering areas in Africa) showed much less migratory restlessness than birds from the Finnish and German populations, which would, in the wild, have much further to fly. Crosses between birds from different populations resulted in hybrids with intermediate degrees of migratory restlessness.

Such techniques can even be extended to crosses between two different species, as Dilger (1962) showed in a classic study of nest-building behaviour in two species of lovebirds. These members of the parrot family breed readily in captivity and, within the genus *Agapornis*, two types of nest-building behaviour are represented. All species tear strips of material from leaves to build with (in the laboratory newspaper forms an excellent substitute) but whilst some, such as the peach-faced lovebird, tuck the strips into their rump feathers and fly back to the nest carrying several pieces at once (Fig. 5.8), other species, such as Fischer's lovebird carry the strips singly in their bill. Dilger crossed the peach-faced with Fischer's lovebirds and watched the nest-building behaviour of the hybrids. For some time the hybrid birds were incapable of building a nest at all because they attempted to perform some kind of compromise between the two collecting methods. They might start to tuck a strip between the rump feathers, but either failed to let go of it or failed to tuck it properly. The end result was usually that the strip fell on the ground and the whole process began again. The only success the hybrids had was when

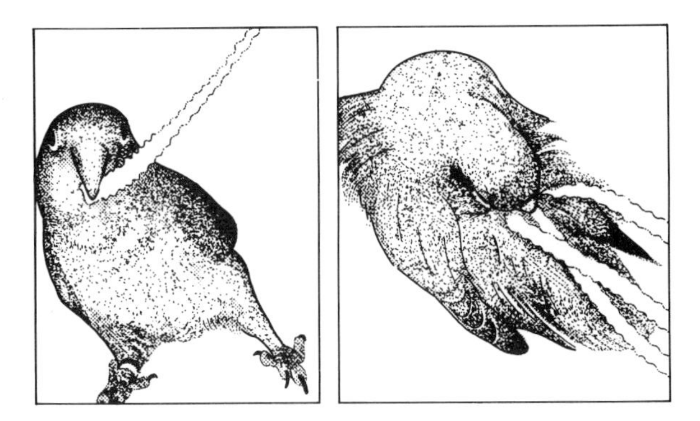

Fig. 5.8 A peach-faced lovebird tearing strips of paper as nest material and then tucking a strip into the feathers of its back to carry them to the nest (from Dilger 1962).

they managed to retain a strip in their bill after the attempted tucking procedure. Dilger found that even after months of practice the birds managed to carry nest material successfully in only 41 per cent of their trials. Two years later they were successful in nearly all trials, but before carrying a strip in their bill they would make a brief turning movement of the head, which is the preliminary to tucking.

Unfortunately, because the hybrids were sterile, it was not possible to take genetic analysis any further. At least this result shows that there is clearly a genetic basis to the nest-building behaviour and further, reveals the remarkable force of an inherited potential to behave in a particular way. Parrots are intelligent birds and they are known to learn readily in other situations but here the inherited predisposition to perform the tucking sequence, outweighed experience for a long time.

In some dramatic instances the hybrid between two species shows behaviour that is not found in either of the parental species, but seems to belong to different species altogether. Neither the yellow-billed teal nor the pintail shows a male courtship pattern that Lorenz (1941) called 'down–up', in which a drake dips his bill into the water and then suddenly lifts his head, raising a plume of water. However, hybrids between yellow-billed teals and pintails do show a well developed down–up display, as do a number of closely related species of duck. The most likely explanation for this is that the block of genes necessary for the down–up display are still present in both parent species, as they are in the related species, but that natural selection has eliminated its performance in pintails and yellow-billed teals by raising its threshold. This threshold change has been accomplished by different sets of genes in these two species so that when the two sets combine in the hybrids, their effect is reduced, and the down–up display is once more performed. We would need to study F2 hybrids and backcrosses to confirm this idea but, unfortunately, as with the lovebirds, the hybrids are usually infertile.

Rather more precise understanding of the role of genes in behaviour can be obtained by selectively breeding animals over a large number of generations (often not possible with interspecies crosses because of reduced fertility or viability). Manning (1961) selectively bred for fast and slow speed of mating over a number of generations (Fig. 5.9). Mating speed is a complex character that involves the interaction of male and female and more than just their sexual behaviour. However, amongst the other changes, the behaviour of both sexes had been altered in a quantitative way. The males from fast-mating lines performed high intensity courtship movements more frequently than those from slow lines. Conversely, females from fast lines were more easily stimulated to accept males of their own or from other lines than are slow mating females.

In *Gryllus* crickets, there are, in nature, two kinds of males—'callers', which sing a great deal and attract females to them and 'satellites', which are relatively silent and intercept the females on their way to the calling males. Calling males attract more females than satellites but also suffer from more parasitism by flies, which are also attracted to the males' songs. Cade (1981) reared males in isolation in the laboratory and fitted each of their containers with a microphone so that he could measure the amount of singing each male did. By measuring the mean calling rate per night from the 7th to the 16th day of life, he showed that there was a bimodal distribution in the number of singing males. He then selected 2–4 males from each end of the distribution and mated them with non-sister virgins and continued this process for four generations. The result

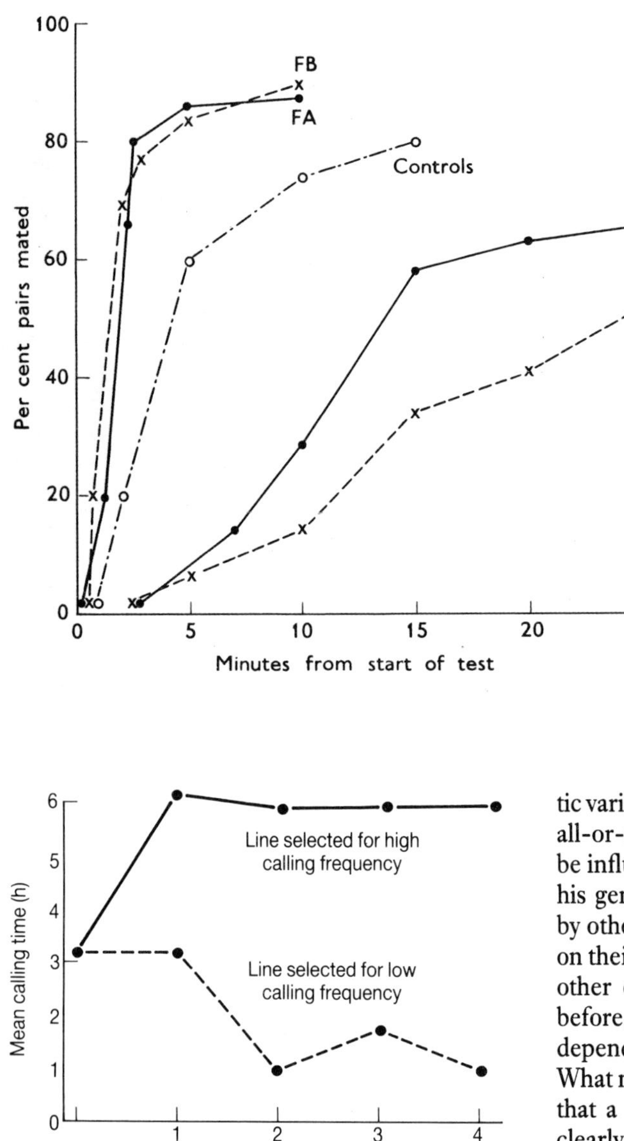

Fig. 5.9 The mating speed of groups of 50 pairs of *Drosophila melanogaster* from two lines selected for fast mating, FA and FB, and two selected for slow mating, SA and SB, compared with unselected controls. These samples are from the 18th selected generation. Some 80 per cent of the fast lines have mated before the first of the slow lines begin (from Manning 1961).

Fig. 5.10 Calling rate/h in male crickets selected for high and low calling rates over four generations (from Cade 1981).

was a divergence in the amount of singing shown by the two groups (Fig. 5.10), showing that there must have been a considerable genetic component to the variation in the original population. Artificial selection, like natural selection, can operate only if genetic variation already exists. However, calling is not an all-or-nothing trait and the amount a male calls can be influenced by the environment he is in, as well as his genetic make-up. 'Callers' that are surrounded by other callers tend to sing less than when they are on their own, and 'satellites' may be quite vocal if no other cricket is singing nearby. As we have seen before, the behaviour an animal actually shows is dependent on both its genes and its environment. What matters, as far as evolution is concerned, is not that a gene always has the same definite effect (it clearly does not) but that at least some of the observed differences in behaviour between individuals can be traced back to differences in their genotype. These differences may be added to and complicated by differences in the environments the individuals have experienced (we know they are). But as long as there are some genetic differences (some 'genes for' behaviour), then natural selection can get to work and we have the raw material for the evolution of behaviour.

KIN SELECTION AND INCLUSIVE FITNESS

Behaviour evolves because some behavioural genotypes—those that make individuals slightly more aggressive, say, or slightly better at attracting females—leave more descendants than their competitors and these descendants, too, carry the advantageous genes. 'Success' in evolutionary terms means leaving offspring that themselves reproduce, but the 'success' of an individual is short-lived and ephemeral. In sexually reproducing species, an individual does not survive for more than one generation. It is possible to argue that it is only our genes that reproduce and leave descendants and our adult bodies are simply the elaborate packaging that protects them. As far as a gene is concerned, the body it happens to be in at a given moment is a useful, if temporary, vehicle for getting itself passed on into the next generation, as Dawkins (1989) puts it. Samuel Butler's aphorism that 'A chicken is an egg's way of producing another egg' can be rewritten as 'An animal is a gene's way of producing more copies of that gene'.

Many people object to this way of looking at animals and feel uncomfortable with the idea of dispelling the sovereignty of the individual in favour of a gene-centred view of evolution. But consider something as basic and fundamental as parental care. We take it for granted that parent animals should feed and protect their young, but parental care itself is the result of strategies genes have for perpetuating themselves. Genes that help to make the bodies they are in more effective at defending their young will perpetuate themselves in the bodies of the protected young. Genetic variation in ('genes for') a tendency to defend young—perhaps mediated through variations in the level of a hormone —will result in a variation in the numbers of offspring that survive to pass on the favoured genes, and so on down the generations.

But there is a twist to this tale. The direct line of parents to offspring is the only way that genes are passed on into the future, but direct parental care is not necessarily the only genetic strategy that will be successful. Helping a brother, sister, or other relative to reproduce may also enable genes to perpetuate themselves. Full brothers and sisters share, on average, half their genes (although the vagaries of Mendelian segregation mean that particular pairs of siblings may have much more or much less than this). So a genetic tendency to help a sister to reproduce could be favoured by natural selection because the sister, being so closely related, has a high chance of having the same genetic tendency. 'Genes for' helping sisters thus help copies of themselves in the sisters of the body they are in and perpetuate themselves through the children of those sisters. Hamilton (1964) showed how genes for 'care of relatives' (not necessarily direct offspring) could spread and Maynard Smith (1964) suggested the term 'kin selection' to describe selection that takes account of other relatives as well as immediate descendants.

There is a very important point that has to be made here about kin selection. Helping a given relative to reproduce will be favoured ('genes for' it will be spread) only if the benefit—the increase in reproductive chances of that relative as a result of the help—more than makes up for the cost—the decrease in reproduction the helper incurs as the result of its action. For example, there is no point in 'helping' a brother if the help does not enable the brother to have any more children and at the same time prevents the helper having several children of his own. Genes for brother-helping clearly cannot spread under these circumstances; parental care genes would do much better. Hamilton generalized the circumstances in which relative-helping of various sorts would evolve into the equation: $rb - c > 0$, where r is the coefficient of relatedness and expresses how closely two individuals are related to each other, b is the benefit and c the cost to the relative-helping genotype. The net benefit minus the cost must be positive and greater than zero for the behaviour to be favoured. To work out whether or not the equation does work out greater than zero, we have somehow to calculate values for the three terms, r, b and c.

r does not usually cause too many problems. From basic genetics we can work out that full siblings and parents and offspring have a 50 per cent chance of sharing a given rare gene ($r = 0.5$); nieces and nephews have a 25 per cent chance of sharing with an uncle or an aunt ($r = 0.25$) and so on.

In practice, it may be more difficult to discover how closely related animals are to each other in nature. They may have to be studied over a long period of time to find out which are the offspring of which adult but, if it is practicable to get small samples of blood, there are now a number of biochemical techniques, such as 'DNA fingerprinting', which can give an exact answer on relationships quite easily. Obviously, it is not just the degree of relatedness that matters, but the number of relatives that can be helped. Hence J. B. S. Haldane's famous after dinner remark 'I am prepared to lay down my life on behalf of four grandchildren or eight first cousins!'

b and c in the equation above are somewhat more problematical. If we observe one animal helping another to rear its young, how do we know that the parent wouldn't have been just as successful without the help? And how do we know what the cost to the helper was in terms of the offspring it would have had if it hadn't been helping someone else? We seem to be dealing with mythical offspring that don't exist (the cost of helping) and extra offspring that do but are indistinguishable from the others (the benefits of helping). We can look at some examples to see how estimates of b and c are arrived at in practice.

In his two original and very important papers, Hamilton (1964) addressed himself to a problem that had long puzzled zoologists. The social insects, the Isoptera (termites) and many of the Hymenoptera (ants, bees and wasps) show extreme altruistic or helping behaviour. There is usually just one reproductive female (the queen) and large numbers of sterile workers. The workers—both males and females in termites but solely females in the ants, bees and wasps—perform all the tasks of the society such as foraging, rearing young, nest construction and defence and do not reproduce at all themselves. The remarkable fact is that in the Hymenoptera this ultimate form of self-sacrifice—the sterile workers do not, of course, reproduce themselves—appears to have arisen independently at least eleven times and perhaps more. Clearly it has not been easy to evolve social life or more different types would be expected to have done so, yet somehow this one order of Hymenoptera seems predisposed to achieve it.

Hamilton drew attention to the significance of r—the coefficient of relatedness—in the Hymenoptera's unique form of sex determination. Ants, bees and wasps exhibit 'haplodiploidy': males are haploid and develop from unfertilized eggs, females are diploid and develop in the normal fashion, from fertilized eggs. All the sperm from one male are therefore identical—a simple copy of the male's own haploid chromosome set.

When the queen bee fertilizes eggs with this sperm, all the resulting daughters receive the same paternal chromosomes and so have half their genes in common (the half donated by their common father). In addition, they share, on average, half the genes inherited from their common mother and so their degree of relatedness, $r = 0.75$ (0.5 from father) plus (0.5×0.5 from mother).

This means that, although the workers are sterile themselves, many of the genes that they share with the young queens (which are also their sisters) will be passed on to the next generation. The sterile workers benefit not because they help other workers that are closely related to them but are a reproductive dead-end, but because they care for the small number of their sisters that will develop into young queens.

While haplodiploidy appears to predispose the Hymenoptera to the high degree of sociality they show, it is clearly not the only factor because, as we have mentioned, a similar degree of sociality is also shown by the termites, which have an ordinary diploid mating system and sterile workers of both sexes have a degree relatedness of only 0.5 to the young reproductives. King and queen termites are long-lived and monogamous. A queen may lay up to 36 000 eggs a day and in some species live for 60–70 years. The king and queen together may have literally millions of offspring in their lifetime, the vast majority of which will never reproduce at all. In order to understand why so many termite workers should be sterile, we have to remember that there are three variables in Hamilton's equation. A high r predisposes towards helping because it increases the effective benefit, but even a relatively low value of r would lead to helping if the benefits were high enough and the costs low enough.

The cost of being sterile to each worker is the loss of those offspring it would have had if it had not been helping the colony. Given that a pair of termites on their own would probably not survive, let alone reproduce even modestly, the costs of helping must also be small (no hope of reproducing means no cost to losing it). The monogamous nature of the termite breeding system means that the workers are guaranteed a long series of full brothers and sisters to take care of. Since their own parents offer no care to them whatsoever, without the workers, the young termites would die. The workers, therefore, make a substantial difference to the survival chances of close relatives, even though the benefit accruing to each worker is only a fraction (because it is shared with the other workers that have helped) of the output of each reproductive.

Nevertheless, it appears that the benefits of helping are significant, the costs low and r at least as high as between diploid parents and offspring. Termites often live in deserts or in very dry regions and can do so only because the mounds (Fig. 5.11) built by the collective labours of millions of workers, enable them to create their own microenvironment. A single pair of termites, removed from this specialized microenvironment, would stand no chance, and this fact effectively reduces the costs and increases the benefits of helping the royal pair in the home colony.

Until quite recently, worker sterility and a caste system of sociality were thought to be found only in insects but Jarvis (1981) showed that at least one mammal, the naked mole rat, also has a very termite-like social system. These mole rats (Fig. 5.12) live in underground colonies in dry desert regions of southern Africa. There is usually just one breeding female and a caste of workers, most of which never reproduce throughout their lives. Instead, they defend the colony and, working like a chain-gang, dig tunnels through the hard earth to find roots and

Fig. 5.11 The ventilation 'chimneys' built by the colonies of some termite species can reach an astonishing size (photograph by Martin Speight).

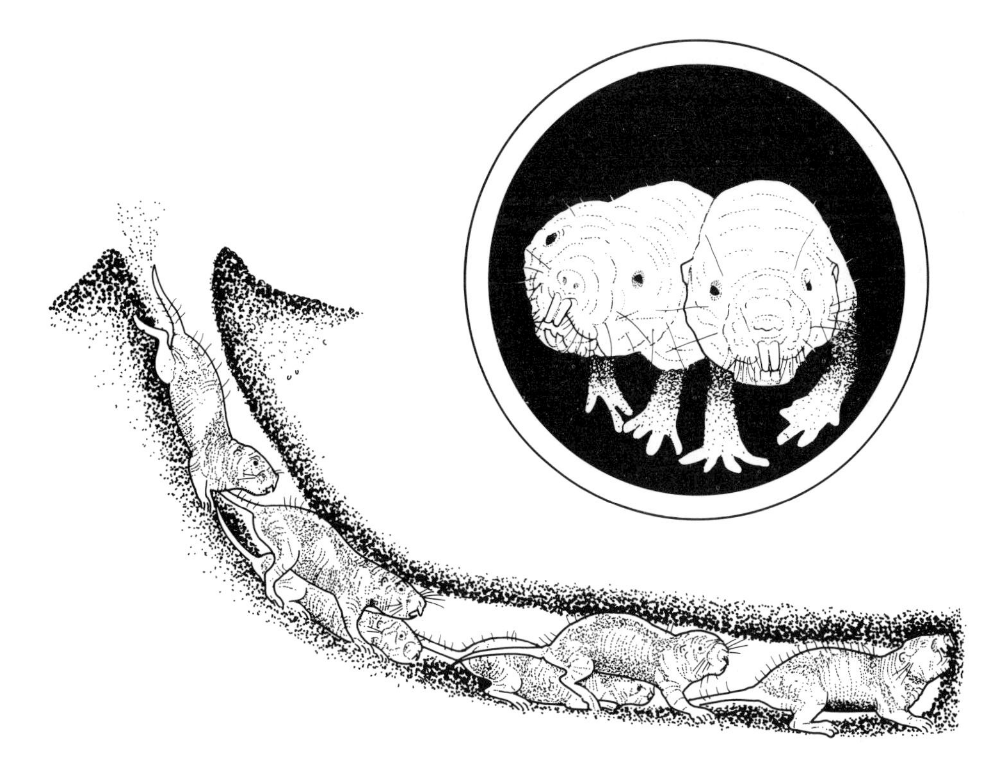

Fig. 5.12 Naked mole rats live underground in colonies of closely related individuals. The workers cooperate to tunnel through the hard earth in search of food (from a photo by David Curl, drawing by Priscilla Barrett).

tubers on which to feed the rest of the colony. The reproductive female does not take part in the food-gathering activities and has food brought to her by the workers. Food is so hard to come by that a pair of mole rats on their own would probably not survive at all. By cooperating with the colony and contributing to the reproduction of others, at least the workers get a 'part-share' of the offspring produced. Monogamy (giving an r of approximately 0.5 between workers and the young mole rats), prolific parents giving lots of young to look after, overlap of generations and a harsh environment making reproduction hazardous without help seem, as with the termites, to have tipped the cost–benefit equation in favour of the ultimate in self-sacrifice, life-long sterility.

Birds offer some interesting comparisons here. None, as far as we know, go to the lengths of being totally sterile, but in over 200 species, the parents are helped in some way by other individuals, often their own young from previous years. Later, these helpers frequently become parents in their own right but the selective advantage to their juvenile helping bears a strong resemblance to the helping seen in termites and mole rats. First, most birds are monogamous, so that the coefficient of relatedness between siblings of different years is similar to that between parents and offspring ($r = 0.5$). Secondly, helpers can contribute significantly to the viability of their siblings. They may help in nest-building, territorial defence and feeding and they are often very important in keeping predators away. Woolfenden (Woolfenden and Fitzpatrick 1984) has shown that in Florida scrub jays, 57 per cent of pairs have more than one helper and some have up to six (Fig. 5.13). The majority of helpers are males and are the sons of the birds they help. They appear to increase the number of chicks the parents can raise to fledgling largely through their active defence of the nest. So the helpers gain through having an increased number of young siblings and they seem to have a further benefit in that they tend to take over part or all of their parents' territory in the future, when they themselves start to breed. The cost of helping, that of postponing reproduction for a year or so, may not be all that great, since there appears to be a shortage of nest sites, making it very difficult for young birds to breed at all. Until a young male can secure a suitable breeding territory, he appears to be better off helping his parents to rear extra siblings than trying, and then failing, to reproduce himself.

Juvenile 'helping at the nest' of this type is much rarer in mammals but has been found in the black-backed jackal, where Moehlman (1979) proposed a similar explanation for why young jackals sometimes help their parents rather than breed on their own. She argued that in the area she was studying, there was an absence of suitable breeding sites, so that many of the young jackals would not be able to find anywhere to breed for themselves. Their best bet, therefore, was to stay and help their parents (Fig. 5.14).

So we can see that kin selection, of which 'helping

Fig. 5.13 Three adult Florida scrub jays at the nest. All are colour-ringed and can be identified as the two parents and a yearling bird from a previous brood. This latter is now helping to rear its young siblings in the nest (from Wilson 1975).

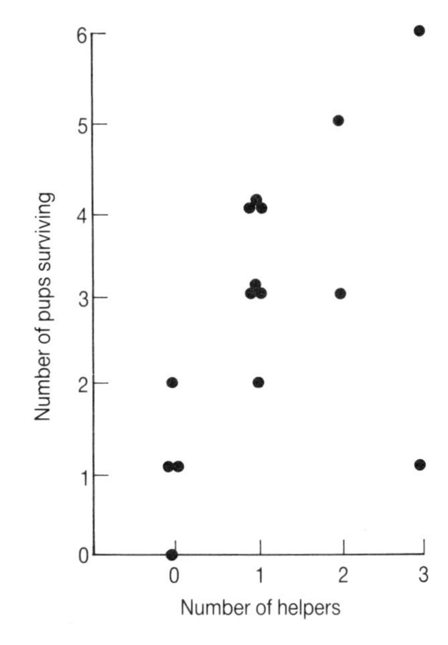

Fig. 5.14 The number of pups produced by pairs of jackals with different numbers of helpers (from Moehlman 1979).

at the nest' is a clear example, is not a separate, special sort of selection, in some way different from natural selection. It is a logical extension of the theory of natural selection as put forward by Darwin. But whereas on 'core' natural selection we can measure success or fitness as the number of offspring reared to reproductive age, with the extension of kin selection, the measure of success is slightly more complex. We have to know not just the number of offspring produced by an individual, but the effects of its behaviour on how many offspring its relatives have and how many it itself does not have as a result of its helping. Hamilton (1964) used the idea of 'inclusive fitness' as a way of calculating the conditions under which a gene might spread, taking into account the effects that bearers of that gene might have on different sorts of relative. The concept of inclusive fitness has been quite widely misunderstood (Dawkins 1989; Grafen 1984). Fitness is about success in leaving offspring. Inclusive fitness is also about success in leaving offspring but, as we have seen, that success may sometimes come about through aiding the offspring of relatives that, for a variety of reasons, may be reproductively more rewarding than an individual's own offspring. Genes that contribute to making their bodies help relatives can, under some circumstances, find themselves better represented in the children of the next generation than those that always encourage self-reproduction.

It will be obvious that, if kin selection is to operate, animals must have some way of selectively directing aid towards animals they are related to. Unfortunately, this process has come to be known as 'kin recognition', a misleading term because it implies that some active process of singling out relatives must be taking place. In practice, directing aid to relatives could be achieved simply by aiding individuals that are physically nearest. In species that do not disperse very far, proximity may be a reasonable guide to relationship, without any other special mechanism being involved. This method is open to exploitation, as when nest parasites such as cowbirds and cuckoos deposit their eggs in the nests of host species. The parents accept these nestlings as their own because they use **location** (i.e. the fact that the young are in the right nest) to label them as kin.

Some of the recent work on more sophisticated ways animals can recognize their kin is reviewed by Fletcher and Michener (1989) and Waldmann *et al.* (1988) and it is now clear that some animals have highly developed abilities to distinguish kin from non-kin, that go far beyond simply responding to the place they are in. Ground squirrels, honey-bees and even colonial sea-squirts are some of the species that have been shown to have this ability. Holmes and Sherman (1982) studied Belding's ground squirrels whose mothers had been fertilized by several different males and where they were, therefore, full and half-siblings in the same litter. Even though they grew up in the same burrow and had even developed in the same uterus, full siblings were less aggressive to each other than half-siblings. If baby ground squirrels were transferred to new litters when very small, they showed reduced aggression to their foster siblings and also to their full genetic siblings in the original litters. It seems that ground squirrels show a combination of mechanisms for kin recognition. Familiarity obviously plays a part, since individuals are less aggressive to individuals they grew up with than to strangers. But there also seems to be some recognition of full siblings even when they are reared completely separately from each other.

Honey-bees, too can discriminate (apparently by smell) between full and half-sisters. In the case of the sweat bee, *Lasioglossum*, Greenberg (1979) showed that the willingness of the bees guarding the entrance to the nest to let other bees in was directly correlated with the degree of relatedness between guard bee and intruder (Fig. 5.15).

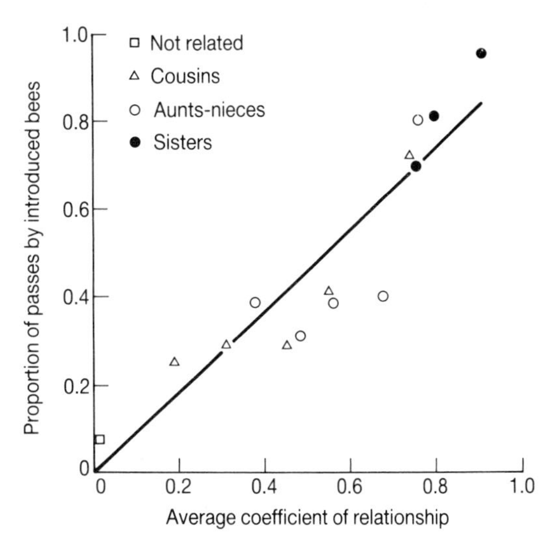

Fig. 5.15 Relationship between the proportion of sweat bees, *Lasioglossum* allowed into a nest by guard bees at the nest entrance and the degree of relatedness of the potential intruder to the guard (from Greenberg 1979).

CONFLICT AND INFANTICIDE

Parents may be selected to put themselves at great risk of injury and even death because, if their actions result in the survival of enough offspring, the genetic tendency to behave in this way will be perpetuated in those offspring. There is, however, a darker side to this process. Parents are not genetically identical to their children, and the genetic interests of the two may not always precisely coincide. Sometimes the interests of the parent may even be best served by killing or at least withholding care from some offspring if, as a result, the parents are enabled to have many more offspring in the future. Owls and other birds of prey often lay more eggs than they can normally rear as an 'insurance policy'. All the chicks are reared if food is plentiful but the smallest one dies in times of scarcity. It is only a small step from withholding food to passive or even active killing. Mock *et al.* (1987) showed that herons and egrets often practice 'siblicide'—active killing of nestlings by their older brothers and sisters, with the parents looking on and not intervening.

If this sounds strange, it shouldn't do. Every time a parent feeds its young it is expending effort and energy that could have been spent on rearing other offspring. If, by withholding that care, the parent could rear two or more other offspring then withholding care might be the most beneficial strategy. As Trivers (1974) discusses, conflict between parents and offspring is most apparent at weaning, when it is in the offspring's best interest to demand a bit more food and in the parent's best interests to conserve resources for future children. For example, Hinde (1977) showed that as baby rhesus macaques become older, the mother rejects her infant's advances and stops it making nipple contact with steadily increasing frequency until at the end of the period of investment, all the offspring's advances are ejected (see Fig. 2.10, p. 33).

In a number of species, males are known to kill infants that are not their own. When a new male lion takes over a pride of females he tends to kill young cubs, particularly those that are still taking milk from the females. Breaking off lactation brings the females more rapidly back into oestrous. The new male can then make them pregnant and replace cubs fathered by other males with his own offspring. Cub-killing by male lions ceases by the time the lionesses are bearing the new males' cubs. A similar explanation—competition between males to sire as many infants as possible—has been put forward to account for infanticide among primates (Struhsaker and Leland 1987). As with the lion, males have been regularly observed to kill infants when they move into a new troop. Showing a rather curious role-reversal, in the wattled jacana, it is the females that kill the young. In these birds, females are larger than males, defend territories against other females and have several males incubating their eggs. Emlen *et al.* (1989) showed that when a female takes over a new territory, she kills young already there, and so hastens the time when the males will be incubating her own eggs.

COOPERATION BETWEEN NON-RELATIVES

It is not just between animals that are related that we find examples of animals helping each other. Mutual grooming or preening are quite common, with one animal grooming parts of the body, such as the head, neck and back, that the animal itself cannot get to. This may occur between relatives, but is also found in animals that are unrelated but very familiar associates. It is particularly well developed in the primates where, as we shall discuss in Chapter 7, friendly contact helps to cement bonds that may have all sorts of other pay-offs within a social group. Both parties gain from the interaction and both can break it off if the other does not participate.

An even more interesting form of mutual help is called **reciprocal altruism**, where one animal helps another but may not then receive assistance back itself until some time later. This 'mutualism with a time-lag' is therefore potentially open to cheating, because one animal could apparently receive benefit itself and then not reciprocate. For this reason, we do not find reciprocal altruism evolving in situations where there is a 'one-off' advantage, with two animals being unlikely to meet again. Rather, it evolves in situations where the same animals associate over a long period of time and where a 'cheat' will be penalized because benefits will be withheld in the future if it does not reciprocate. It is most highly developed in animals that have the capacity to remember which other individuals are reciprocators and which are cheats, such as the chimps described by van der Waal (1989) or the baboons by Packer (1977). What is of interest is that their repeated, long term cooperation leaves both parties to the pact better off than either would be alone.

Wilkinson (1984) has described a remarkable example of reciprocal altruism in vampire bats, which feed on the blood of mammals, particularly domestic animals such as cattle and horses (Fig. 5.16). A bat will inflict a tiny (3 mm) wound on its host with its sharp teeth. Its saliva contains anti-coagulants and the bat flicks its tongue rapidly in and out to take the blood. These bats are active during the darkest hours of the night and return to communal roots in hollow trees or caves by day. Sometimes a bat will return to the roost without having fed and, if it fails to find food for three successive nights, it may starve to death. But a bat that has not found food itself will often be fed by another bat regurgitating a blood meal to it. Starving bats are more likely to be fed than well fed ones, and although bats feed their kin, they also feed unrelated bats, particularly those that have fed them in the past. Regurgitation of blood meals was seen only between bats who regularly (over 60 per cent of sightings) associated with each other. The bats fulfil the criteria for successful reciprocal altruism. They associate with particular individuals over long periods of time and remember benefits given to them. This means that they can detect cheating individuals that do not reciprocate. The bats are also sensitive to whether another bat is starving or well fed and are able to help a starving bat without too great a cost to themselves. A starving bat loses weight at a high rate, but a well fed one at a much lower rate, so that the giving of a blood-meal from a well fed to a starving bat gives the recipient more time until starvation than the donor loses.

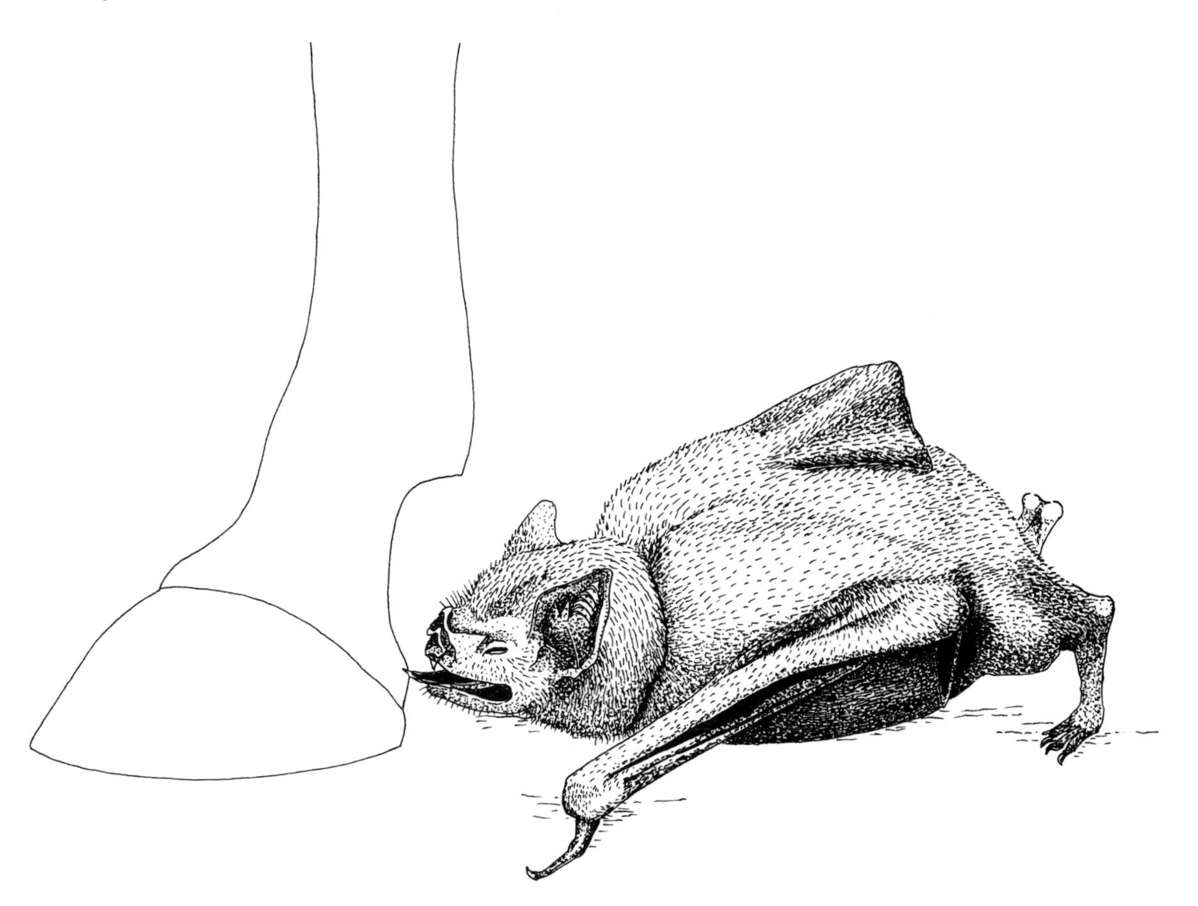

Fig. 5.16 Vampire bats feed on the blood of large mammals. They make a small incision with their teeth and then lap the wound with their tongues. Their saliva contains an anti-coagulant that prevents the blood from clotting (drawing by Melissa Bateson).

Finally, failing to find food on its own appears to be something that can happen to any bat at some time, so all bats benefit from the 'insurance policy' of being fed by others in time of need. But those that donate food will be in a more favourable position to receive it when they themselves are in need of it. Cooperation, in the long run, is selfishly the best policy.

SPECIES ISOLATION

So far in this chapter, we have looked at the raw material of evolutionary change—genetics, kin selection, cooperation and competition. Eventually, over the course of time, the gradual accumulation of small genetic changes may lead to the formation of a new species. It may be that part of a population moves off into a new area, where it gradually becomes so different from the parent population it left behind that there are now two distinct species, recognized by the fact that they cannot interbreed. With some rare exceptions, it is generally thought that some sort of geographical barrier—such as a mountain range or a stretch of sea—is necessary to keep the populations separate for long enough to allow two distinct species to evolve. Then, if they do come back into contact again, they may be so different they will not recognize each other as 'the same'. Hybridization is usually a hazardous process on genetic grounds because hybrids, although they may show some signs of 'hybrid vigour' are usually sterile or less fertile than the offspring of within-species matings. Mules (horse × donkey) are a good example of this. Even if they are fertile, they are rarely as successful as either parental type. The latter each have a set of genes that has been selected over many generations as the best for their own environment. The hybrids inherit a compromise set of genes that does not equip them so well in either parental environment. Dilger's lovebirds (p. 108) showed how confused hybrids can be in their behaviour. In addition, their chromosomes may also be incompatible, so that there is a mechanical breakdown at meiosis.

Consequently, there is a strong selective advantage to choosing a member of one's own species as a mate. This is particularly true for females; males can usually mate several times in their lifetimes but some female insects, for example, mate only once. Accordingly, it is common to find that discrimination is stronger by females than by males. In fruit-flies, for example, females are more selective than males and it is often their active rejection of foreign suitors that prevents interspecific hybridization.

Sexual isolation may be defined as a behavioural barrier to hybridization between species or populations. It forms one aspect of the more general phenomenon of reproductive isolation and is one of the most important ways in which behaviour affects the evolution of animal populations.

The degree to which two related species will meet under natural conditions varies greatly. Even if they coexist in the same general area, two species will rarely live in exactly the same part of the habitat: competition drives them into specialization. In Britain, chiff-chaffs and willow warblers inhabit the same woods, but for feeding the chiff-chaff moves amongst the high trees while the willow warbler feeds in the lower branches of trees and bushes. Clearly they are specializing, and this conclusion is strengthened when we observe that in the Canary Islands, where the willow warbler is absent, the chiff-chaff occupies the willow warbler's niche as well.

In most animals, isolation by habitat, or part of habitat, is reinforced by the evolution of highly specific signals that enable animals to detect their own species, sometimes from a considerable distance. There is little doubt that diversification of such signals is a direct result of selection of dicriminating own from other species. Blair (1955) describes how two species of frog, *Microhyla olivacea* and *M. carolinensis*, have a wide range in the United States, the former more to the west and the latter more to the east. At the extremes of their ranges, where only one species is present, their assembly calls are quite similar. More centrally, where both species overlap and use the same ponds for breeding the calls have diverged and are quite distinctive. Hybrids are occasionally found in nature and Blair has shown that, apart from the calls, there is little other barrier to hybridization.

Distinctive calls are a common way for animals to recognize members of their own species. Not only frogs, but crickets (Pollock and Hoy 1979), fruit-flies (Bennet-Clark and Ewing 1969) and birds use sound in this way. Searcy and Marler (1981) used synthetic bird songs to find out exactly what it is about the songs of their respective males that female song sparrows and swamp sparrows use to distinguish between them. Their synthetic songs consisted of either song sparrow or swamp sparrow 'syllables' presented in either song sparrow-like or swamp sparrow-like order. For the females of both species, the song had to have both the right syllables and be in the right order, but male swamp sparrows seemed to show no particular preference for the temporal pattern of their own species. As we mentioned earlier, 'fussiness' by females but not by males is not surprising in view of the fact that females usually have a greater investment in each offspring and so choosing the wrong species as a mate has more serious consequences. In the red-winged blackbird, females clearly distinguish between real red-winged blackbird song and that produced in imitation of it by the mockingbird, whereas males seem to be completely taken in by the deception (Searcy and Brenowitz 1988).

We have already seen (Chapter 3) that fireflies use vision to recognize species. Male fireflies signal to females by producing flashes of light that are extremely distinctive at night. Females respond only to the pattern of flashes that is given by the male of their own species (Lloyd 1965, 1975).

We often find such mutual adaptation between a male's signal and a female's response. In fireflies, Lall *et al.* (1980) showed that there is a good match between the colour of light emitted by the male and the colour sensitivity of the female's eyes. A comparable coevolution, this time within one species, is described by Ryan *et al.* (1990) for the cricket frog. They studied two US populations of this species 2500 km apart—one in New Jersey and the other in South Dakota. Not only were the calls of the male frogs different in the two populations (dominant frequencies were 3.56 kHz and 3.77 kHz, respectively) but so was the frequency that most excited the auditory systems of the two sets of females. The basilar papilla—the inner ear organ that is used in the reception of calls responded maximally to 3.52 kHz in New Jersey females and to 3.94 kHz in South Dakota females. Such differences between populations, particularly where there is geographical separation, pave the way for the eventual evolution of completely new species.

THE PHYLOGENY OF BEHAVIOUR

Over long periods of time, new species will evolve and others become extinct. New structures and new patterns of behaviour appear and we can ask how they arose and what they evolved from. The theory of natural selection assumes that everything about animals and plants, however extraordinary, evolved gradually from something that was there before and did not spring fully formed like a phoenix from the fire. In considering the last of Tinbergen's four questions about behaviour—that of phylogeny—we are therefore enquiring about the evolutionary origins and antecedents of behaviour that we see today. We want to know how, back in the mists of evolutionary time, behaviour patterns evolved.

Although we can readily speculate about the origin and evolution of general behavioural capacities, such as aggressiveness or learning ability, it is easier to be specific about more proscribed behaviour patterns, such as calls or courtship displays, which are shared by a group of related species. We might call this the study of behavioural microevolution.

The task is somewhat similar to finding out how anatomical structures evolved but it is more difficult. We could ask how the vertebrate ear ossicles evolved and deduce from the fossil record that they evolved from bones of the lower jaw. With behaviour we are hampered by the fact that (with one or two notable exceptions, such as fossilized dinosaur tracks) we have no fossil record of behaviour. But we do have two other methods for tracing the phylogeny of behaviour. One is to look at closely related species to see exactly how they have diverged: it is often possible to identify the nature of evolutionary changes in the past from the pattern of diversity in the present. The other is to look at genetic changes that take place in a population within a manageably short space of time, such as a few weeks or months.

One sort of evolutionary change that attracted a great deal of attention from early ethologists is 'ritualization' (Huxley 1914). Here, a behaviour that initially does not serve any communicating function becomes elaborated and exaggerated over evolutionary time to serve as a signal. Huxley studied the various courtship displays of the great-crested grebe, one of which involves both partners rearing

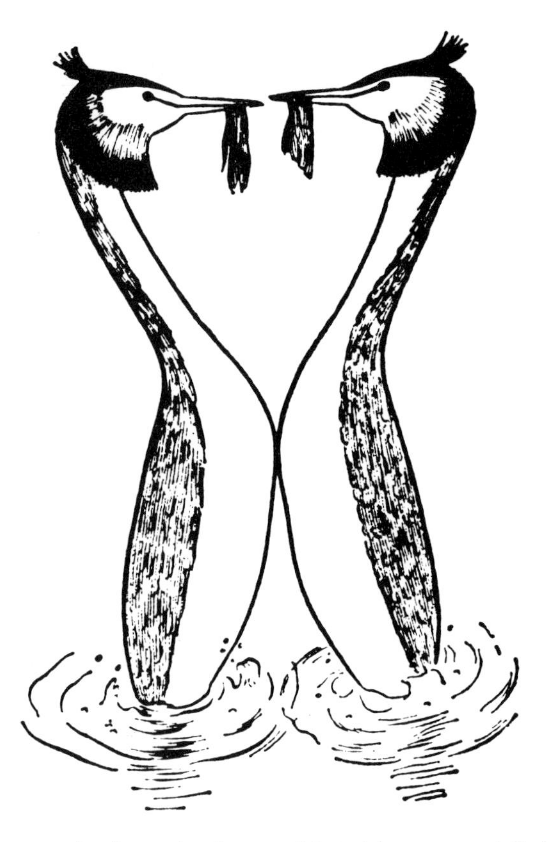

Fig. 5.17 Male and female great crested grebes performing one of their elaborate mutual displays. This one involves presentation of nest material (from Huxley 1914).

out of the water and presenting nest material to each other (Fig. 5.17). This display has been derived (in an evolutionary sense) from elements of the grebe's nest-building behaviour and in becoming adapted to its courtship function, it has become more elaborate and 'ritualized'.

If we compare closely related species or different strains of the same species, we can sometimes see how their behaviour has diverged from that of common ancestor. In some cases, a behaviour pattern may remain more or less the same but the frequency with which it is performed changes; in other cases, the form of the behaviour itself alters.

For example, on page 109 we discussed how artificial selection can lead to changes over a few generations in the calling behaviour of male crickets. The calling behaviour remains the same but how much calling a male does in a night can be either increased or decreased by selective breeding. Here selection acts on the frequency of the behaviour. On the other hand, Bentley and Hoy (1972) have shown that the form of behaviour can be changed by quite small genetic differences. They showed that, in backcross hybrids between two closely related species of Australasian crickets, *Teleogryllus oceanicus* and *T. commodus*, some hybrid songs contain an extra pulse of sound, produced by one extra impulse to the wing muscles, which in turn gave an extra 'scissor' movement to the wings that produce the song (Fig. 5.18)

Hunsaker's (1962) comparative study with lizards of the genus *Sceloporus* shows most beautifully how form and frequency changes move together during microevolution. Whenever they meet other lizards, males perform rhythmic head-bobbing movements, which act as identifying signals. Each species has a characteristic pattern of bobbing (Fig. 5.19), which is produced by rhythmic contractions of the muscles that extend the front legs, thereby raising and lowering the head and shoulders. The pattern of microevolutionary change is particularly well indicated because only two groups of muscles are involved,

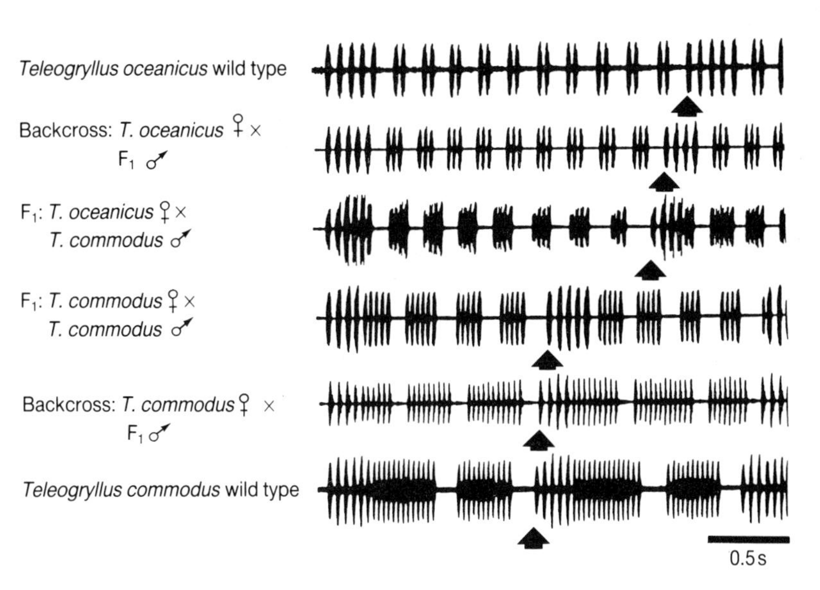

Fig. 5.18 Songs of two species of cricket, *Teleogryllus oceanicus* and *T. commodus* and of various hybrids between them. Arrows indicate the beginning of a song 'phrase' (from Bentley and Hoy 1972).

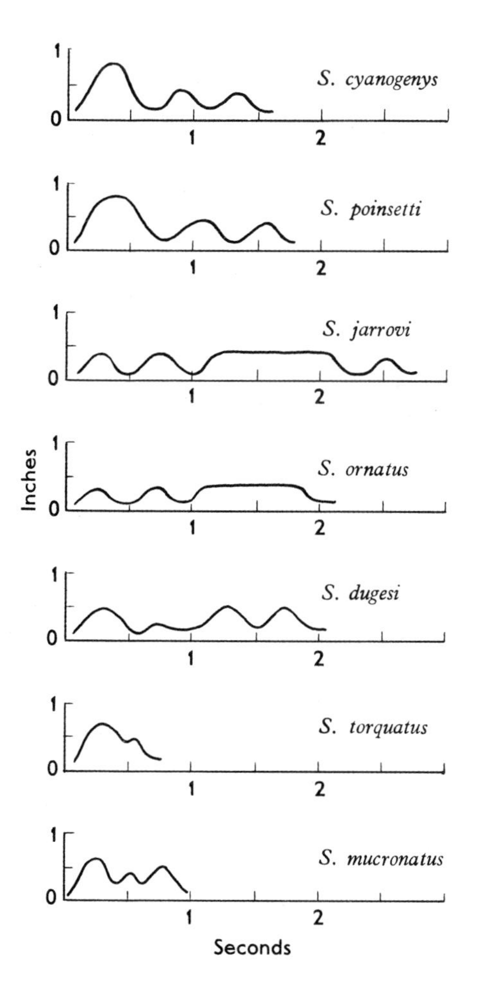

Fig. 5.19 The specific head-bobbing movements of some *Sceloporus* lizards. The movements of the head are represented as a line with height on the vertical axis and time on the horizontal axis. Changes to amplitude, speed and length of movements are all clearly shown (from Hunsaker 1962).

and yet we see evidence of change in amplitude, speed and length of movement from species to species.

The overall theme of this chapter has been the interplay between genes and behaviour. The behaviour of animals is not only shaped by natural selection: it is itself a powerful influence on evolution. We have already discussed (p. 104), the powerful influence that culturally transmitted behaviour may come to exert on selection for genetically based traits, and obviously behaviour such as parental care and helping at the nest will have a direct effect on the gene frequencies of the next generations, as does behaviour that leads to an individual choosing a mate from its own rather than a different species. The end result is the most extraordinary diversity in the species living today, in form, colour and behaviour.

6 LEARNING AND MEMORY

Throughout this book we have made frequent reference to learning as it affected some particular aspect of behaviour under consideration. Now attention must be focused on learning itself. Learning and memory go together because, whilst the former involves changing behaviour as a result of experience, its effects can be put to use only if the results of the experience can be stored in some way and recalled the next time they are needed.

We shall examine learning in a wide range of very different animals and Thorpe's (1963) book, although quite old, remains very useful because it deals with the whole animal kingdom—learning in molluscs and insects as well as in birds and mammals. Thorpe defines learning as '. . . that process which manifests itself by adaptive changes in individual behaviour as a result of experience.' This definition draws attention to two important features. First, learning normally results in **adaptive changes** and, as we discussed in Chapter 2, learning and instinctive behaviour are both ways for equipping an animal with a set of adaptive responses to its environment. Normally both are found in combination and logically they have much in common. In one case we have the selection of individuals whose genes operate best during development so that, as these animals are most successful, adaptive behaviour evolves in a population of animals. In the case of learning, individuals select and retain the best responses over the course of their own lifetime.

The second important point arising from Thorpe's definition is that, strictly speaking, learning is a 'process' that cannot usually be observed directly; we measure what has been remembered as a result of learning. Because we can communicate so easily with human subjects, they are in many respects better material for learning studies than animals. For instance, in the laboratory, it is possible to test our memory in two ways: 'recall', i.e. by reciting or writing down a list of nonsense syllables that we have previously learnt and 'recognition', i.e. when presented with a set of nonsense syllables, including those we have learnt, we record the syllables we recognize. Recognition is always an easier task than recall because the situation provides stimuli that, as we say, 'jog our memory' and help the process of recall. If we have trained a rat to run through a maze we cannot ask it to draw a map of its route on a piece of paper. The only way to test it has learnt and retained is to put it back in the maze and observe its behaviour. If the rat makes mistakes we have no means of knowing whether it failed to learn adequately or learnt but failed to recall.

This difficulty brings us to the question of learning mechanisms. What happens in the nervous system when an animal learns something? When a young gamebird crouches on the first exposure to the sound of its parent's alarm call, it must utilize pathways already present in its brain whereby a particular auditory input has easy 'access' to the motor system controlling crouching. However, new pathways must be established when a rat learns to press the bar of a Skinner box because, prior to learning, the bar evoked no special response. We know too that mere presentation of the bar to the rat is not enough, there must be some sort of reward as a result of doing so. Some part of the learning mechanism records the result of the rat's reaction and if this result is 'good' it increases the probability that the reaction will occur again the next time the situation is presented. Further, we know that somewhere in the nervous system there is stored a more or less permanent record of the learning that can be 'consulted' or recalled on future occasions. The investigation of the behavioural and physiological basis of memory has been an active area of research for decades. Recently there have been some real advances and we shall return to this subject after looking at the phenomenon of learning itself.

There was a time during the first half of this century when a huge amount of effort was directed towards the study of animal learning. Psychologists hoped that animals would serve as models for human beings in this respect and the majority of their work concentrated on two convenient laboratory species, the rat and the pigeon. By investigating every variation of learning task, of reinforcement (reward or punishment) and of cue they hoped to generate 'laws of learning' that would enable us to predict performance and outcome no matter what subject or what situation. The *Handbook of Psychological Research on the Rat* by Munn (1950) is a fascinating record of this 'heroic age' of experimental psychology, the diverse models and the theories of learning that it produced. It has to be said that, for the most part, they attract little attention now. The classification of learning into distinct types, the proscribed learning situations in which they were studied and the rigid models into which all animals and all learning were expected to conform, have been replaced by a much more biologically based approach. This recognizes that an animal's learning abilities must have evolved to suit its own special requirements, just as any other aspect of its behaviour. We should not expect learning in pigeons to resemble that of honey-bees, except in a very general way. Further, learning is now seen as a way in which animals attempt to identify key aspects of a fluctuating environment; to detect its regularities and ignore the distracting 'noise' that is not important for them. They will integrate their learning with the inherent biases of perception and response that all animals bring to the world. Encouragingly, these moves towards a more truly comparative and biological approach have come both from psychologists, dissatisfied with the old learning theories and from ethologists, who observe the role of learning in their animals' natural lives. Articles by Mackintosh (1983), Roper (1983) and Rescorla (1988) form a good background to this new synthesis. Books by Seligman and Hager (1972), Hinde and Stevenson-Hinde (1973) and Dickinson (1980), provide more examples and a more detailed analysis of some of the changing attitudes towards learning.

SENSITIZATION AND HABITUATION

If animals are to learn to change their behaviour to meet a new situation, then clearly anything new appearing in their environment—and we do not mean just a visual stimulus—will have to be taken note of and its importance assessed. The easiest way to get some estimate of importance is to note what happens just before or just after it appears. An alert animal, for instance one that has had its appetite aroused by the smell of food or has just fled from a predator's attack, is particularly sensitive to such stimuli. To take a laboratory example, a rat that has just received a small electric shock to its feet jumps in alarm to any novel stimulus—a flash of light, a tap on its box—stimuli that would normally evoke little or no response. Evans (1968) studied the marine rag worm *Nereis*, which live in tubes they construct in the sandy floor of the sea, stretching out the front part of their body to forage on the surface near by. Evans kept his *Nereis* in laboratory aquaria in dim light. Under such conditions he found that retracted worms sometimes emerged from their tubes to a flash of light and in one experiment 21 per cent of worms did so. However, if he took a second group of unexposed worms and fed them just once in the dim light conditions, then over 60 per cent of them emerged from their tubes to a light flash. The arousal following feeding had made them much more sensitive.

This phenomenon—a period of high responsiveness following arousal by rewarding or punishing experiences—is, reasonably enough, called **sensitization**, and it is widespread in such situations. Why then, do not animals respond with alacrity to all stimuli, because opportunities to have become sensitized to them must often exist? The solution lies in a second, and in many ways, opposite effect of stimulation. If repeated several times in the absence of any significant accompaniment—no more foot shocks to the rat, no more food for *Nereis*—then the animal gradually ceases to respond. From being initially sensitized it calms down and eventually ignores the stimulus. This waning of responsiveness is called **habituation**.

If we now think back to the definition of learning—adaptive changes as a result of experience—then perhaps we should regard both sensitization and habituation as learning. Habituation, as we shall see, fits quite well but sensitization really does not. It is too short-lived and too indiscriminate—for a short time the animal is hyperresponsive to a variety of events in the outside world and pays more attention to them. This is probably an essential preliminary to any learning but not the process itself. Soon one of two things must follow. If the novel stimuli that initiated sensitization are not repeated then the effect wanes quickly. If they are repeated enough times then the animal goes on to learn. What it learns depends on the results of the situation. Do the stimuli signal pleasant or unpleasant events—food, water, a predator, pain? If so then the stimuli that give the best prediction of these rewards or punishments (together we can conveniently call them **reinforcements**) will be picked out for continuing attention. We say that the stimulus and the reinforcement have become **associated**—association learning is a major category for discussion later (p. 126). There is an alternative result of repeated stimulation—that nothing happens, there is no reinforcement. The stimulus then captures less and less of the animal's attention. Birds soon come to ignore the scarecrow that put them to flight when it was first placed in a field. We say that the birds habituate and that habituation can be regarded as a simple form of learning. It is relatively long-lasting and it is 'stimulus-specific', i.e. only the stimuli that are repeated without reinforcement are affected—the animal remains alert to others.

These qualities will be clearer from an example. Once again the relatively simple behavioural repertoire of *Nereis* has provided good material for studies by Clark (1960a,b). As already mentioned, the worm burrows in mud or sand on the floor of brackish estuaries. Its head and anterior segments protrude from the tube to feed on the surface around the burrow. During its bouts of feeding a variety of sudden stimuli will cause the worm to jerk back into its tube. In the laboratory, Clark could easily get the worms to live in glass tubes in shallow basins of water. He found that jarring the basin (mechanical shock), touching the head of the worm, a sudden shadow passing over and a variety of other stimuli would all cause retraction into the tube, but the majority of worms emerged again within a minute. If these stimuli were repeated at 1-min intervals the proportion of worms responding fell off until none of them were retracting—they had habituated. Clark found that habituation occurred more rapidly if stimuli were given close together. For example, with a bright flash of light it took less than 40 trials at half-min intervals, but nearly 80 trials if the interval was 5 min. The speed of habituation also depended on the nature of the stimulus—mechanical shock, shadow, touch and light flash each produced their characteristic rates of habituation. Further, habituation was to a large extent stimulus-specific. Figure 6.1 shows how the waning of retraction to repeated mechanical shock is independent from that to a moving shadow.

There are a number of other processes that may be confused with habituation because they also lead to a reduction in responsiveness. Results such as those illustrated in Figure 6.1 eliminate any possibility that the waning of response is due to motivational changes or to muscular fatigue, but it is often more difficult to eliminate sensory adaptation. Many sense organs eventually stop responding to repeated stimulation. We cease to be aware of our clothes within a few moments or so of putting them on because the tactile receptors in the skin cease to respond. In *Nereis*, Clark could also eliminate sensory adaptation as an explanation for the waning. For example, the worm soon ceased to retract when touched by a probe, but clearly still detected the stimulus because it then attempted to seize the probe with its jaws. Sensory adaptation is usually a short-lived phenomenon; a few minutes without stimulation is usually sufficient for complete recovery. We sensorily adapt to the feeling of our clothes rather than habituate to them and we retain the term 'habituation' for a more persistent waning of responsiveness, which must be a property of the central nervous system and not the sense organs.

Clark could detect some recovery from habituation within an hour and the worm's retraction response was completely recovered and back to full strength within 24 h. This is not very long and, if we

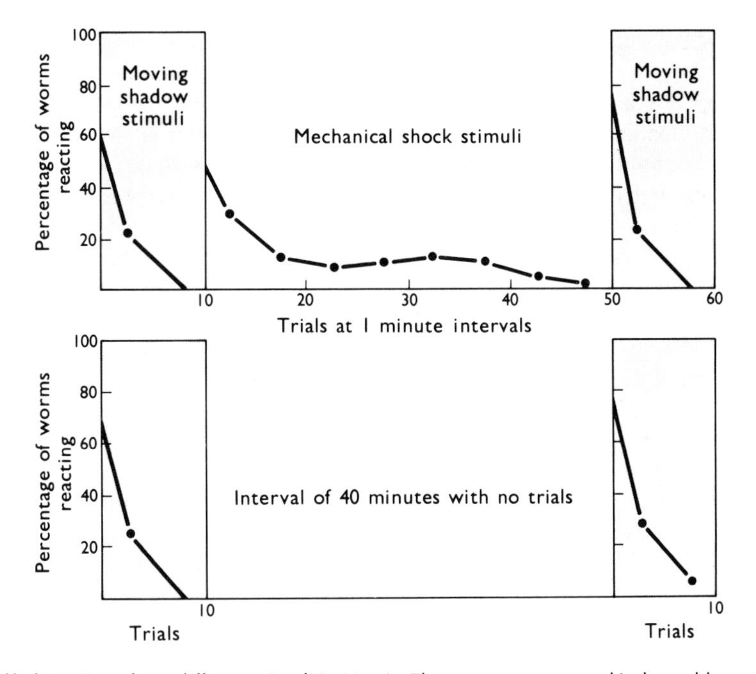

Fig. 6.1 The rates of habituation of two different stimuli in *Nereis*. The response measured is the sudden retraction of the worm into its tube and trials are given at 1-min intervals. The shaded areas record the responses of a group of 20 worms to a moving shadow. Within 10 trials they have all ceased to respond, but switching to a mechanical shock stimulus (unshaded areas in upper graph) brings back the response in half the worms, and it characteristically habituates more slowly, taking more than 30 trials. The recovery of the response to moving shadow is complete after 40 min, whether the habituation trials to mechanical shock intervened (upper graph), or if the worms were simply left alone (lower graph). Clearly, habituation is quite independent for these two very distinct types of stimulus (from Clark 1960a).

think in terms of a memory span, not impressive, but this is to ignore the role that habituation plays in the life of animals like *Nereis*. It is, in fact, a vital process for adjusting an animal's behaviour to the minute-to-minute events in its environment. Habituation enables it to concentrate on important changes there and ignore the others. Small prey animals cannot spend too long skulking in shelter—they must normally be out feeding. Although it is best at first to retreat rapidly to sudden shadows, not every shadow means a bird or a fish overhead: it might just as well be a floating piece of seaweed. In general, frequent shadows are more likely to be seaweed than predators and it is certainly adaptive to cease responding when a repeated stimulus has no attendant consequences. Just as certainly it is dangerous to ignore shadows from then on. With the next incoming tide the seaweed will be in different places and a shadow could now signal a predator's approach. The same cautious testing out must begin again.

This is not to imply that all habituation is so short-lived. Newly hatched chicks begin to feed by pecking at any object that contrasts with the background. Responses to inappropriate objects rapidly habituate and do not return—rather the growing chick begins to learn positively which objects represent food.

In nature we must expect to find sensitization, sensory adaptation and habituation almost inextricably combined, so that the time course and persistence of diminished responsiveness depends on a variety of factors. Habituation processes are shown by all animals, but predators and prey animals, for example, are unlikely to respond in the same ways because the costs and rewards will be so differently balanced.

ASSOCIATIVE LEARNING

Habituation is generally regarded as a simple form of learning because it involves the waning of a response that is already there. We tend to think of learning as a process whereby we acquire new responses and new capacities. For this reason, the basic characteristics of associative learning are immediately familiar to us. A previously neutral stimulus or action has sufficiently important consequences to be singled out from other such events. After some repetitions followed by the same consequences, a long term association is built up between the event and its result and the animal's response changes accordingly. Having found sugar solution there previously, a honey-bee picks out the blue dish from an array of dishes laid out by an experimenter. Rodents exploring new territories quickly learn the shortest routes to shelter for use when a hawk or an owl swoops.

The best known type of associative learning situation is the **conditioned reflex**, inseparably linked with the name of the great Russian physiologist, I.P. Pavlov, whose school was active around the turn of the century. Everyone has heard of Pavlov's dogs who salivated to the sound of a bell! So far has his influence run, the type of learning he studied is often referred to as **classical conditioning**. Pavlov's (1941) influence on behavioural work and neurophysiology in Russia is still very great but, perhaps because the theory he developed attracted little favour here, his influence in the West has been less. Pavlov and Sherrington were working at the same time but from completely different viewpoints. Sherrington studied the organization of reflexes in the isolated spinal cord of dogs and cats, having deliberately cut off influences from the higher centres. Pavlov worked with intact animals and considered that, just as simple reflexes are a property of the spinal cord, so conditioned reflexes are the particular property of the higher centres of the brain especially the cerebral hemispheres. Pavlov's aim was to study 'the physiology of higher nervous activity', but most of his experiments were, in modern terms, pure experimental psychology. Indeed, Pavlov was really one of the founders of experimental psychology; he was applying objective techniques to the study of learning years before J.B. Watson, whose book, *Behaviorism* (1924), had such a huge influence on American psychology.

Pavlov's classic experiments with dogs often involved the 'salivary reflex'. Dogs salivate when food is put into their mouths and Pavlov could measure the strength of their response by arranging a fistula through the cheek from the salivary duct, so that drops of saliva fell from a funnel and could be counted. A hungry dog was placed on a stand, restrained by a harness and every precaution was taken to exclude disturbances. In this position, it could be given various controlled stimuli, such as lights, sounds or touch, and meat powder could be puffed into its mouth through a tube (Fig. 6.2). A standard quantity of meat powder caused the secretion of a certain amount of saliva. Now Pavlov preceded each ration of powder by, say, the sound of a metronome ticking. At first, this stimulus caused no response, save perhaps that the dog pricked up its ears momentarily. However, after five or six pairings of metronome followed by food, saliva began to drip from the dog's fistula soon after the metronome started and before the meat powder arrived. Eventually the amount of saliva produced to the metronome alone was the same as that elicited by the meat powder.

The dog had learnt to respond to a new stimulus, previously neutral, which Pavlov called the **conditioned stimulus** (CS). The salivation response to the CS is the **conditioned response** (CR). Prior to learning, only the meat powder or **unconditioned stimulus** (UCS) produced salivation as an **unconditioned response** (UCR).

Pavlov found that almost any stimulus could act as a CS provided that it did not produce too strong a response of its own. With very hungry dogs even painful stimuli, which initially caused flinching and distress, quite soon evoked salivation if paired with food. The CR is formed by the association of a new stimulus with a reward and in the same way a CR for withdrawal can be formed by associating the CS with punishment. An electric shock to the foot causes a dog to lift its paw; if a metronome is paired with the shock, the dog soon raises its paw to the sound alone.

Pavlov carried out exhaustive tests on the precision with which a particular stimulus was learnt. He found that if a dog was conditioned to salivate when a pure tone of, say, 1000 Hz was sounded, it would also salivate when other tones were given but to a lesser extent. It **generalized** its responses to include stimuli similar to the conditioned one and the more similar they were the more the dog salivated.

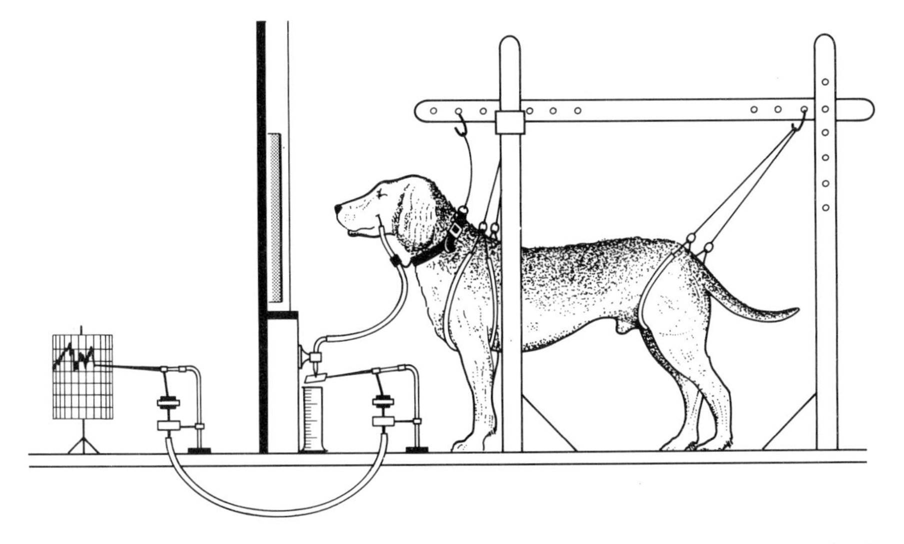

Fig. 6.2 A typical experimental set-up in Pavlov's laboratory. The dog is restrained on a stand facing a panel and, under experimental conditions, is well insulated from external disturbances. Tactile, visual or sound stimuli can be presented in a carefully controlled fashion. Tubes run from the fistula in its cheek, which collects saliva as it is secreted. A simple arrangement of a hinged plate, onto which the saliva drips, and a measuring cylinder where it collects, enables the amount of saliva to be measured together with the intensity and duration of its secretion. Not shown is the device for blowing a controlled amount of powdered meat into the dog's mouth to reward the salivary response.

The opposite process to generalization is **discrimination**. Dogs naturally discriminate to some extent or they would salivate equally to all sounds, but their discrimination becomes refined after repeated trials when only one particular tone is followed by reward. We can accelerate discrimination if, as well as rewarding the right tone, we slightly punish the dog when it salivates to others. This **conditioned discrimination** method has been of enormous value for measuring the sensory capacities of animals. After training to one particular stimulus —it may be a colour, brightness shape, texture, sound, smell, weight, etc.—we then test to see how far the animal can discriminate this stimulus from others. We present it, together with another stimulus of the same type, and reward only responses to the former, perhaps giving slight punishment for incorrect responses. The two stimuli are made increasingly similar until there comes a point beyond which the animal can no longer learn to discriminate between them. This marks the limit of its sensory capacities as measured by its behaviour.

To give but three examples from many hundreds, this method was used by von Frisch (1967) in his classical studies of the colour vision of bees, it was also used to examine the touch of sensitivity of the octopus (Wells 1962), and the chemical senses of fish (Bull 1957).

Conditioned reflexes of the type investigated by Pavlov have been observed in many different animals, from arthropods to chimpanzees. For example, birds learn to avoid the black and orange caterpillars of the cinnabar moth after one or two trials revealing their evil taste. They associate this with the colour pattern and generalize from cinnabar caterpillars to wasps and other black and orange patterned insects. Because predators generalize it is advantageous for different distasteful insects to resemble one another—the phenomenon of Müllerian mimicry (Fig. 6.3).

In nature we rarely observe a conditioned reflex in as 'pure' a form as in the laboratory. When foraging, bees do not just learn to associate a colour with the nectar reward, they also learn the position of the group of flowers with respect to their hive, and learn what time of day the nectar secretion is highest, directing their foraging trips accordingly. Pavlov, despite his scrupulously controlled environment, found that his dogs learnt more than one particular response to one particular stimulus. A hungry dog familiar with his laboratory would run ahead of the experimenter into the test room and jump up on to the stand wagging its tail with every sign of expectancy. Animals, then, don't just hang around waiting for a stimulus to signal reinforcement, they try actively to put themselves into the situations or perform

Fig. 6.3 An example of Müllerian mimicry: (a) is a wasp, showing its striking black and yellow striped 'warning' pattern; (b) is a group of cinnabar moth caterpillars, showing a similar pattern. Two very different sorts of insects have converged on the same pattern to warn potential predators that they are noxious. Each derives a certain degree of protection from bad experiences predators may have had with other similarly coloured prey (photographs (a) by Paul Embden and (b) by Mike Amphlett).

actions that lead to reward or escape from punishment.

This introduces another familiar form of learning that is easiest to study in a simplified laboratory situation where particular actions can be reinforced. Thorndike, one of the pioneers of American experimental psychology, investigated this type of learning with cats using a series of 'problem boxes' of the type illustrated in Figure 6.4. For example, we may have a box with a door that can only be opened from the inside by pulling a loop. A cat is shut in and tries hard to escape, it moves around restlessly and after a time—by chance—it pulls the loop and the door opens. The second trial may be a repetition of the first and also the third, but soon the cat concentrates more attention on the loop and eventually it moves swiftly across the box and pulls the loop as soon as it is confined. Thorndike gave the descriptive name **trial and error** to this type of learning. The cat learns to eliminate behaviour that led to no

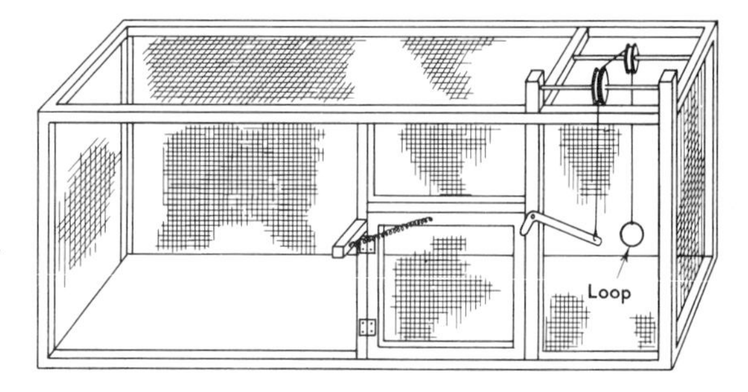

Fig. 6.4 One of Thorndike's problem boxes. A cat is confined inside the cage and must learn to pull the string loop to open the door (from Maier and Schneirla 1935).

reward and increases the frequency of behaviour that is rewarded, but in the early stages there is little system to its activity—the first reward is obtained by pure chance.

The famous Skinner box, called after B.F. Skinner, who used it extensively for the study of learning, is basically a problem box of a convenient form in which an animal learns by trial and error that pressing a bar yields a small reward. Because the animal's own 'spontaneously generated' behaviour is instrumental in its gaining a reward, such learning has been called **instrumental conditioning** (Skinner also used the term **operant conditioning**), but it is no different in principle from trial and error.

Skinner, whose book *The Behavior of Organisms* was published in 1938, was a very influential figure in learning theory and, because his school has concentrated largely on instrumental conditioning whilst others have worked on classical conditioning, there has been a tendency to regard the two situations as leading to two rather distinct types of learning. At first sight they do seem clearly distinguishable. With classical conditioning the animal comes to associate a novel stimulus with reinforcement and its response—the CR in Pavlov's terminology—was originally a UCR, there from the outset. So we have something added on to the perceptual, sensory end of the responses' control but the response itself is unchanged. With instrumental conditioning, or trial and error, there is no novel stimulus to be learnt but a new response that leads to reinforcement and, in turn, the CR. Something has been added to the motor response end of the system.

Close examination of how the animal behaves in the two situations shows that this distinction may be more apparent than real, at least in some cases. The clearest examples come from pigeons, familiar subjects for instrumental conditioning studies because they can readily learn to peck at a key on the wall of a Skinner box to obtain a food or water reward. If, as is commonly the case, key-presses are recorded automatically, then it looks like very conventional trial and error—the pigeon learns this new response and can use it to obtain food or water, depending on circumstances. However, there is real value in bringing an ethological approach into the Skinner box. A pigeon key-pecking for a food reward does not behave in exactly the same way as when pecking for water. Close observation of the head and bill reveals that, when hungry and pecking for a food reward, the pigeons' eyes are partly closed and its bill open and, when pecking for water, its eyes are fully open and the bill almost closed (Fig. 6.5). These are precisely the contrasting features that distinguish a pigeon pecking to pick up food grains when feeding from one that dips its bill into water to drink (pigeons, unlike most birds, suck up water like mammals). In other words, the pigeon that pecks instrumentally is treating the key 'as if' it were food itself in one case, as water in the other.

Pavlov regarded the conditioned stimulus as coming to replace the unconditioned one, i.e. the dog treats the metronome sound 'as if' it were food. It seems that pigeons in the Skinner box respond in a similar way, and such a conclusion certainly forms a bridge between classical and instrumental conditioning and forces us to look at the learning situation more from the animal's point of view. This is not to say that any distinction between classical and instrumental conditioning is meaningless, but we have to recognize that animals will come to any

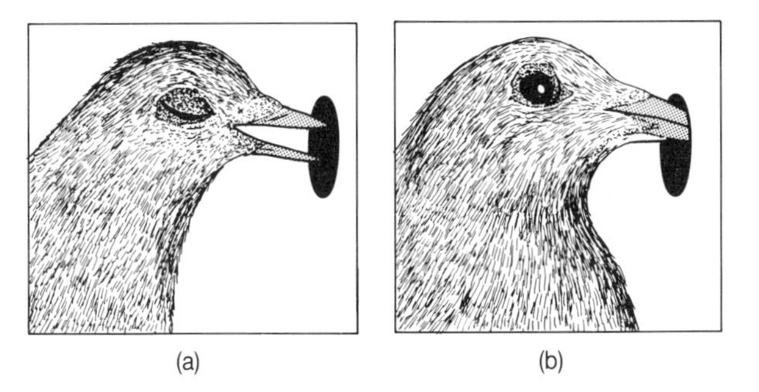

Fig. 6.5 Trained pigeons in a Skinner box 'pecking' a key to obtain a reward. In (a) the reward is food and the bird's head moves sharply towards the key. As it reaches it, the bill is opened as if to seize a food item and, in a characteristic protective reflex, the eyes are almost closed. In (b) the bird is 'pecking' for water reward and the bill, almost closed, is pressed against the key more slowly whilst the eyes remain fully open (drawn from photographs in Moore 1973).

learning situation with some built-in predispositions that will affect the way they respond.

It was once widely believed that, given the right opportunity, animals could learn to associate any stimulus they could perceive, or any response that they could perform, with any reinforcement—it was the association that counted. In practice, this assumption has proved quite inadequate. In Chapter 2 (p. 43), we have already discussed some evidence that animals may have inherent biases to learn particular things. In conventional learning situations we usually find that they most readily learn responses that form part of their natural behaviour to the reinforcement. Cats will quickly learn to lick or bite at a lever to get food; it is very difficult to train them to turn a treadle wheel with their paws for the same reward. Exactly the converse happens if the reinforcement is escape from electric shock. Now the turning of a treadle is easily associated with escape, but licking a lever cannot be. We can understand these biases in terms of the normal way the cat responds to food or runs away from unpleasant situations.

Problems with species-specific biases are well illustrated by avoidance conditioning. One familiar form of laboratory avoidance conditioning is the 'shuttle-box'. A rat is put into a box whose floor is a grid that can be electrified to deliver mild shock to its feet. The box is divided into two halves by a partition with a gap for easy passage between them. Each half can be electrified separately. On a signal (a buzzer or light usually) the rat has a few seconds to run through into the other half to avoid electric shock. This is not an easy task for rats because during the early trials they are being asked to take refuge in another part of the box where they have also received a shock just previously. It takes them some time to accept this as a refuge, but at least they have the right natural bias—they run in response to shock. Bolles (1970) discusses the varying fortunes of psychologists who have used other species in a shuttle-box situation. For many animals it is a nearly impossible task because their natural response to foot shock, as to any alarming stimulus, is to crouch or freeze. It is no use expecting a hedgehog to learn to run away from approaching danger—hence the carnage on our roads.

Sometimes the biases animals show are so strong that, when placed in unnatural learning situations, they will perform patterns that actually delay their getting reward. Breland and Breland (1961), pupils of Skinner, tried to put their techniques to good use by training animals to perform various eye-catching tricks for TV commercials. They knew it should be possible to train hungry pigs to drop money into piggy banks or to get chickens to ring bells, provided these actions were associated with food. It worked, up to a point, but they could not eliminate 'undesirable' side actions. The pigs would root with their snouts on the way to the bank, the chickens scratched and pecked at the ground. Spending time on this behaviour was certainly inefficient in operant conditioning terms. It was never reinforced and it delayed the arrival of food. The Brelands, recognizing the instinctive behaviour patterns that chickens and pigs use in feeding, came to accept that animals do not always simply associate response and reinforcement. They mischievously entitled their paper *The Misbehavior of Organisms* (note Skinner's title); Seligman and Hager (1972) give many other such examples.

The fact that animals do not always learn to respond so as to minimize the delay between an action and its reinforcement was most surprising to conventional learning theorists. Contiguity—the close association in time between stimulus or response and reinforcement—was considered essential. Figure 6.6 shows the familiar picture from Pavlov's work—CS and UCS must be close, or must overlap, if learning is to occur. If the CS ends too early, so that there is a delay before the UCS signals the reinforcement, then no association is formed. Skinner found, in general, that delays of more than about 8 s between a response like bar-pressing and its reinforcement greatly slowed learning. In practical terms the deleterious effects of delayed reward can often be overcome by introducing a **secondary reinforcement**. Suppose a rat learns that a reward is delivered when a light comes on in the Skinner box (light and reward must overlap, as discussed above), then it will learn to press the bar to switch the light on. Light becomes a secondary reinforcement or a 'bridging stimulus' between the response and the primary reinforcement—food. Bridging stimuli are useful for training animals, as in a circus when it is often difficult to reward immediately after the response is made.

However, just as there were exceptions to the rule about it being possible to link any response to any reward, so we find that immediate contiguity is certainly not an invariable requirement for learning. There are certain types of associative learning that

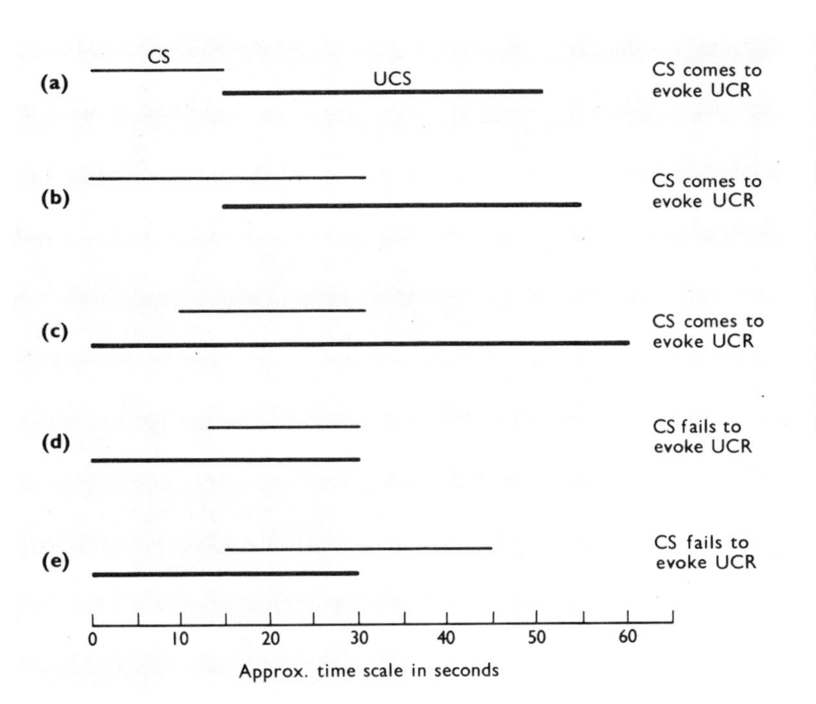

Fig. 6.6 The effect of the sequence of stimuli upon the formation of a conditioned reflex. In each case the upper, thin line denotes the duration of the conditioned stimulus (CS) and the lower, thick line that of the unconditioned, reinforcing stimulus (UCS). The results are given on the right; note that the CS must not end with or persist beyond the UCS if a positive conditioned reflex is to be established. If it does so, then the CS will remain neutral and may even tend to inhibit the response to the UCS (from Konorski 1948).

consistently occur when reinforcement is delayed for a matter of hours. Barnett (1963) describes how wild rats only nibble at small amounts of any novel foods that appear in their territory. If it proves edible, they will gradually take more on successive nights until they are eating normally. If it is poisonous, and they survive, they avoid it completely on subsequent occasions. This type of behaviour is highly adaptive and makes poisoning rats no straightforward task. The interesting feature for our discussion is the delay that must ensue between a rat tasting poison bait (always made superficially palatable with sweet substances) and any subsequent ill-effects. Few rat poisons take less than an hour to produce effects. Laboratory findings have confirmed this ability. Not only will rats learn to avoid tastes associated with sickness that sets in at least an hour later, if deprived of the vitamin thiamine they will learn to choose a diet containing it, although many hours must elapse before they can feel its benefits. Rozin and Kalat (1970) provide an interesting review, in which they link the specializations of learning that are involved in such 'specific hungers' with the rat's ability to avoid poison baits.

It is only when a new taste acts as the CS that reinforcement can be so delayed. In one ingenious experiment, Garcia and Koelling (1966) supplied rats with a drinking tube containing saccharin-flavoured water so arranged that when they licked the tube, bright lights flashed on. After these sessions the animals were irradiated with X-rays or, in other experiments given an injection of lithium chloride, both treatments made them sick about an hour later. Subsequently, the rats avoided saccharin taste but did not avoid flashing lights. Conversely, if they were given flashing lights plus immediate electric shock to the feet whenever they licked the tube, they subsequently avoided the light but still licked at saccharin. Rats are in some way 'prepared' to associate taste with sickness after a single trial and a long delay, but visual stimuli and sickness are not so connected. Conversely light and electric shock are easily associated if they occur close together, but taste and shock are not.

However, visual stimuli and sickness can be associated in birds, and again this exemplifies how natural selection has been able to shape the way learning operates to suit the requirements of very different animals. Birds are visual hunters, in contrast to rats and many other mammals and if delayed sickness is produced with lithium, they associate the appearance of novel food with this effect, and avoid it later (Martin and Lett 1985). In the wild, they learn rapidly in this way to avoid poisonous caterpillars for example.

Hitherto, although its timing can clearly vary with

circumstances, reinforcement itself seems central to the learning process. Certainly, in the situations covered by standard Pavlovian conditioning or trial and error there can be little doubt that it is a crucial factor. If Pavlov ceased the supply of meat powder, his dogs rapidly ceased salivating to the stimuli he was presenting, a process he called **experimental extinction**. Extinction also follows in the Skinner box situation but, for reasons that are not clear, it usually takes much longer than with classical conditioning. Once a rat has learnt to press the bar for food, the proportion of reward to presses can be reduced to as low as 1 in 100 in some cases, and the rat will go on pressing. If rewards are stopped altogether it is a long time before the response finally extinguishes.

Pavlov realized that an extinguished CR did not just disappear and leave the animal as it was before conditioning started. In the first place, if we simply leave the animal alone for a few hours and then give it the CS again, the CR returns, i.e. it shows **spontaneous recovery**. This recovery is not back to the original level and the response extinguishes more rapidly, but this process of a pause followed by spontaneous recovery can be repeated several times.

A second way of reviving an extinguished response is to give a novel stimulus along with the CS. A dog that has had its conditioned salivary response to a bell extinguished, salivates again if a light flashes as the bell is sounded. Similar results have been obtained with rats in Skinner boxes. Pavlov called this process **disinhibition** because he regarded extinction as another new learning process that inhibited the original CR. Neutral stimuli presented with the CS early in the original acquisition of the CR often 'inhibit' it temporarily and reduce its strength. Similarly, perhaps the neutral stimulus disinhibits an extinguished CR by inhibiting the new learning that takes place during extinction.

In the heyday of rat and pigeon experimental psychology, reinforcement was a central concept and was often equated with 'drive-reduction'. Animals were said to have needs for food, water, etc. and would learn tasks that reduced these needs or drives. Without drive reduction, learning would not take place. This immediately raised the question of how many drives there are and how distinct they are from each other—questions that were touched on in Chapter 4. Reduction of 'anxiety' or some similar concept was needed to explain avoidance conditioning (p. 130) because once a rat has learnt the shuttle-box problem it gets no more conventional reinforcement, i.e. foot shock.

The drive-reduction hypothesis also runs into trouble with what we might call exploratory learning. Thorpe (1963) and Munn (1950) called it **latent learning**. Rats given an opportunity to explore a maze at will, without any inducement from hunger or any other drive, turn out to learn it quicker than other naïve rats when both groups are later running the maze for food. No biologist is surprised by this result—animals learn the vital features of their home ground in this way. Initially, however, it caused great controversy amongst the adherents of drive-reduction, for what drive is reduced during exploration? We could equate it with anxiety if we wished, but there seems little point to the exercise. Monkeys will learn to press a bar to move a shutter aside, giving them a few seconds' view into another room with a toy train set in action. Reduction of boredom seems the best hypothesis! It is better to accept that conventional reinforcement is not the only route to learning and recognize the adaptiveness of the result and the biases—sometimes highly specialized ones—that may have become built in to the learning mechanisms, whatever they may be, as a result of natural selection.

SPECIALIZED TYPES OF LEARNING ABILITY

If we look beyond the familiar rat and pigeon, the prime subjects for laboratory studies of animal learning, we must expect to find learning systems whose features are specially adapted to match their very different life histories. The honey-bee provides an excellent case in point. Learning is particularly crucial to worker bees during the second half of their brief 6 weeks of life. This is the time when they act as foragers, making regular excursions outside the hive to collect nectar and pollen from flowers. Von Frisch (1967) and Lindauer (1976) give good accounts of the role learning plays in bees and the types of features in their environment that they learn. Menzel and Erber (1978) review experiments on the learning process and its underlying mechanisms.

Honey-bees learn with remarkable rapidity and

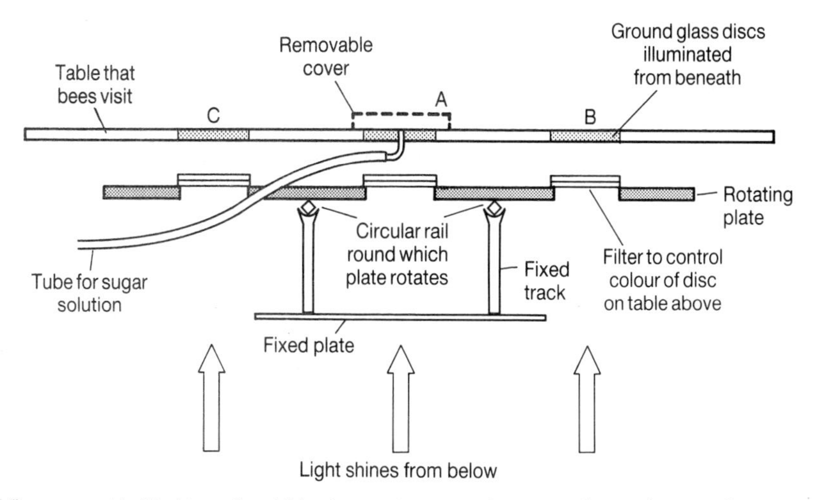

Fig. 6.7 Artificial flower provided in Menzel and Erber's experiments on learning in bees. There are three ground glass discs on the table and these can be illuminated from below through interference filters giving pure blue or yellow light. Which colour appears at which disc can be controlled and changed quickly by rotating a plate underneath the table, which carries the filters into position under the discs. The central disc (A) is for training; sugar solution is provided through a hole in its centre. During training it is lit whilst discs B and C are not. For testing, A is covered and B and C are lit giving the bees a choice of colour close to where they have been rewarded. Responses to B and C can be counted and related to the colours received during training at the central disc (simplified from Erber 1975).

retain remarkably well the effects of a single association between a colour and food reward. They make marvellous subjects for experiments on their learning ability and will readily learn to visit a dish of sugar solution on a table. The experimenter can have the dish lit from below with light of any wavelength —forming in effect an artificial flower whose colour can be switched at will (Fig. 6.7).

It takes a bee a minute or two to fill its crop with sugar solution, so we can easily arrange to change the colour of a 'flower' during its visit. In one set of experiments Menzel and Erber (1978), using naïve bees as subjects, changed flower colour between blue and yellow at various stages and then gave the bees a choice of these two colours, to see which colour they associated with the sugar reward. The results were quite dramatic and are illustrated in Figure 6.8. The association period is very brief,

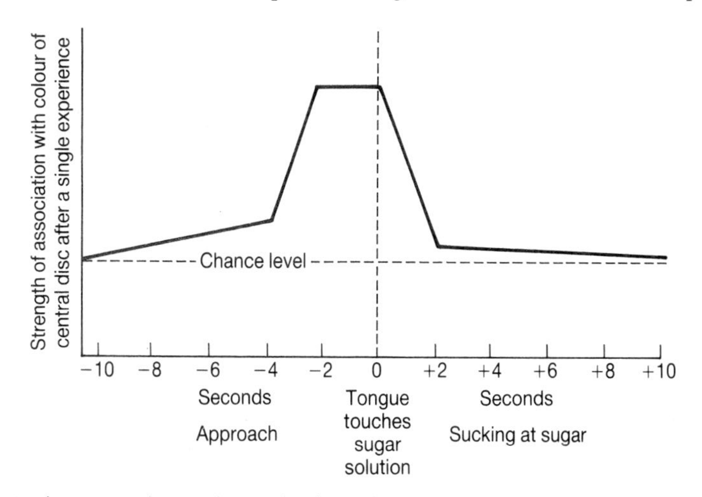

Fig. 6.8 Strong association between colour and reward is dependent on exact timing. As the bee approaches the disc its colour has no significance until about 4 s before the bee begins to suck sugar. Then there is a sharp peak of strong association, which has disappeared within 2 s following this onset of sucking. Apart from this short time, the colour the bee sees during the approach or for the whole time it takes to fill its crop, is of no importance (further explanation in text; from Menzel and Erber 1978).

between 4 and 5 s only, and changes made outside this period are ignored. The timing of the period is also very precise and we can best measure it from the exact moment at which the visiting bee's proboscis touches the sugar solution, i.e. the moment when reinforcement begins. Then the colour stimulus must be present within 3 s before and 0.5 s after the start of sucking if an association is to be made. Sucking has to continue for 2 s for the bee to retain the memory, but the colour signal can be removed after 0.5 s—it will still be associated with the reward and the bee will fly to this colour on the next visit.

So if the 'flower' is, say, yellow as the bee approaches, switches to blue 3 s before it alights and begins to suck, switches back to yellow half a second after sucking starts and remains yellow for all the time it is filling its crop—perhaps a whole minute —and is still yellow as it leaves, the bee subsequently associates blue with reward. This is extraordinary, if we think in terms of conventional vertebrate learning. There we might well assume that the enduring association would be with the colour most recently and most persistently associated with the reward (note, for example, the response of rats given vinegar after saccharine flavour preceding induced sickness, which is described below. However, for the specialized honey-bee learning system the brief flash of blue at the beginning is the significant feature, this degree of contiguity is enough. Menzel and Erber found that a single association of this type is retained for about 6 days but that given three such rewards, then bees retain the association unchanged for up to 2 weeks.

Among the most striking learning specializations of vertebrates is that associated with food storage. When the chance arises, a number of birds and mammals collect more food than they can eat immediately and hide it. Thus they can take advantage of a rich food source whilst it is there and reduce the degree to which they have to share it with others. Some animals make a single larder to which they return regularly but the most interesting, from our point of view, are those species that hide food items dispersed singly around their territory (see Sherry 1985 for more details). For example, marsh tits will store several hundred seeds a day in winter, secreting them in tiny crevices under tree bark or stones, etc. Careful observation in the field and experiments in aviaries have shown that the birds, although not recovering each seed, can recall dozens of dispersed hiding places for one or two days with extraordinary precision (Cowie *et al.* 1981; Shettleworth 1983). This is a very remarkable development of spatial learning and memory, the more so because close relatives of the marsh tit, e.g. the great tit, show no such ability. Evidence is now accumulating that food-storing species have particularly developed those parts of the brain concerned with memory formation (see p. 147), and this suggests quite rapid evolution of brain and behaviour to meet a particular ecological requirement.

If we survey across the animal kingdom we shall expect to find many more such specializations. They show the diverse forms that associative learning can take, but it is nevertheless valuable to look for common features. There is, at a behavioural level, a real equivalence between classical conditioning in a honey-bee, an octopus and a dog—it requires the same logical process.

It is in this area that some of the most valuable interactions between experimental psychologists and ethologists have occurred. Approaching from rather different viewpoints, both have begun to investigate exactly what it is that animals are learning. As we have seen, we can reject any idea that what they form is a simple, almost idealized association between an event and a reinforcement that occur in some regular sequence. It is more profitable to start from the notion put forward early in this chapter (p. 123) that what animals must do is learn which events provide the best prediction of important outcomes. We must also remember that animals will come to all new situations with some inherent biases to pick on certain events as signals and to respond in particular ways.

Mackintosh (1983) and Rescorla (1988) cite a number of interesting examples of how this operates in practice. Thus, in the taste aversion situation, described on page 131, rats learn to avoid saccharin, whose flavour has been associated with sickness some time later. If, between tasting saccharin and being given the injection of lithium, which leads to sickness, the rats are given vinegar-flavoured water to drink, they subsequently avoid vinegar but not saccharin. The physical and temporal association between saccharin and sickness remains the same but the rats—sensibly enough we may feel—choose to pick on the taste of vinegar as the key event. Because it occurred more recently, nearer to the reinforcement, it is taken as the better predictor.

A rather more complicated example shows another side of the same process. Two groups of rats were given experience of electric shocks in a Skinner box where they were obtaining food by bar pressing. In one group—we may call it group A—the shocks were paired with a distinctive light stimulus; in the other group, B, there was no light. Now both groups were given the light plus a sound tone stimulus together with the shock and this combination of light plus tone was certainly effective. The rats signalled that they had learnt, so to speak, because when light plus tone was given, now without any shock, their rate of bar pressing for food showed a sharp decline presumably caused by increased anxiety about the possibility of shock. But had the two groups learnt the same thing? The test comes when they were given just the tone by itself. Group B showed just as strong an effect on bar pressing as with light plus tone. Group A did not; these rats paid little attention to the tone without the accompanying light. A's previous experience had told them that light was the best predictor of shock, so although the contiguity of tone and reinforcement were the same for both groups, they interpreted the situation differently.

Now, it may be said that such results are just what we would expect. We probably mean by this that the rats are behaving as we would. But this is really the key point; the rats are not behaving like associating automata, they are behaving more intelligently. This leads us directly to a further question:

ARE THERE HIGHER FORMS OF LEARNING IN ANIMALS?

Thorpe's book, *Learning and Instinct in Animals* (1963), includes an examination of what he, following others, calls **insight learning**. He defines it as 'the sudden production of a new adaptive response not arrived at by trial behaviour or the solution of a problem by the sudden adaptive reorganisation of experience'. We can readily see the drawbacks of such a definition (e.g. how do we know a response wasn't arrived at by trial and error? etc.), which is not to say that, as humans, we do not recognize the phenomenon. Everyone can recall occasions when the solution to a problem has 'come in a flash', perhaps as the climax to several minutes of concentrated thinking. It is obviously going to be very difficult to demonstrate conclusively that there are similar processes going on in animals.

All we have to go on is what animals actually do in learning situations but this is no reason to despair. As Dickinson (1980) discusses very well in his book, *Contemporary Animal Learning Theory*, mental process can be inferred from behaviour, even if it is difficult to do so. In practice, animal workers have used the term 'insight' when they observe animals solving problems very rapidly, too rapidly for normal trial and error. At least, too rapidly for the animal to carry out actual trials, but there is the possibility that it is 'thinking' about them and trying them out in its brain. This would imply that the animal can form ideas and 'reason' and studies on animal reasoning seem doubtfully distinct from those on insight. Most accounts of insight or reasoning in animals are highly heterogeneous, little more than a rag-bag of tantalizing examples with very few instances of sound, experimental investigation. This is to a large extent due to the nature of the phenomenon and certainly does not reduce its importance. Some of the first attempts by experimental psychologists of the 1930s to get some hold on insight and reasoning, involved 'detour' experiments in maze learning.

The point at issue relates to two contrasting views of how animals learn. One, particularly associated with the learning theory of Hull, envisaged a rat learning a maze as building up a chain of stimulus–response (S–R) associations starting from the goal box and reward. The rat associates a corner, or a left turn with the arrival at the goal—the first S–R association. As it explores further it learns that a previous corner or turn leads it to the first S–R point, and so it 'attaches' the second S–R to the first. Gradually, by a process often called **chaining** the rat puts together a chain of S–R associations that can then be used, in reverse order of their acquisition, to guide it from the start box through the maze. Hull therefore envisaged maze learning as essentially a set of simple associations, with one leading, rather automatically, to another.

The second view of the process is linked with the work of Tolman. It can be stated much more simply, if less precisely than Hull's, because Tolman considered that rats build up a mental picture or cognitive map of the whole maze during the exploration and learning phase. They could then use this to find their way through avoiding turnings that they recall lead to blind alleys, choosing the path that leads most quickly to the goal box.

Figures 6.9 and 6.10 provide two examples of the

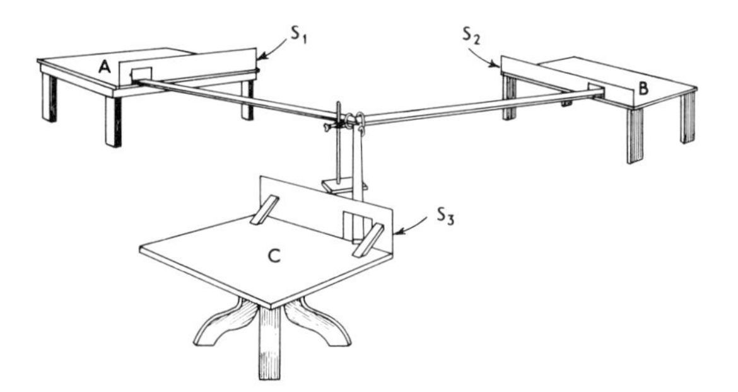

Fig. 6.9 One type of apparatus used by Maier to test 'reasoning' in the rat. The pathways are 8 feet long and the small tables vary in size, shape, and character. S1, S2 and S3 are wooden screens placed on the tables to obstruct vision from one to the other. After exploring the three tables and runways, the rat is fed, let us say, on table A. It is then, let us also assume, placed on table C. After reaching the joint origin of the three paths, the animal now has a choice between A and B. If it chooses A, it is credited with a correct response. Exploration precedes each test and the rat is started from different tables from test to test. In a group of such tests, a score 50 per cent would occur by chance; some rats score much better than this (from Maier 1932).

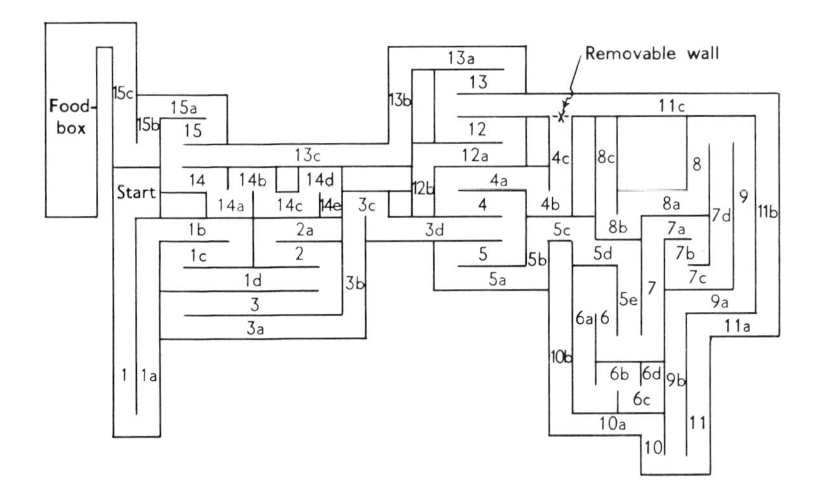

Fig. 6.10 A complex maze used by Shepard to test 'reasoning'. After rats have learned the maze, the section indicated by X is removed, thereby causing a previous blind alley to become a short cut. Having discovered the change whilst running along 11c, and exploring from there a little into 4c, some rats entered 4 (and thence 4a, 4b, 4c) instead of 5 on the next trial (from Maier and Schneirla 1935).

detour experiments mentioned above, whose aim was to test reasoning in the rat. If suddenly presented with a new situation in a maze, or given the opportunity to take a short cut, how will it respond? Although it does not perhaps constitute a rigid test of the two theories given above, most would agree that Hull's 'S–R' view of the learning process would not leave much scope for the rat to take short cuts, especially of the type offered in Shepard's complex maze (Fig. 6.10). The fact that some rats do change their path, and very rapidly, strongly suggests that Tolman's view is more sensible and that rats can show signs of reasoning and insight.

Detour studies of this type (more examples are given by Munn 1950; Thorpe 1963) often have three features in common. First, they all include a longish period of exploration prior to any testing, so that the animals will already have learnt a good deal about the general features of the situation. Secondly, performance is judged by speed of solution. If the rat makes mistakes or has to explore further it is assumed that a simpler, trial and error form of learning is being used, not insight. Thirdly, it is an interesting fact that rats and other animals that have been similarly tested are highly variable in their apparent insightfulness; it is often only 20–30 per cent who respond rapidly and we do not know what factors of prior experience or genetic constitution contribute to these differences.

As mentioned earlier, a diverse series of observations have often been lumped together as evidence of insightfulness or reasoning in animals, but sometimes we may confuse capability *per se* with intellectual ability. For example, some remarkable findings on the capacity of pigeons to 'form concepts', as we might say, is often discussed in this context. Using a quite simple Skinner box situation, which projects photographs of scenes above the key at which the pigeon must peck, Herrnstein *et al.* (1976) have shown that pigeons can learn to recognize photographs of human beings, water, trees (or other such categories) very much as we can. Thus, given a reward for pecking the key when it recognizes 'human being' in the picture, it responds to a person's face in close up, or a whole figure, or distant figures seen in a landscape. Water can be in a lake or stream

Fig. 6.11 Natural concepts in pigeons. Trained to expect reward when they saw a picture that included a tree, pigeons responded to both pictures on the left, i.e. they discriminated against those on the right—creeper on a cement wall and celery (drawn from photographs in Herrnstein *et al.* 1976).

setting, in a puddle or a goldfish bowl, and so on. The pigeon classifies these visual images together so as to extract the common feature associated with reward (Fig. 6.11).

We can understand why at first sight such results seem amazing. We feel they must represent advanced learning capacities and we are reluctant to endow the pigeon with such intellect. In fact, we are probably misled in this view by our anthropocentrism, which tends to equate human capacities with intelligence. Pigeons, evolved for flight and intensely visual animals, have a sensory system adapted to rapid recognition of features in a rapidly changing environment. As they swoop over the landscape they must respond to trees in the distance as a massed feature, or close to and singly as they dive into the branches to alight. Viewed this way the fact that the pigeon can classify diverse images on its retina into certain common categories, though no less wonderful, is more readily seen as a matter of sensory processing than as intellect *per se*. We may note that Hollard and Delius (1982) have shown how quick pigeons are to recognize those rotated shapes that form a component of human intelligence tests (Fig. 6.12). But again, pigeons have far more practice than we do at seeing shapes below them apparently rotate as they fly around over their home areas.

If we are to acknowledge true insight in animals it must come from the identification of truly novel

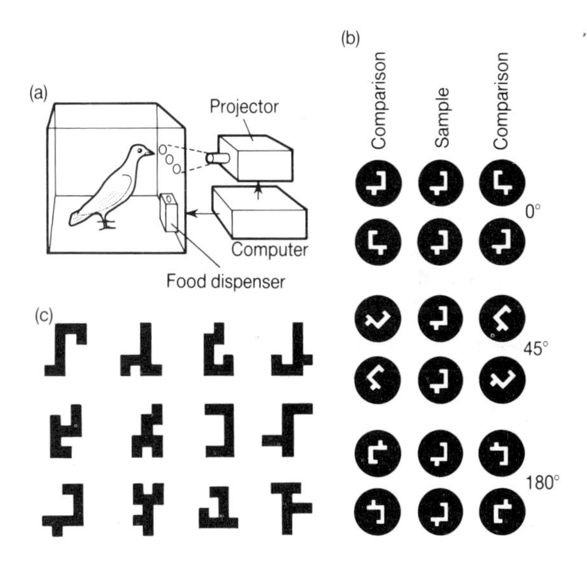

Fig. 6.12 Test to measure the ability of pigeons to recognize complex shapes even when rotated from the familiar. (a) Shows the experimental procedure, a Skinner box with three keys. Trained to respond to a shape presented on the central key the pigeon is now given three shapes on the three keys—see the rows in (b)—with a reward for pecks to the side key whose shape is identical to the sample. When the wrong shape is a straight mirror image of the sample, as in the top two rows of (b) the task is quite easy. It becomes (for us) much more difficult when the comparison has to be made with shapes rotated from the sample orientation. (c) Shows some of the shapes with which pigeons were tested in this way. They are remarkably quick and efficient at such discriminations (from Hollard and Delius 1982).

associations, such as we referred to in our original definition. It is not surprising that many examples come from primates, and perhaps the most famous is the work of Köhler (1927) on chimpanzees, described in his book, *The Mentality of Apes*. Loving bananas, they would learn to overcome hurdles of increasing difficulty to get at them. If presented with a bunch too high to reach, they would pile up boxes to make a stand for themselves or fit two sticks together to pull down the bananas. Often they arrived at his solution quite suddenly, although they benefited by previous experience of playing with boxes and sticks and showed considerable and obvious trial and error when actually building a stable pile of boxes. Köhler's chimpanzees were using knowledge obtained in one context (something of the properties of sticks and boxes) and applying it in another. There surely is no question but that apes and other primates can show true reasoning on occasions. We may recall the examples of deliberate deception within wild primate groups mentioned in Chapter 3, page 66. Many dog and cat owners will cite examples of reasoning in their pets; this is possible, after all we have just described experiments which suggest that rats have some reasoning ability, but we must take care to exclude other explanations.

Before we try to extend our discussion of insight learning to consider the nature and implications of reasoning ability in animals, it is important to consider why we should assume that primates will be more able to reason than dogs, and dogs more than rats. We are, in effect, suggesting that learning ability and 'mentality' have shown trends during the evolutionary history of the vertebrates.

THE COMPARATIVE STUDY OF LEARNING

Comparative psychology has a long history (Dewsbury 1984) and, in the early stages, a wide range of animals was worked upon, even if later the rat and the pigeon came to dominate. Psychologists have always tended to concentrate upon studies of learning and recently those with an ethological background have also come to study the diversity of learning abilities, so we have a reasonable body of data to search for evidence of evolutionary changes.

There are all kinds of inherent difficulties in making such a survey. The first and most obvious is our essential vanity concerning human intellectual powers, which often leads us to look for progression upwards amongst animals leading towards human beings on a very detached, high pinnacle. Any biologist must expect there to be other pinnacles on alternative pathways. Honey-bees are not to be judged on the same criteria as monkeys. Nor is it sufficient to rely on laboratory studies of learning alone, we must try to get some estimate of the role that learning plays in the natural life of different animals and also try to distinguish between simple and more advanced types of learning which, as we have seen, is not always easy.

One approach has been to examine the correlation between brain development and learning ability. Dethier and Stellar (1970) include a good introductory survey of nervous system structure through the various animal groups, and use it as a basis for considering how far behavioural complexity can be linked with brain development. The prevalence of learning, the capacity to process information and the general complexity of behavioural responses are greater in mammals and birds than they are in fish and reptiles and we can reasonably associate this difference with the evolution of a large brain—encephalization as it is termed (Jerison 1985). It is not, of course, just brain size that is important; whales and elephants have larger brains than ourselves but smaller, though considerable, learning powers. It is less easy to construct a series among the invertebrates, which are much more heterogeneous, but advanced insects and cephalopods have the largest brains of their respective phyla, Arthropoda and Mollusca, and also the greatest capacity for learning.

The most dramatic aspect of brain evolution within the vertebrates is the growth of the cerebral hemispheres, especially their cortex, which is greatest (reaches a climax) in the primates. However, we can reject any simple equation of cerebral cortex with learning ability. In the past birds have often been underestimated because, although their brains are relatively large, those parts homologous to the mammalian cerebral cortex are small. But birds have evolved along a line separate from the mammals for over 200 million years. They have evolved another type of brain structure and their learning ability is in some respects second only to the primates; see, for example, the remarkable account of the capacities of a grey parrot by Pepperberg (1990). A converse example has already been mentioned on page 21—amongst insects the Diptera and the Hymenoptera both have brains that diverged greatly from the ancestral insect type, but they exhibit very different degrees of learning ability.

It is clear that brain structure alone is inadequate as a guide to learning abilities and to study the evolution of learning we need to compare how different animals perform on particular behavioural tests. In selecting representative animal types for our evolutionary analysis, it is all too easy to refer to 'higher' and 'lower' animals. Within the vertebrates, for example, we often find the sequence fish, amphibian, reptile, bird, mammal quoted as an evolutionary scale of increasing complexity of behaviour and increasing learning capacity. The construction of such a scale from living representatives of each class ignores the actual course of evolution. We have just mentioned the completely separate histories of the birds and the mammals. All the living vertebrates are equally distant in time from their common ancestors and all are specialized for their particular mode of life. We would be naïve to expect the learning capacities of a modern teleost fish (e.g. a goldfish, commonly used in learning studies) to reflect accurately those of the ancestral fish from which the teleosts and other vertebrates diverged some 400 million years ago. Hence the construction of a valid phylogeny of behaviour is fraught with difficulties because behaviour does not fossilize. Hodos and Campbell (1969) provide a vigorous critique of the phylogenies that have sometimes been constructed by comparative psychologists.

A further problem for comparative learning studies is that of devising truly comparable situations for testing different animals that vary so widely in their sensory capacities and manipulative ability. The procedures needed to measure discriminative conditioning in an octopus, a honey-bee and a rat have to be very different and we can no longer be sure that problems are of equal difficulty or that the animals 'see' them in the same way. Motivation and reinforcement present further problems; the level of motivation often affects the rate of learning and may, indeed, determine whether the animal learns at all. How can we equate levels of hunger motivation in a rat and a fish? The latter may live for weeks without food, the former only days. It is just as difficult to equate reinforcement between different animals. A small piece of food may be an excellent reinforcement for a hungry mammal, but mean much less to a fish and less still to a worm. It is perhaps easier to equate punishment—all animals 'dislike' electric shocks. Even here there are difficulties because shock or fear affects the behaviour of animals in such diverse ways. As we discussed earlier (p. 130) animals come to learning situations with a good deal of built-in bias. To equate the effects of punishment we require some knowledge of each animal's natural responses in fearful situations. Whenever possible, one must try to use a reinforcement that is directly relevant to the animals we are testing. Escape into a darkened area may be best for many small invertebrates. Schneirla (Maier and Schneirla 1935) and Vowles (1965) both found that the best reinforcement for maze learning by ants was to get back to their nest.

In the face of all these difficulties there are those who question the validity of any comparisons of intellectual ability between different animals. McPhail (1985, 1987), for example, takes up a position that we might call constructive provocation and argues that, '. . . there are, in fact, neither quantitative nor qualitative differences among the intellects of non-human vertebrates.' Few would agree with this but recourse to common sense should not be our only reply; we must try to devise valid comparative tests.

Amongst the vertebrates, a good series of learning tests ought to be able to record stages in the evolution of 'intelligence'. Speed of learning—a quality much admired in human schoolchildren —might seem to be one useful measure. However, a brief survey of the literature shows that speed alone does not tell us much. Ants and rats, for example, show very comparable speed when first learning a fairly complex maze. Gellerman (1933) describes in detail experiments in which two chimpanzees and two 2-year-old children were learning that a food reward was associated with a white triangle on a black square and not with a plain black square. One child learnt in a single trial, but the other took 200 trials and both the chimpanzees took over 800 trials to reach the criterion of 19 correct trials out of 20. On a comparable test most rats would learn in 20 to 60 trials, although admittedly they would usually be mildly punished for wrong responses as well as rewarded for correct ones. Discrepancies of this kind abound both within and between species and we have no reliable evidence that speed of initial learning for simple associative problems varies within the vertebrates, or even between them and the advanced invertebrates.

However, speed is only one aspect of learning, we might also ask what is learnt. For instance, Figure 6.13 shows that, although the chimpanzee may take longer to learn the triangle discrimination outlined above, it learns more about 'triangularity' than the rat. One aspect of 'intelligence' is the ability to strike a reasonable balance between generalization and discrimination in tests of this type. Similarly, if we consider more complex forms of learning then we can at least note some gradation of ability within the vertebrates.

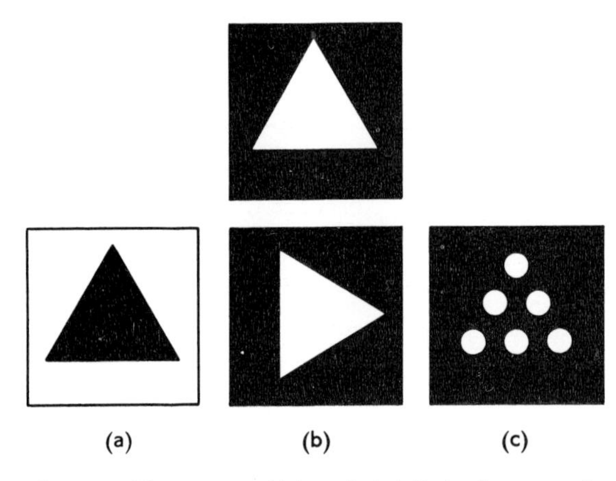

Fig. 6.13 The concept of 'triangularity'. Trained to respond to the top figure, a rat makes random responses to any of the lower figures. A chimpanzee responds to (a) and (b), but makes random responses to (c). A 2-year-old human child recognizes 'a triangle' in (a), (b) and (c) (from Hebb 1958).

Testing for what have been called **learning sets** has been useful in this respect. If an animal can form a learning set it means that it can learn not just a problem, but something about the principle behind it and then can steadily increase its learning speed when given a series of similar problems. Harlow (1949) has described the basic technique with primates. A monkey is presented with a pair of dissimilar objects—a matchbox and an egg-cup, for example. The matchbox, no matter where it is placed, always covers a small food reward, the egg-cup never has a reward. After a number of trials, the monkey picks up the matchbox straight away. Now the objects are changed, a child's building block is rewarded, a half tennis-ball is unrewarded. The monkey takes about the same time to learn this, again the objects are changed, and so on. After some dozens of such discrimination tests the monkey learns each discrimination much more rapidly, though viewed as an individual problem it is just as difficult as the first one. Eventually, after 100 or so tests, the monkey presented with a pair of objects lifts one, if it yields a reward he chooses it for all subsequent trials. It has learnt the principle of the problem or, in Harlow's terminology, it has formed a learning set.

This is one type of learning set based on successive trials of discrimination. Perhaps a simpler version is the 'repeated reversal' problem. Here we train the animal to select object A in preference to object B. Once learnt, object B is now rewarded and A unrewarded; when this first reversal is learnt, the reward is again given with A, and so on. If the animal gets progressively quicker at learning each reversal this again implies it has learnt a principle.

The ability to form learning sets was once regarded as a capability of the more advanced mammals only, but we now know this is certainly not the case. Warren (1965) summarizes the comparative data to that date that suggested that all the vertebrates with the exception of fish shared it. In fact, later work (Mackintosh *et al.* 1985) has shown that goldfish can form repeated reversal sets and so can one of the most advanced invertebrates, the octopus (Mackintosh 1965).

Although, as we have already seen, the speed of learning a simple discrimination does not vary systematically between different animal groups, the speed with which learning sets are acquired does change dramatically. Figure 6.14 compares rats and goldfish in the 'easier' repeated reversal situation. Both learn the original discrimination and its first reversal—always the most difficult—with about the same number of errors, which suggests that both animals find the problem roughly 'equal' in difficulty. However, thereafter rats improve much more rapidly than goldfish. Figure 6.15 shows more comparisons with the more difficult discrimination sets, where each successive problem involves new objects. Here we find rats performing much more slowly than dogs, cats, or primates and nobody has yet succeeded in demonstrating that fish can ever improve their rate of learning in this situation.

Many of the uncertainties with which such comparisons are afflicted still remain. Nevertheless, with caution we may conclude that cats and dogs do reveal more signs of intelligence than rodents and

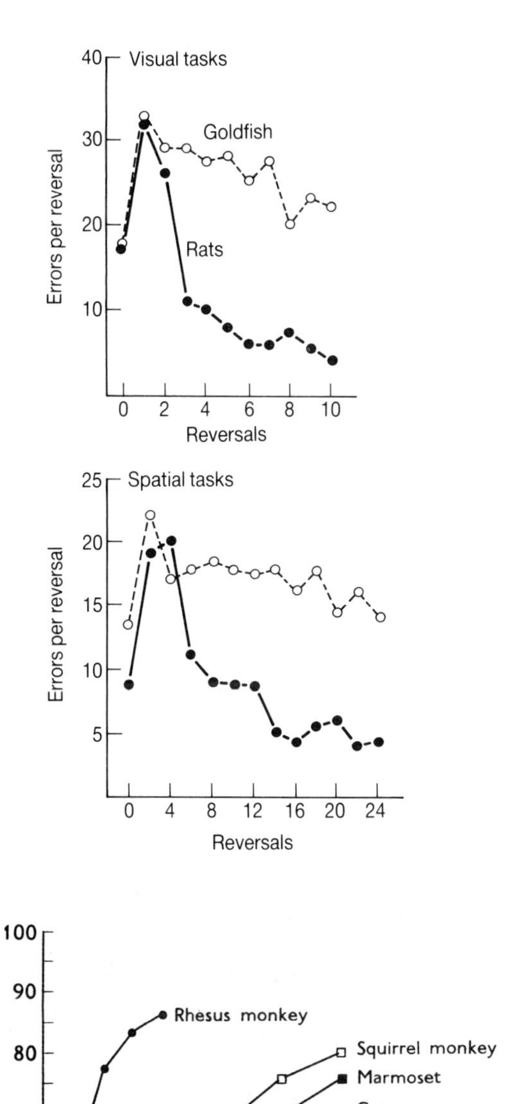

Fig. 6.14 Repeated reversals for learning visual and a spatial discrimination by goldfish and rats. The graph plots the number of errors to achieve the criterion of all responses to the rewarded object. The point 0 represents the original task, note how scores both for it and for the first reversal (always the most difficult) are very comparable for both animals. Thereafter, with repeated reversals, the rat improves much more rapidly than the goldfish (from Mackintosh *et al.* 1985).

that the superiority of primates is clear. However, it seems safest to regard it as a quantitative superiority; we have no evidence that they possess any abilities that are not foreshadowed elsewhere among their mammalian relatives.

One cannot help feeling vaguely dissatisfied by this conclusion. Especially for the primates, the rather circumscribed artificiality of many learning experiments does not seem to do justice to the incredible flexibility and ingenuity of animals such as chimpanzees. Here we are undoubtedly influenced by their similarity to ourselves, particularly because they can manipulate objects as we do. We may underestimate the intelligence of other animals because they lack good hands and good eyesight. Because their structure and environment is so different from our own, we have only recently become aware of the remarkable intelligence of dolphins.

Fig. 6.15 The rate at which various mammals can form the more demanding discrimination learning sets in which each problem presents them with completely novel objects. Thus, at each change the animal's choice on the first trial has to be random, but if it has learnt the principle behind the problems, trial 2 should be correct. Note how long it is before the scores of rats or squirrels on trial 2 become better than chance or 50 per cent. Many monkeys reach almost 100 per cent within 400 problems (from Warren 1965).

CAN ANIMALS THINK AND REFLECT ON THEIR ACTIONS?

This issue brings us back to questions that we raised when introducing the topic of insight learning, and attempts to go beyond the description of how animals behave in learning situations. It is surely impossible to keep pets or to observe animals carefully in captivity or in the wild without wondering whether they 'think' or have a 'mind' and 'consciousness'. These words are in inverted commas because, whilst none of us doubt that we possess them, they are not easy to define exactly and it is even harder to suggest how we might set about proving them to exist in animals. After all, we have no proof positive that they exist in any human being other than ourself; it is just reasonable to assume it because we are built the same way and behave the same way. As we mentioned at the outset, the information we have that bears on these issues is not very systematic and often frankly anecdotal. We may consider three examples as illustrations.

The first comes from casual observations made over a century ago by the great biologist, Alfred Russell Wallace, who shared with Darwin the original formulation of a theory of evolution by natural selection. He spent much of his life on collecting trips in the Far East and, whilst staying on the island of Borneo, he had opportunity to observe a captive orang-utan. Around its cage foraged some of the local village's domestic chickens. The orang showed great interest in them and once or twice tried to capture one, but they evaded the ape's long arms. One day Wallace saw the orang scatter grain from its food dish outside its cage, dropping some grain just outside the bars. It then sat quietly as the chickens foraging nearby found the grain and followed it right up to the cage. Whereupon the orang flashed out an arm, captured a bird and killed it!

Our second example comes from an interesting article on the behavioural capabilities of sheep-dogs and the way they are trained:

> 'Dogs experience particular difficulty when faced with recalcitrant ewes with lambs, one such ewe which had split off from the main flock refused to be moved and faced the dog square on, stamping its hooves. The dog returned to the main flock, cut off several sheep, and brought them over to the stubborn ewe. The ewe promptly joined this group and the dog was able to move them all back to the main flock.' (Vines 1981)

The last example, records an aspect of honey-bee behaviour, which a number of people have noted rather than studied. In the kind of experiments we have already described in connection with honey-bee communication (p. 69) it is often necessary to train honey-bees to forage at a dish some hundreds of metres from the hive. It is possible just to put out a dish of sugar solution and wait, but more often one begins by placing a colour-marked dish on the landing board of the hive. Once some bees have begun to feed, the dish is moved a few centimetres back; usually some bees will still find it and continue to feed. The dish is then moved further out, and further. At first, moves of more than a metre or two disrupt the foragers and they have considerable difficulty, but as the day wears on it becomes possible to move the dish in much bigger steps—10 m, 20 m, or more. At this stage, people have reported a remarkable phenomenon. Sometimes as they move the dish out to its next position, they find bees flying around, already there searching. Markings show that these are not new recruits alerted by dancing and searching vaguely in the same direction, they are the regular visitors, 'anticipating' the next position of the dish.

All these examples raise a host of intriguing questions. As is so often the case, the descriptions of the orang-utan and the sheep-dog are one-off cases. We might see similar behaviour again, but it is the essence of the situation that the observations are not repeatable. This does not make them any less interesting or important. Some may argue that to have observed such behaviour at all is of profound significance for our view of animals. Others will immediately suggest caution. How are we ever to know their significance? Perhaps the orang spilt its food completely by accident—apes are not tidy feeders. Wallace's—and our—assumption of insight and imagination may be completely misplaced. So we have to face difficulties both with the accuracy of such observations and with their interpretation. Similarly for the sheep-dog; what prior experience had it? Vines seems to suggest that dogs often have trouble with individual sheep. It might have had an opportunity to copy this strategy for dealing with them from older dogs during its training. If so, would this make any difference to the way we interpreted the behaviour? The honey-bee example raises some different questions. The behaviour in itself is not very complex but it is the fact that it comes from an insect that makes it so extraordinary. We are usually unwilling to entertain the possibility that insects think about the future, or deduce a rule that tells them the food dish will next appear in that direction.

At various times in the history of our thinking about animals they have been regarded as little better than automata, at other times people have not hesitated to regard them as fully sentient beings like ourselves; Boakes (1984) and Walker (1983) both provide interesting reviews of these changes. For most of this century, with the rapid rise of science, the predominant view has not regarded it as profitable to speculate on the nature of animal thinking. Rather, as we have seen, attention has been focused on analysing what animals actually do, both in natural and laboratory situations, and deducing the causal factors, the function and so on. Most serious workers have always had at the back of their mind a rule proposed around the turn of the century by Lloyd Morgan, one of the founding fathers of comparative psychology, now referred to as 'Morgan's canon': 'In no case may we interpret an action as the outcome of the exercise of a higher psychical faculty, if it can be interpreted as the outcome of one which stands lower in the psychological scale' (Morgan 1894).

It must be said that Morgan's canon has served us well. It has enabled us to approach the analysis of behaviour sensibly and to avoid the superficial anthropomorphism that led to many absurdities in the past. Around the turn of the century, in Germany, the eccentric Baron von Osten seriously believed that many animals, and certainly horses, could perform mental arithmetic—all they needed was the right training. His best pupil, 'Clever Hans' became world famous through being able to give the answer to sums by beating the ground with his hoof the correct number of times. In fact it turned out that he was responding to subtle unconscious cues from his owner. If the Baron did not know the answer himself, the horse got the wrong answer too.

Not long before, another Morgan—Lewis—studied beavers in the northern USA and believed that they revealed an extensive understanding of hydraulics by the construction of their dams and channels. Both of these classic examples of anthropomorphism are described more fully by Sparks (1982). However far we can progress without invoking concepts of mind thought or foresight, there is now a considerable body of examples accumulated —such as those just given—where it is much more difficult to avoid them. We may struggle to explain the sheep-dog's response, in terms of a chain of S–R units with appropriate reinforcements, but such a procedure totally lacks conviction. As Walker (1983) puts it: 'sometimes explanations can be too simple to be sensible'.

We are indebted to Griffin (1984) for reopening the whole issue in a compelling way. Probably a majority of animal behaviour workers would now accept that, no matter how difficult they may be to study, it is no longer possible to deny the existence of true thought processes and even consciousness of some type in some mammals and birds. One of the most powerful arguments for this conclusion is an evolutionary one. It is hard to accept that mind and consciousness in human beings have just arisen *de novo* without any precursors in animals that were ancestral to us and probably very similar to the non-human primates that we observe now. It is not difficult to envisage the advantage of possessing even some slight ability to anticipate the consequences of one's actions. Such abilities would be selected for and perhaps one of the bases for them was increasing brain size. Humphrey (1976) has argued that the complexities of social life, especially in the primates, was one of the most powerful selective forces for the emergence of thinking ability.

Even if we accept this argument we must not abandon Morgan's canon. For example, we should not accept the idea that honey-bees have such capacities before eliminating every other possibility. In explanation of the example we cited above, Griffin suggests that, in their normal lives, honey-bees sometimes encounter a situation when a food supply does extend out in one direction as, for example, when the shadow of a hill, or tall trees moves off a flower crop with the rising sun. As the flowers are warmed by the sun, they begin to open and secrete nectar and bees can now extend their feeding range, step by step as the shadow retreats. If we can accept this, then it could be argued that the ability to move out along the line of food dishes may be a manifestation of an inborn ability. This most certainly does not explain how the bee's nervous system can develop or operate so as to organize such behaviour, but it need not require that the bees consciously reflect on what they are doing. Some kind of unconscious more automatic rule-of-thumb may be working, as we suppose it does in the bees' remarkable dance communication discussed in Chapter 3.

Similar rule-of-thumb explanations may suffice for many animals, but not all. Premack and Woodruff (1978) and Savage-Rumbaugh *et al.* (1978) describe a series of experiments with chimpanzees to try to test not just for anticipatory thought, but also for the recognition that another individual outside themselves will share their own 'internal experience' —a fair definition of consciousness in some ways. Thus, one chimpanzee was shown videos of a human being interacting with certain objects, all known well by the chimp. Various situations that implied a need were portrayed—the person was seen shivering and huddled with cold in a room with an unlit stove, or struggling to get out through a locked door. The chimp was then offered a series of photographs of objects, one of which offered a solution, e.g. the burning wick of the stove, a key for the door. Could the chimp anticipate the need of the person separate from itself and choose for them as it might choose for itself? The evidence suggests it can. One chimp chose the 'correct' picture seven times out of eight. (Fascinatingly, it chose 'neutral' photographs much more commonly with videos of a person it disliked!)

Such studies are difficult to arrange and carry out. In the discussion following the papers referred to, various commentators point out difficulties of interpretation. It is essential to be rigorous in eliminating all other explanations but it is also essential to keep up such investigations not just with primates but with some birds, such as crows, with carnivores and, if possible, with dolphins and elephants, which common observation would tell us are extremely intelligent animals. The new approach to communication in animals, especially the primates, outlined in Chapter 3, is also giving us insights into the complexity and richness of their behaviour. However admirable has been our policy to reject subjective explanations of our observations, it may have led us to underestimate some of our animal relatives in the past. It is most important that we no longer continue to do so because everywhere, in the Western world at least, there is increasing concern for the welfare of animals, both domestic and wild. It is the responsibility of behavioural scientists to provide a sensible and practical basis for our efforts (Dawkins 1980; Fraser and Broom 1990).

THE NATURE OF MEMORY

Learning is nothing without memory. We must be able to store the results of experience and recall them to our benefit later. It is surely one of the most remarkable properties of the nervous system that it can retain some representation of past events for almost a lifetime—tens of years in some cases.

The nervous system operates by transmitting impulses along defined pathways and probably the process of learning involves heightened activity in those channels that record sensory impressions and their outcomes. However, it seems unlikely that heightened activity *per se* could constitute memory, that a memory could be stored in the form of a continuous train of nerve impulses running for years around the same pathways. Such ideas of 'self-reverberating circuits' have been entertained at times in the history of psychology, but can now be abandoned. Perhaps the neatest disproof was provided by Andjus and his colleagues (1955). They managed to cool rats down to 0°C for periods of up to an hour, at which temperature all electrical activity in the nervous system ceases. When warmed up again, these rats retained their old memories of events prior to cooling as well as normal animals.

If, on the other hand, the establishment of memory involves structural changes of some kind in the nervous system, so that some channels are facilitated, then we can readily understand why function is restored when electrical activity begins again. It is now generally accepted that memory storage must be represented in a physical form and we have enough evidence to understand, in some cases, how such changes are brought about and what is the nature of the store.

Obviously the study of memory mechanisms will involve neurophysiological and biochemical methods, because we must investigate the fine-scale operation of neurons and the synapses between them. Yet it remains the case that the most crucial evidence for memory still has to come from behavioural observations. It is what the animal actually does that provides our knowledge of what it has learnt, stored and now recalls and from this knowledge we go on to deduce mechanisms and then check them physiologically and biochemically.

Just as with learning itself, so the study of memory mechanisms has benefited from the comparison of different types of animal. Valuable evidence has come from certain molluscs, the honey-bee, a few bird species and a few mammals. Human memory itself has been a rich source. Although invasive experiments are impossible, there are many cases where the effects of brain damage can be studied. People, unlike animals, can tell you what they can or cannot recall, although we must recognize that their recollections are not always reliable.

Different types of memory

Introspection tells us that not all events are stored in the same way in our memory. We look up a phone number to make a call, but may not be able to recall it a couple of minutes later. On the other hand we can recall the numbers of close friends after months or years. Repetition and the greater significance attached to certain events must make a difference to the way they are stored. Perhaps all events go into a **short term** store but only events of some consequence are held in a **long term** store.

The idea that short term and long term memory are somehow distinct is supported by some remarkable psychological and physiological evidence. People suffering from concussion or other severe shock are often unable to recall the events that led up to the accident, but their memory of events in the more distant past is unaffected. Further, if their recent memories recover (and they often do) they do so roughly in order—the most distant first—until gradually almost all memory is recovered. Commonly, though, the few moments before the accident can never be recalled. This phenomenon of **retrograde amnesia** can also be reproduced in animals.

It suggests that the process by which events are moved into a long term memory store is more labile and sensitive to disturbance than the store itself, once in place. More evidence to support this conclusion comes from studies in rats and chicks using a range of drugs that affect the way the nervous system functions. Drugs can act in both directions. Thus some powerful stimulants of neural activity, such as low doses of strychnine or picrotoxin, can actually enhance the early stages of memory formation if they are given just before or just after learning a new task. Conversely, if drugs which inhibit protein synthesis, e.g. puromycin, are given at the same stage, memory of the new task does not consolidate and fades rapidly, but long-term memories already established are not affected (Andrew, 1985, 1991; McGaugh, 1989).

It has been particularly useful for such studies to develop tasks that can be learnt very rapidly in an appropriately all-or-none fashion. Learning a maze, for instance, is a complex task that takes time. How can we ever be sure, at a given moment, how much a rat has learnt? We can get more precise timings for memory processing from situations where animals can recall accurately after a single learning trial. Rats learn to step across a barrier to avoid electric shock after one such experience. Similarly, chicks refuse to peck at a coloured bead after a single experience when the bead is coated with the intensely bitter substance, methyl anthranilate.

Studies using this kind of task suggest that, at least in the chick and the rat, there are three types of memory with different time courses that we may call short term, intermediate term and long term (Andrew 1985). The evidence for this comes from two sources. First, there are drugs that affect memory specifically at each of three time phases, and are ineffective earlier or later (Fig. 6.16). Secondly, corresponding to these times of drug sensitivity, there are remarkably precise timed fluctuations in the ability to retrieve the memory of an event after it has occurred. Figure 6.16 indicates these fluctuations: chickens have pecked at a coloured bead coated with methyl anthranilate. Now different batches of them are tested at increasing delays after this unpleasant experience. Retrieval, revealed by their avoidance of beads of the same colour is, at first, good. Then, at 14–15 min after the event there is a fading of retrieval. This is not conventional 'forgetting' because a minute or two later retrieval returns, only to take a second dip 55 min after the original event; it also recovers (and more permanently) from this fall. Similar evidence comes

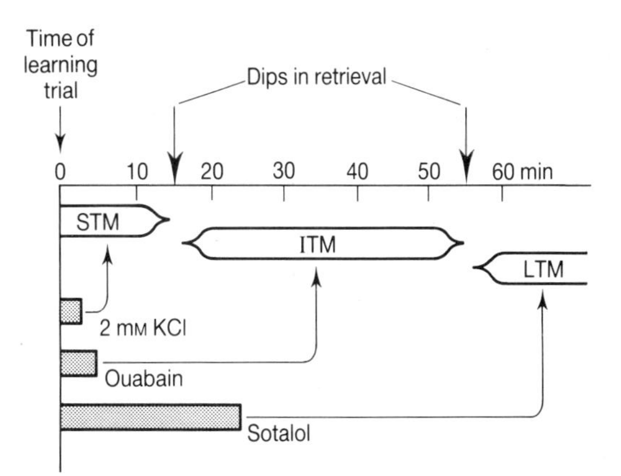

Fig. 6.16 A scheme for the three stages of memory storage constructed from experimental evidence in chicks. As time since the learning trial increases the chick retrieves memory of the event first from short term memory (STM) then intermediate term memory (ITM) and finally long term memory (LTM). The dips in retrieval at 15 and 55 min represent periods when the chick is switching from one store to the next. The three chemical agents listed each affect memory of one type only and lead to subsequent amnesia when administered during the period indicated by the extent of the bars at the left. This is supportive evidence for a three stage system (modified from Andrew 1985).

from rats, which also show two dips in retrieval before long term memory is established, although the exact timing is different. Even honey-bees, with their very specialized type of learning system (see p. 133), show some parallels to the vertebrates in their memory processing. Their retrieval shows a sharp fall around 3 min after a learning event and then recovers to a sustained level by 10 min (Erber 1981).

Such precision in the timing of the ability to recall events suggests that animals retrieve their memories from different stores in turn. It is the process of shifting from one store that is now fading, to the next, which is now forming, which appears to manifest itself as a dip in retrieval. We still have to speculate about the nature of such processes, nor can we be certain how the different phases of memory interact. Is each a necessary precursor of the next, playing a part in its formation, or does each form and fade in parallel, perhaps interacting and passing information across as it fades and the next comes to play?

In whatever way they interact, each phase of memory must be carrying a representation of the events that have been learnt. Long term memory can be shown to be in place after about an hour. As mentioned above, there is good reason to believe that its establishment involves structural changes in the brain, which subsequently facilitate neural transmission along new pathways (we will discuss the nature of these changes below). The processes leading to the setting up of short and intermediate term memories probably involve nervous activity alone. This is why they eventually decay and why they are vulnerable to physical shock and to drugs.

The anatomy of memory

We can now identify regions of the mammalian brain involved in these stages of memory formation (Morris 1983; Mishkin and Appenzeller 1987). The original evidence came from human subjects with particular types of brain damage. Last century the Russian neurologist, Korsakov, described patients who were able to recall distant events normally but who had permanently lost all recent memories following a head injury, a stroke or severe alcoholism. A common feature of such patients was damage to neural structures at the base of the forebrain, especially the thalamus and the mamillary bodies of the hypothalamus (Fig. 6.17). Since Korsakov's original descriptions, damage to other closely associated brain regions has been shown to have the same effect, notably the hippocampus, on the inner side of the base of the cerebral hemispheres, and the amygdala, which lies anterior to it (Fig. 6.17). Bilateral damage to any of these structures can lead to a complete inability to store new memories, and recall beyond a few seconds may be impossible. There is no better description of the behavioural consequences of such damage than the essay, *The Lost Mariner*, in Sacks (1986). He provides a vivid account of the extraordinary and tragic existence of an intelligent person who has lost all ability to memorize events and has only a distant past to which he can relate.

This type of human evidence has been amply supported by work with rats and primates. Numerous studies using specific brain lesions and drugs are building up a picture of a series of memory circuits involving the hippocampus, amygdala and thalamus. Short term and intermediate term memory must involve persistent neural activity in these areas. It is significant that they are very much at the hub of a series of fibre tracts connecting with the hypothalamus and its control of motivational systems (Chapter 4) and with those parts of the cerebral hemispheres that process sensory information. Thus they are in a position to combine information on the sensory nature of an event—visual, auditory, tactile, etc. —with its outcome in terms of reward or punishment.

Mishkin and Appenzeller (1987) give more detail and also discuss evidence of diversity in the memory system of monkeys so that, for example, visual recognition memories and spatial memories may be recorded, both in the short term and the long term, using different circuits and stores. Morris and his collaborators (Morris *et al.* 1982; Morris 1989) have shown elegantly that rats also utilize different memory systems for these two types of task. They have rats swimming in a water tank to reach an escape platform. In the first situation the rats learn to recognize which of two platforms visible above the water surface will give them a safe place to stand (the other platform sinks beneath their weight); this is a visual discrimination task and they soon learn to swim straight to the correct platform. In the second situation, rats learn to swim towards a platform that is constant in position but which they cannot see because it is just below the surface and the water has been made opaque by adding milk. This is a spatial learning task, but note that the rats still use their eyes

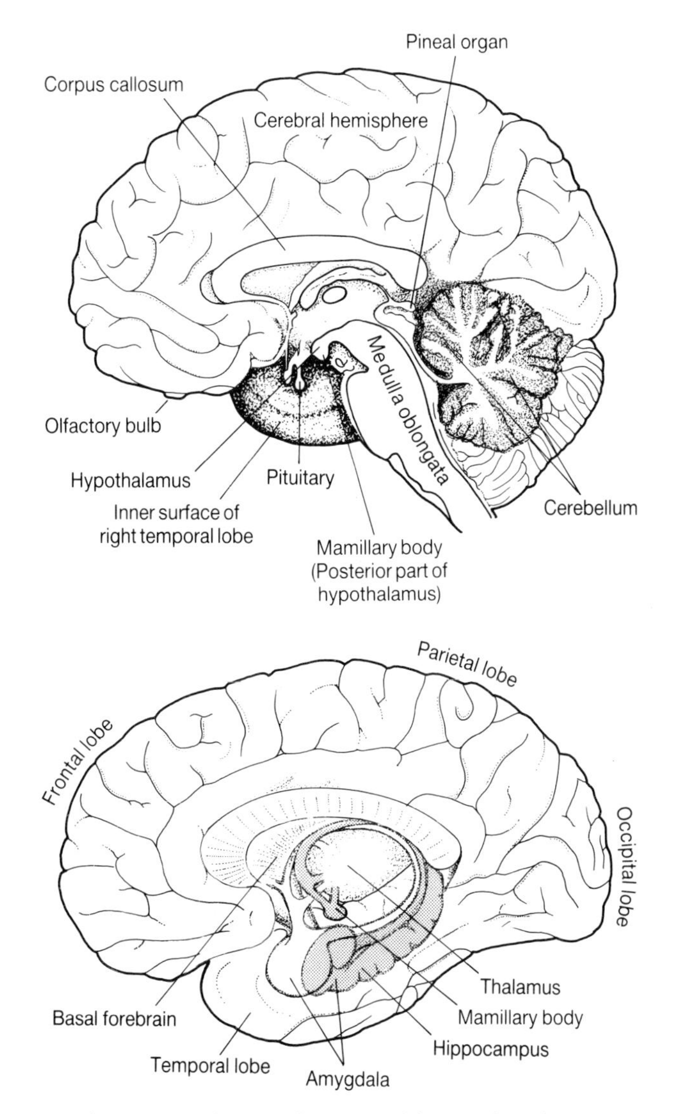

Fig. 6.17 The human brain viewed (top) in straight saggital section and (bottom) from the same aspect but with brain stem and cerebellum removed so as to reveal more of the structures involved in memory formation. Further explanation is given in the text.

to some extent because they have to take account of landmarks around the tank, using this information to navigate in space as they swim to the unseen platform. Lesions in the hippocampus prevent spatial learning, but not visual discrimination learning and from this, and other evidence, we can deduce that the hippocampus is not just involved in short term memory formation but in the organization of spatial memory generally.

Other evidence supporting this conclusion comes from the food-storing birds, whose behaviour and remarkable learning abilities we discussed on page 134. Birds have had a long history separate from mammals and it is necessary to be cautious in homologizing their brain structures. Nevertheless, there is some evidence to suggest that the avian hippocampus is homologous with that of mammals and does function in memory. Krebs *et al.* (1989) have examined the brains of a wide range of bird species and find that the food-storing species have a hippocampus significantly larger in proportion to the rest of their brain than that of relatives that do not store food. Their learning specializations have apparently required neural specializations to match.

Long term memory storage

We may turn finally to consider the nature of long term memory itself. The diversity of memory mechanisms in mammals mentioned above might prepare us for accepting that the areas of the brain where long term memories reside may also be diverse. The long term representation of an event stored in the brain is sometimes called an **engram**. In 1950 the psychologist Lashley published a paper entitled *In search of the engram*, reviewing years of work studying memory storage by rats; the engram proved to be elusive. In one set of experiments rats learnt a fairly complex maze and then had parts of their cerebral hemispheres removed. In the course of many experiments every part of the hemispheres, and many other structures, were tested. The overall result was that the degree to which the rats lost their memory seemed to depend not on which parts of the brain were removed, but only on how much. But is the engram really that diffuse?

Maze learning is certainly complex and will involve visual, spatial and tactile cues at least. Their processing and subsequent storage probably involves several separate pathways and interactions between them. Consequently, we might expect memory of a maze to be disrupted by lesions at several different sites. Remembering may be a process that reflects the term's origin—the opposite of *dis*membering—a putting together again of information gathered from several sources. More precisely defined, learning tasks may give a clearer picture of storage.

One of the best examples of this approach comes from the work of Bateson, Horn, Rose and colleagues, who have used visual imprinting by newly hatched domestic chicks as a learning and memory system for behavioural, anatomical and biochemical studies. Horn (1985, 1990) provides full reviews. The great advantages of imprinting are the rapidity with which young chicks learn, the fact that they are innately biased to approach conspicuous objects and require no period of deprivation followed by reward (as with rats being trained to run a maze) and lastly that imprinting occurs very early in life. Thus it may be possible to study the anatomical effects of visual learning on a brain that has been little 'marked' by previous experience.

In many of their experiments chicks were hatched in the dark, and then, when about 24 h old, were put into a small running wheel (Fig. 6.18(a)) close to an attractive visual stimulus—often a flashing red light. It was possible to vary exposure time and to measure the chick's effort to approach the light by recording the number of wheel turns. Subsequently, the extent of imprinting could be measured by giving chicks a choice of two objects—a discrimination test between familiar and unfamiliar—and seeing which they preferred or which they would strive hardest to approach (Fig. 6.18(b)).

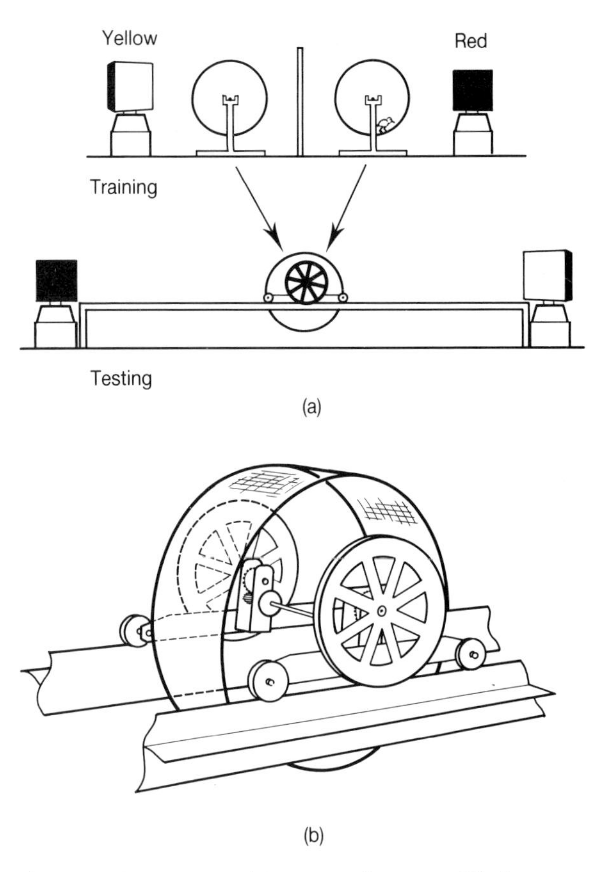

Fig. 6.18 Apparatus used to measure the extent of filial imprinting in young chicks as a preliminary to for studying its neural basis. In (a) chicks are placed in running wheels close to attractive objects—in this case flashing lights of different colours. Length of exposure can be varied and the number of turns of the wheel gives a measure of how much time the chicks spend 'following' the object. Then individual chicks of both groups are placed in a wheel running on a rail track between two objects (one familiar the other infamiliar) and asked to choose between them. This apparatus (b) measures how much effort they will put into approaching the model of their choice. By a clever arrangement of gearing, as the chick in its wheel strives to approach a model, it is carried away from it along the rails. Eventually it gives up and the distance at which it comes to rest is proportional to the degree of its effort to approach ((a) from Horn 1985; (b) from Bateson and Wainwright 1972).

Using a variety of techniques, these researchers have been able to identify a region of the chick's forebrain—the intermediate and medial part of the hyperstriatum ventrale (IMHV)—that is crucial for storing a representation of the imprinting experience and for mediating discrimination. Other parts of the forebrain are not so involved. It is of special interest that differences were found in the way the left and right hemispheres and their component IMHVs operated. Andrew (1985) describes other evidence that the two sides of the chick brain act complementarily during memory formation and the same may be true for mammals.

Using biochemical techniques and electron microscopy Bateson, Horn and Rose have been able to go further and show some of the growth and microstructural changes involved in forming the store. The association of events that sets up a novel pattern of neural activity definitely brings about synaptic changes at specific sites, which facilitates activity at these same sites again; see Horn (1985) for a review.

Similar patterns of synaptic changes have also been found in mammals. Greenough and Bailey (1988) describe work on rats using a variety of learning situations and techniques which reveal alterations to the number and nature of the connections formed between neurons as a result of experience. There are also parallels from invertebrates. Bruner and Tauc (1966) first showed that habituation of a protective withdrawal response in the sea hare *Aplysia*, a large marine mollusc, is associated with a depletion of neurotransmitters (see p. 3) at crucial synapses. Kandel and Schwartz (1982) have extended this work and record synaptic changes with simple associative conditioning also.

In conclusion, we can certainly now identify some of the short and long term events associated with memory formation and storage and it is even possible to make some generalizations across very diverse groups of animals. It is perhaps our own human memory system that leaves us with the most tantalizing questions. Even allowing for the garnishings of our imagination, the richness of association in human memory is amazing. How can it all be stored? Is there any limit to the amount stored? Luria (1975) in his study of a Russian mnemonist describes a person who could memorize long lists of nonsense syllables, meaningless mathematical formulae, strings of numbers and, without prior warning, repeat them back again many years later. Luria could find no limits to his storage capacity and no evidence that he ever forgot anything. Such an astonishing memory was as much a burden as a blessing. Selective forgetting will always help us, and our animal relatives with the management of our lives.

7 SOCIAL ORGANIZATION

In this final chapter, we draw together the threads of evolution, learning, cooperation and kinship that we have touched on already, and see how they underlie one of the most striking of all features of animal lives—their tendency to be social. Virtually all animals exist in pairs or groups for at least part of the time and, for some, sociality is the dominating feature of their whole lives. It is easy to lose sight of the individual when looking at huge flocks and herds and to forget that the cohesiveness and coordination shown by the group as a whole is, in fact, the result of natural selection acting upon individual animals and the genes that they carry. A genetic tendency to group with others and to interact with them in certain ways will, in the right circumstances give the animals bearing it an advantage and their genes will spread. What we see as social interactions between animals, and describe as 'social organization', is the net result of such selection. As we discussed in Chapter 5, the social environment that results can itself impose new selective forces on the individuals involved and can affect both the direction and pace of their subsequent evolution.

The term 'social organization' refers to populations or groups and not to individuals, and defines how members of a species interact with each other. In some instances—the various social insects, for example—social organization is fairly rigid and species-specific. In many vertebrates, on the other hand, it is a much more dynamic phenomenon and may vary with changing conditions. Certainly, the use of the term is not restricted to highly social animals. Tigers, which usually live and hunt alone in large territories, avoiding contact with others except for breeding and honey-bees, which spend their entire life in a dense colony, both provide examples of social organization, even though they are very different.

Indeed, as we will see, social organization among animals takes very diverse forms. In elephants, the females may live in the same family unit for 40 or 50 years. They clearly know and react to each other as individuals and the stability of their relationships suggests that their groups could be called 'societies'. On the other hand, the organization within many flocks or birds or schools of fish is much less complex, although individuals may stay together for months. And when we look at a swarm of water fleas gathered in some rich food area, or a mass of fruit-flies collected on some rotten fruit, then clearly the word 'society' would be quite inappropriate. Fruit-flies and water fleas form only 'aggregations' because they are attracted to a common food source, not specifically to each other. However, even they show basic social response because they react to one another's presence by spacing themselves out so they do not touch.

ADVANTAGES OF GROUPING

All animal groups, whether aggregations, flocks, schools or what we may wish to call true 'societies' result in the individuals that are part of them being better off than they would be on their own. However, as we discussed in Chapter 5, it is not always easy to discover exactly why (in an adaptive sense) the animals benefit, because being 'better off' implies that we can compare their survival and reproductive success in a group with that in some other situation, such as being on their own; group living is often so beneficial that animals on their own are usually not around to make the comparison possible.

Discovering the advantage of group living, then, often means putting together evidence from several different sources. Specifically, we can use experimental evidence (artificially creating groups or placing individuals on their own), naturally occurring variation (within a species some animals may be more or less likely to group than others) and comparisons between species that have adopted solitary or social lifestyles.

Allee and co-workers (1938) used simple experiments to show how even loose aggregations can benefit the individuals that comprise them. They showed that water fleas cannot survive in alkaline water, but that the respiratory products of a large group of them are sometimes sufficiently acid to bring the alkalinity down to viable levels. Thus, a group can survive where a few individuals could not. Fruit-fly cultures do not do well if there are too many eggs because the resulting larvae are undernourished, but they fare equally badly with too few eggs. This is because reasonably large numbers of larvae are needed to break up the food, encouraging the growth of yeast and making the food soft enough for all the larvae to feed easily. It is thus advantageous for a female to lay her eggs close to those of others because her own offspring will benefit.

Flocks of birds and schools of fish exemplify groups that are much more than simple aggregations because there is often a high degree of social interaction between individuals. Even here, physical factors may still count, as in the case of Emperor penguins, which huddle closely together as they stand incubating their eggs during the Antarctic winter. Heat is conserved and birds on the outside move more than those in the centre, leading to mixing and a reasonable distribution of shelter.

One of the most obvious advantages of a cohesive group whose members respond to each other's behaviour is protection against predators (Fig. 7.1). With a number of animals on the alert, the approach of a predator is less likely to go undetected and one alarm signal will suffice for all. Lazarus (1979) showed this effect in action by watching how effectively red-billed weaver birds spotted a goshawk flying overhead when they were on their own or in groups with other birds. He found that solitary birds often failed to respond at all, whereas where there were two or more birds, the hawk was much more likely to be seen and responded to. Powell (1974) used not a hawk but a hawk model to study the antipredator responses of starlings when they were feeding. He too found a more rapid response to danger when there were several birds, rather than just one. He also found that, in groups of 10, individual starlings spent significantly less time in surveillance than did individuals on their own. In other words, starlings in the larger flocks increased the amount of time they spent feeding (through not having to keep looking around) but the combined

Fig. 7.1 The defensive formation of a group of musk oxen on the Canadian tundra. When a predator approaches, they bunch with the older animals at the front, facing the threat (photograph by D. Wilkinson from Information Canada Phototheque).

efforts of many pairs of eyes nevertheless enabled them to be more effective than single birds in detecting predators when they do appear. Elgar (1989) reviews over 50 studies that show that birds and mammals spend less time in vigilance and more time in feeding the bigger groups they are in.

In meerkats—socially living mongooses—vigilance is undertaken by particular individuals, which take turns to go to a high look-out point such as a tree and keep watch for predators while the others feed (Macdonald 1986; Fig. 7.2).

Fig. 7.2 Vigilant meerkat, keeping watch for predators while the rest of the group feed (photograph by David Macdonald).

Even when a predator has been spotted, being in a group offers further advantages to the individual because the group can take concerted evasive action, confuse the predator by having individuals scattering in all directions or actually physically attack the predators. Figure 7.3 illustrates the evasive action of a flock of starlings, and very similar behaviour is shown by some fish which bunch together at the least

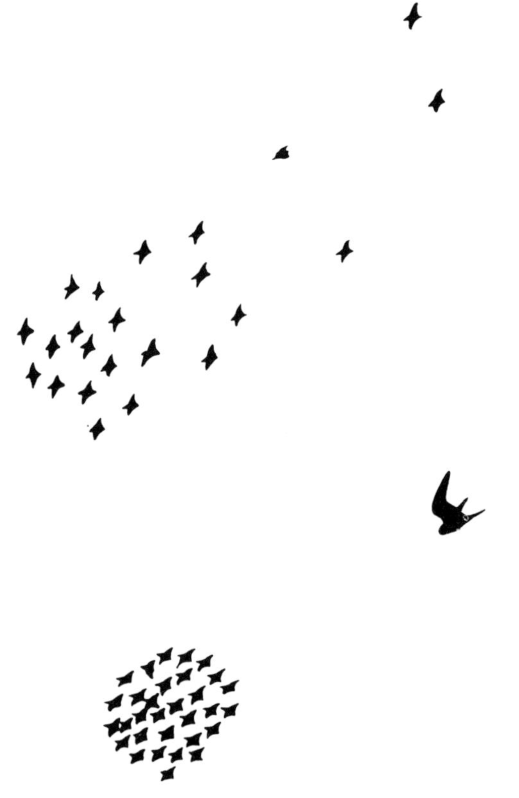

Fig. 7.3 The response of a flock of starlings to the approach of a bird of prey (from Tinbergen 1951).

alarm. Predators rarely attack an individual in a close group and their most common strategem is to make swoops towards the group, which may cause them to scatter, allowing the predator to single out an isolated animal. Hamilton (1971) in his graphically titled paper *Geometry for the Selfish Herd* showed, theoretically, that if each animal in the group attempted to put at least one animal between itself and the predator then tight formations would be expected.

Colonial nesting birds—gulls and terns, for example—may provide formidable opposition to an invading predator, such as a fox, by mobbing it, even hitting it with their feet. Even though each bird is responding individually to defend its own nest, the proximity of other birds all doing the same thing means that their combined efforts can be much more effective than a single bird on its own. As a result, as Göttmark and Andersson (1984) showed, the nesting success of gulls in a large colony is considerably greater than that of those that nest singly or in small groups (Fig. 7.4).

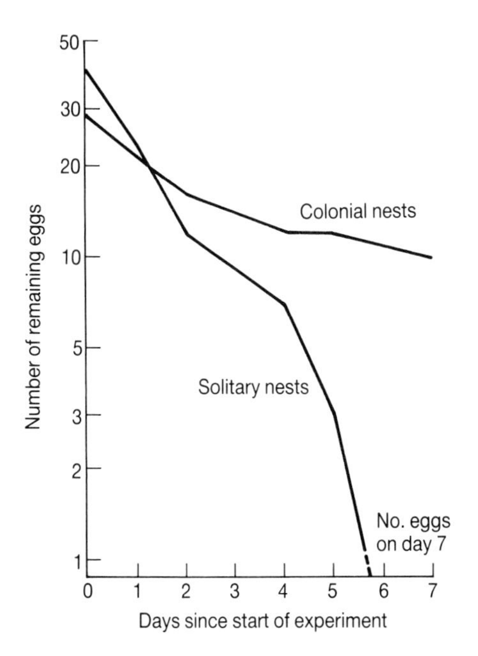

Fig. 7.4 The temporal pattern of predation on experimental eggs laid out near (2 or 5 m) nests of colonial and solitary pairs of common gulls. Predation was much higher near the solitary nests and by day 7, there were no experimental eggs at all left near the solitary nests. Eggs placed near the colonial nests were protected by the massed mobbing of the gulls in the colony (from Gotmark and Andersson 1984).

Protection from predation is only one of the advantages gained by living in groups. Another major factor is utilizing food sources found by other animals. We have already mentioned the fact that birds in flocks can spend more time foraging and less time being vigilant because they rely on the vigilance of others. They may also steal food from other individuals. Krebs *et al.* (1972) showed that when one member of a tit flock finds a food item, the others rapidly alter their searching strategies and concentrate their attention both on the general area and the type of niche in the trees where the food was found. Sometimes this may be to the short term disadvantage of the bird that finds the food but, by staying in the flock, it gains a longer term advantage of being able, in its turn, to utilize food found by other birds and the increased vigilance the flock provides. At other times, when the food source is so rich that there is enough for all there is no disadvantage at all to the finder individual. Brown (1986) showed that cliff swallows follow birds that have located a rich source of insect food. Birds returning to the colony after a successful foraging trip were likely to be followed on their next trip out by birds that had previously been unsuccessful. But the food source—a 'plankton' of insects—was so abundant that the finders were at no disadvantage if they were followed. In yet other cases, it may even be of advantage to the finder to have other individuals around. Gannets are often found fishing in groups and Nelson (1980) argues that a group of birds diving together so confuse the fish that they are more easily caught. Göttmark *et al.* (1986) showed a similar effect in black-headed gulls. Group fishing led to more successful feeding for each individual than a bird could accomplish on its own because, with more gulls around, the fish were more vulnerable and more likely to be caught.

Lions, hyaenas and Cape hunting dogs are all examples of predators that hunt together. Their strategems may involve some individuals driving prey towards others hidden in cover or, as with the hunting dogs, taking it in turns to run down an antelope to the point of exhaustion. Spotted hyaenas are to be found in different sized groups depending on the size of prey they are hunting. Kruuk (1972) showed that when hunting zebra the mean number of hyaenas in a group is 10.8, when hunting adult wildebeest the mean is 2.5 and when hunting gazelle fawns they hunt in groups with a mean size of 1.2 individuals. The interesting thing is that the hyaenas ignore the 'wrong' sort of prey altogether. So if they were in a small hunting group, they completely ignore zebra which, on another occasion and in a larger group, they would regard as highly prized prey.

However, we should not assume that, just because animals hunt together, they are necessarily cooperating to bring down prey. Packer (1986) argues that although lions pursue prey in a group, they may not actually be helping each other. Rather, the advantage of group hunting for the lions lies in the fact that, after the kill, there will be more individuals to keep scavengers and other potential thieves away from the carcass. In other words, the hunting may not be cooperative (more individuals do not necessarily lead to greater success) but the defence of the kill is. Extra individuals are tolerated because they help keep away vultures and hyaenas, even though they eat some of the meat themselves.

It will be clear from the preceding discussion that group living confers some clear advantages but, at the same time, some undoubted disadvantages. Being near other individuals means increased competition for food, increased risk of disease transmission and greater conspicuousness to predators as well as greater risks of cuckoldry, mixing and cannibalism of young, and so the balance between the advantages an individual gains as a result of being part of a group and the disadvantages it inevitably suffers from the same source is a very fine one. Hoogland and Sherman (1976) give a long list of disadvantages suffered by bank swallows as a result of their habit of nesting communally in sand banks. These range from increased chances of picking up fleas to having nest burrows collapse as a result of other birds nesting close by. Nevertheless, the birds benefit from nesting in a colony because they derive safety from the massed attacks of all the other birds on potential predators. This is more than enough to outweigh the evident disadvantages of being close to other birds. The behaviour that will be favoured by natural selection—whether it is territoriality or extreme clumping—will be that that favours the reproductive interests of the individual in the long run. We now discuss some of the types of social groups that have evolved as a result.

TYPES OF SOCIAL GROUPS

Groups of animals differ widely in the complexity of their organization and the types of interactions individual animals have with one another. It is no good simply describing a group by its size. The sex ratio, the degree of differentiation into roles, relationships with other specific individuals in the group, kinship and other factors all affect the costs and benefits of group living. A look at some contrasted types of groups from very different animals will illustrate what this means in practice.

The caste system of social insects

If we look across the vast group of the insects as a whole, we have some of the most completely solitary animals of all. Many of them lack even the most basic of social contacts, that between parent and offspring. The short life span of many insects coupled with their strictly seasonal reproduction often means that parents have died long before their offspring have emerged from larval life. In Chapter 2, we described the life history of the mason wasp, which has only brief contact with another member of her species when she mates. Apart from that she is totally solitary; she builds cells and provisions them with food for young she will never see.

It is obvious that one essential for the development of parental care is a reasonably long life span so that there is some overlap between the generations and contact between young and old individuals. The aggregations formed by insects such as cockroaches and earwigs contain such overlapping generations. The life span of a cockroach may be a year or more, three quarters of which is development when it passes through a series of nymphal stages, becoming more like the adult insect. During this period, cockroaches of all ages live together in a loose aggregation near sources of food and shelter. Some species of cockroach incubate their eggs inside the females body and bear live young, which remain in contract with the mother for some hours after birth. This contact may be important for the survival of the young because, not only are they extremely vulnerable to cannibalism when first born, but they may pick nutrients from the mother's body surface.

We know that this nutritional factor is also of major importance in the termites (which are related to the cockroaches) because, feeding largely on wood, they rely on symbiotic protozoa living in their gut to digest cellulose. These protozoa are acquired by the young termites when they feed on fresh faecal matter from the adults, and they can be transmitted only in this way.

We have already described the evolution of the highly organized insect societies found in the termites (Isoptera) and the ants, bees and wasps (Hymenoptera). Here we find such remarkable adaptations to social life that they are called not just social, but **eusocial**: their societies are divided into 'castes' such as soldiers, workers and reproductives, which are morphologically different from one another, and that each have a separate role to perform in the life of the colony. There is a huge literature on these social insects, which have attracted man's attention for centuries. For an introduction to the whole literature, Wilson's masterly survey (1971) is unmatched and Butler (1974), von Frisch (1967) and Lindauer (1961) are all excellent accounts of that most studied of all social insects, the honey-bee.

The term 'caste' is well suited to describe the division of labour within insect societies. It implies a rigid, limited role in society largely determined by upbringing, which is what seems to be the case here. One of the most important factors determining caste is what the insects are fed when young. In bees, wasps and termites, all eggs laid by the queen are potentially equal, but most larvae are fed a restricted diet and develop into workers (Fig. 7.5). There is evidence that when queen ants are laying rapidly, their eggs are 'worker biased' and develop accordingly no matter how the larvae are fed. But for most of the time their eggs are also equipotential and only richly fed individuals develop into the reproductive castes.

Pheromones are also an important determinant of caste. They are secreted by the insects themselves and coordinate development and social behaviour. The development of worker termites is controlled by pheromones produced by the king and queen. Similarly, the queen honey-bee secrets a pheromone ('queen substance'), which both suppresses the ovaries of the workers and prevents them from rearing new queens. The queen is always surrounded by attendant workers who lick her body, subsequently offering food to other workers and with it, the pheromones from the queen.

Fig. 7.5 Honey-bee larvae in cells (photograph by Paul Embden).

The level of the pheromone must be kept up and once the source is cut off its concentration rapidly drops, which happens if the queen becomes ill or dies. The effectiveness of the incessant food sharing in circulating queen substance is shown by the fact that some workers in the brood area of the hive exhibit changed behaviour within an hour or two of the colony losing its queen. They begin construction of 'emergency' queen cells in which some of the youngest larvae, destined in the normal course of events to become workers, are fed royal jelly throughout their larval life and become queens, which will eventually fight one another to replace their mother. As colonies grow, the dilution of queen substance below a critical level is one of the factors that leads to swarming in honey-bees.

The honey-bee is exceptional because it reproduces its colonies by swarming—in most other social insects new colonies are founded by a single queen (or a pair in termites). The queen begins the construction of the nest and rears the first batch of workers herself. These then take over the tasks of extending the nest and bringing food and the queen usually stays in the nest laying eggs from this point on. The tasks performed by the worker castes vary greatly in detail, but in most colonies they cover the main categories of foraging, rearing the young, nest construction, attending the queen and guarding the colony. In termites and ants this last task is sometimes the sole responsibility of a special soldier caste, which has enlarged jaws and other weapons.

A honey-bee worker lives for about 6 weeks as an adult and her activities are roughly synchronized with her physiology. Thus, she spends the first 3 days cleaning out cells, then begins feeding the older larvae on a mixture of pollen and honey. From about the 6th–14th days of her life, the worker feeds 'royal' jelly (secreted from the pharyngeal or 'nurse' glands on her head) to the younger larvae and any queen larvae in the hive. (Royal jelly is fed to all larvae for a brief period early in their development, but those destined to become young queens are fed royal jelly throughout.) The worker then gradually changes her behaviour from feeding larvae to cell construc-

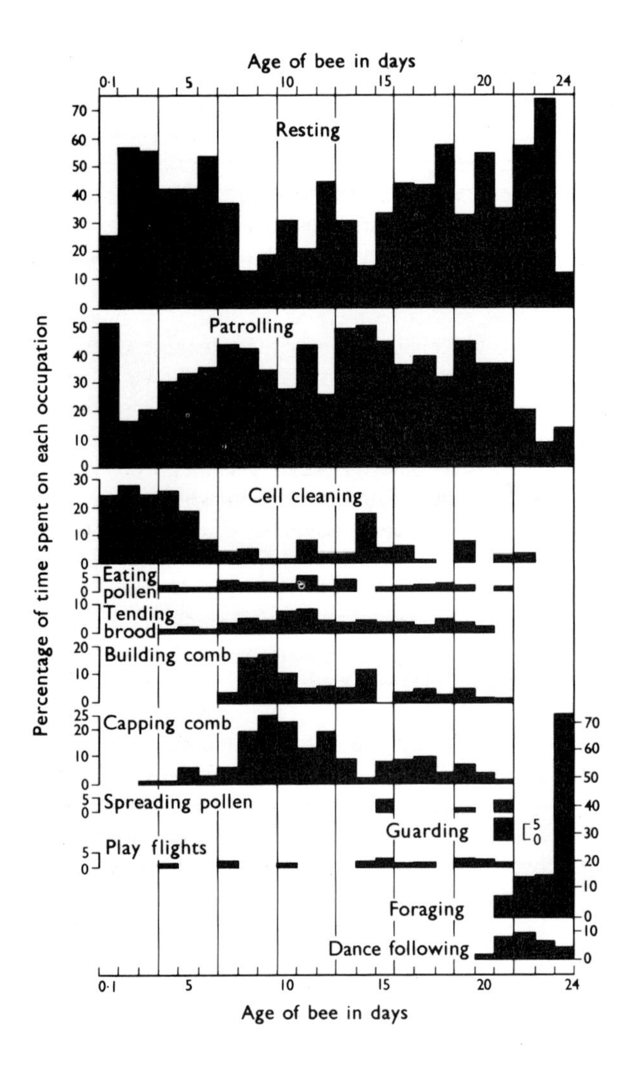

Fig. 7.6 Lindauer's complete record of the tasks performed by one individual worker honey-bee throughout her life. The records are classified according to the type of task. One can recognize the age-determined succession of cell cleaning, brood care, building, guarding and foraging. Note, however, the large amount of time spent in patrolling the interior of the hive and in seeming inactivity. During such periods the worker may be acquiring information on the situation in the colony and adjust her behaviour accordingly (from Lindauer 1961, with permission Harvard University Press. © President and Fellows of Harvard College.)

tion as her pharyngeal glands begin to regress and the wax-secreting glands on her abdomen become active. From the 18th day, she may be found guarding the hive entrance or venturing outside for a few brief orientation flights. From 21 days onwards, the worker is primarily a forager, bringing back nectar, pollen and water and will remain so for the rest of her life—about 2–3 weeks. Figure 7.6 illustrates this sequence of behaviour through the brief life of a worker.

This is the general sequence of events but it can be modified to suit the needs of the colony, depending on the flower crop, temperature, age of the colony or other factors. Modification is possible because of the sophisticated communication systems that exist between the members of a honey-bee society. We have already mentioned the importance of food-sharing as a method of communication, both as a way of circulating pheromones and as a means to keep each worker directly informed of the state of the food supplies within the colony. In addition, the 'dance-language' of honey bees (p. 66) is the most sophisticated example of communication because of the detailed information it conveys about the distance and direction of a food source. Very close attention is paid to returning foragers by other workers. As she moves up on to the comb, the forager is contacted many times and she can gauge the needs of the colony by the eagerness with which her nectar load is accepted. On the other hand, if the food she discovered was not particularly profitable, she may not dance at all. She is quite likely to stay in the hive and subsequently be attracted by the dances of more successful foragers.

Although other social insects lack the detailed communication of the honey-bee, their foraging is not random. Incoming worker ants or stingless bees (*Meliponinae*) may lay scent trails and other workers become aroused to go out and forage by the highly excited behaviour shown by a forager who has discovered a rich food source. Thus, social insects tend to forage as a team that directs its effort where it will be most profitable.

The beautiful adaptiveness of the social insect colonies and the control it gives them over their environments are based on a relatively simple series of responses to other workers in the colony and to the nest itself. There is no suggestion that they are intelligently working out what to do. To a certain extent, the social organization of the insects is flexible, as when honey-bee workers change their normal sequence of tasks in response to a sudden requirement in the colony. However, the degree of flexibility is limited. Consistency of social organization within species is highly characteristic of the insects and we shall not find it so well marked in other groups.

Territory in the social organization of vertebrates

In contrast to the social insects, the social organization of vertebrates is not rigidly species-specific. Thus, the answer to the question, 'What is the social organization of the house mouse?' is not fixed. It depends upon the nature of the food supply, the density of the population, its age, sex structure and a number of other factors.

An important key to understanding the social organization of vertebrates is to remember a point made earlier: that group living brings costs as well as benefits. Other animals are competitors for food, for mates and for nest sites, and we have already seen, in the section on aggression (p. 97), that animals have well-developed mechanisms for 'seeing off' unwelcome intruders. Not surprisingly, therefore, some animals adopt a solitary way of life or live in monogamous pairs. Many defend territories to keep other animals away from them. Some, like the great tit, are both territorial and gregarious at different times of the year. During the breeding season, great tits are strongly territorial and vigorously defend the area around their nest against other great tits. But during the winter they group into large flocks both with members of their own species and of others such as blue tits and nuthatches. This means that when they have hungry young to feed in the spring, they defend a private supply of insect food and, in the winter, when food is scarcer, they gain the advantages of locating food and protection against predators that being part of a flock gives them.

Defence of a territory of some sort is in fact a common feature of the social organization of many vertebrates (as well as some invertebrates). A typical case is represented in Figure 7.7, which shows the spatial distribution of willow warblers. Each male defends a substantial area, which will include food for himself and eventually a mate and young, and he will rarely leave the territory during the breeding season. As will be seen, each territory impinges closely on those of other males and, where habitat permits, nearly all the available ground will be occupied. In birds, territorial defence is usually achieved by song and visual displays. In mammals, which may also have densely packed territories, the territorial boundaries are often defined by scent posts marked with urine, special glandular secretions or faeces.

Defending a territory against constant intrusions is clearly costly because an animal has to expend time and energy seeing off rivals and may make itself vulnerable to predators while doing so. Brown (1969) introduced the idea of 'economic defendability', pointing out that animals should go to the time and trouble of defending a territory only if the resource they were defending (the food in it, for example) was worth defending and, indeed, could physically be defended. A very scattered food source might take so long to defend that the animals would lose more energy than they would gain by doing so. As we saw in Chapter 1, pied wagtails defend a riverbank territory only when there is enough insect food washed up by the river. A male wagtail's tolerance of a second male on his territory is also dependent on whether there is enough food for both of them. Gill and Wolf (1975) also applied the idea of economic defendability to the territories defended by the tiny golden-winged sunbird, which is a nectar-feeding bird found in East Africa. They calculated the energy used by a sunbird defending its territory against intruders and also the energy available in the *Leonotis* flowers they were defending (Fig. 7.8). If flowers are defended they become a better food source because the nectar levels rise if they are

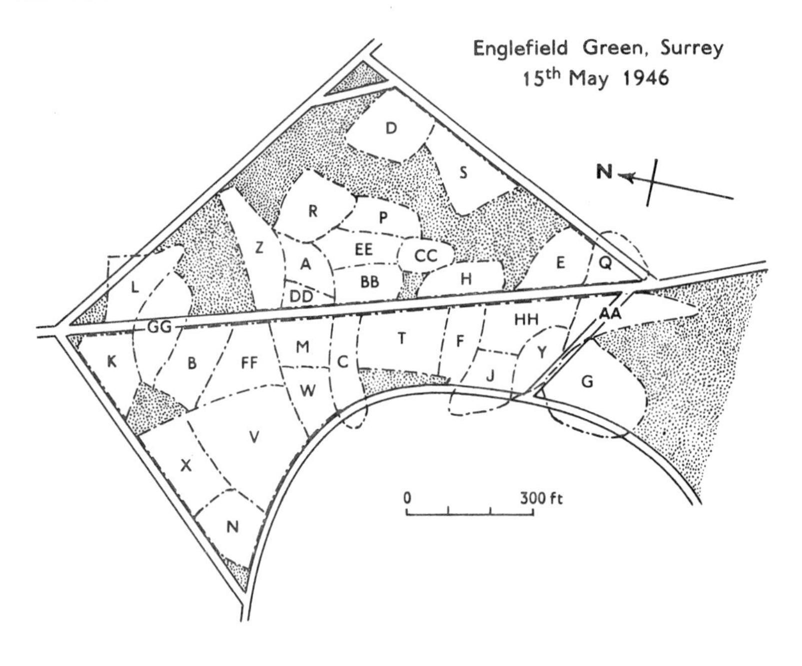

Fig. 7.7 A territory map of a population of willow warblers in birch woodland. The size of territories varies from under 1000 square yards to more than 5000 square yards. The shaded areas are unoccupied (from May 1949).

not depleted by other birds. Gill and Wolf found that the energy the birds expended in keeping other birds out of their territory was more than compensated for by the increased levels of nectar in their 'private' flowers. Defending a territory was, therefore, energetically worthwhile.

It is clear, however, that such considerations of energy gained and lost do not apply in every case of territorial behaviour. The territories of ground-nesting seabirds such as gulls, terns and gannets are about 1 m across and contain no food at all. They are simply a small defended area around the nest and are important because of the cannibalistic nature of other gulls in the colony, some of which specialize in eating the eggs and chicks of other members of their own colony. Yet another function of territory is seen in the 'lek' system of birds such as the sage grouse, ruff and prairie chicken and in mammals such as the Uganda kob and the hammerhead bat. Here the males gather in a tight group or lek and, within the group, each male defends a small territory. There is no food in the territory, but the sight and sound of many males all displaying together attracts females from a large area. The females then appear to choose between males on the basis of their display. In the case of the sage grouse, Gibson and Bradbury (1985) showed that it is the males that display for the longest and in the most vigorous way that are chosen by the females. However, once she has mated the female sage grouse has no further use for the terri-

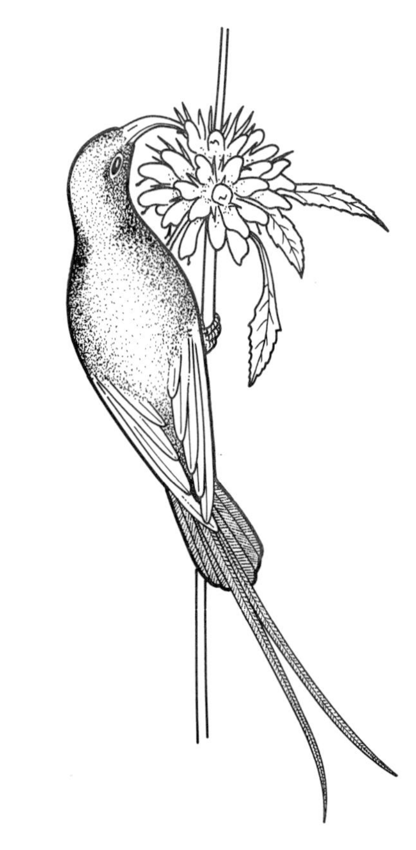

Fig. 7.8 The golden-winged sunbird defends nectar-filled flowers against intruders if it is energetically worth while (from Alcock 1984).

tory and leaves the lek area to rear her young alone. In such polygamous species, the male's contribution to his offspring is purely genetic and he will continue to try to attract other females without giving any of them any assistance with rearing the young.

A very different attitude to a male's territory is shown by female pied flycatchers. In this species, the males arrive from their wintering areas to their breeding grounds in northern Europe about a week ahead of the females and set up large breeding territories. The males that arrive first tend to be the oldest males with the blackest plumage. They also have the pick of the available territories, so the fact that they are more successful at gaining a mate than late arriving males could be because females either like their territories or are attracted to some characteristic of the males themselves.

Alatalo *et al.* (1986) devised an ingenious way of separating these two factors in the females' choice. They made use of the fact that pied flycatchers readily nest in artificial nest boxes and can be persuaded to nest in different areas of a wood depending on whether there is a nest box there. They forced males to defend particular territories by restricting the number of nest boxes available at any one time, and only putting up more when the first ones had been occupied. When the females arrived, they noted the order in which the different males attracted a mate. This time, success in obtaining a mate was not correlated with how early a male had arrived or with his age or the blackness of his plumage or any other characteristic, but with the quality of his territory. Males that had a low density of birch trees in their territories and nest sites high up in trees with thick trunks giving safety to the nest were the ones that obtained mates first (Fig. 7.9). Normally, older, blacker males would choose such territories themselves, but when they were experimentally denied the opportunity to do so, the females could be seen to be responding primarily to the territory, not to its owner. From the female's point of view, this makes sense. If the success of her brood is dependent on cover and food supply, it is not surprising that these form the main criteria of choice.

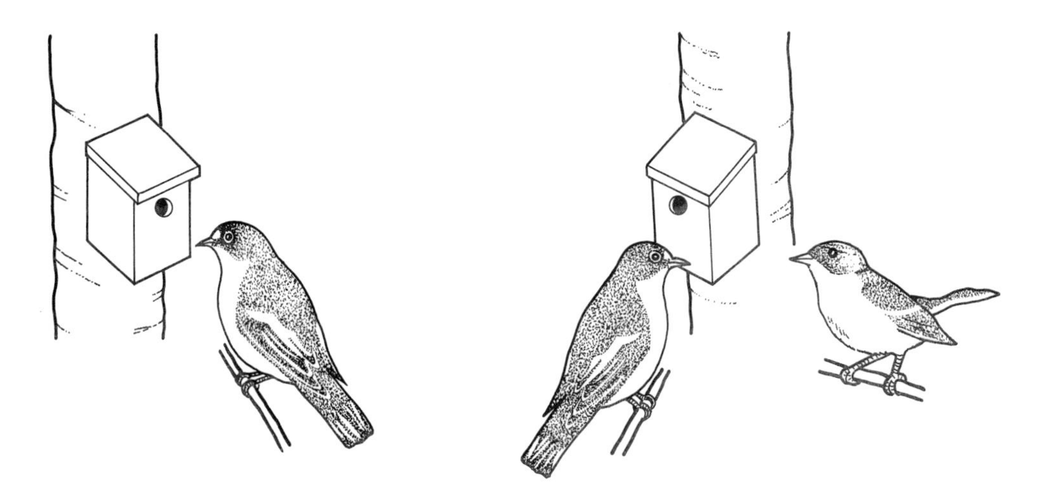

Fig. 7.9 A male pied flycatcher defends a territory against other males. The female chooses a male on the basis of the quality of his territory and the suitability of the nest site it contains.

MATING SYSTEMS AND SOCIAL ORGANIZATION

In general, we find a regular relationship between the type of social organization and the role adopted by the two sexes in the reproduction, i.e. the type of mating system. Animals with 'lek' type territories are polygamous, whereas those that defend large territories tend to be monogamous or to have one male with 2–3 females. This is true not just in birds but in other animals such as antelope. Jarman (1974) looked at a wide range of different antelope species and showed that diet, body size, mating system and territory were all related. Small antelope, such as the tiny duikers and dik-diks, live singly or in pairs. The male defends a territory that provides him and his single mate and offspring with food (the young growing parts of a wide range of food plants) all year round.

Larger antelopes, such as waterbuck, gazelles and impala, live in larger groups of 6 to over 100, feed on a wide range of species and do not stay in the same area all year. Some males form bachelor herds, others defend territories that do not, however, provide enough food for one male, let alone for a female and young. Nevertheless, the territories are defended vigorously against other males and a territory-holding male will attempt to herd any females entering his territory and to mate with them. The mating system is promiscuous—the male mating with many different females as they enter his territory and the females wandering from territory to territory apparently in search of food, and mating with different males as they do so (Jarman 1974). Uganda kob males form leks, or groups of displaying males, to which females are attracted (Leuthold 1966; Fig. 7.10). A proportion of the males thus obtain exclusive mating rights without having any permanent bond with the females, simply by being able to defend a small territory devoid of any substantial resources at all.

Emlen and Oring (1977) pointed out that one of the major factors determining the type of mating system shown by a given species will be the ability of some individuals (usually a proportion of males) to control access of others to potential mates. Thus, polygamy will evolve where some males can defend a group of females and this will in turn be dependent on the distribution of resources, such as food, that the females need. It would, however, be a mistake to see food as the only factor. Crook (1965) in a pioneering study of the weaver birds (Ploceidae) argued that forest species nest solitarily in monogamous pairs and are primarily insectivorous, whereas species living in grassland or savannah nest colonially, are polygynous (one male to several females) and eat seeds. He believed that monogamy was favoured when food was scarce and that polygyny arose when food was plentiful and the help of the male was not essential to the rearing of the brood. But, as Lack (1968) pointed out, in birds other than weaver birds, this relationship does not seem to hold and Haartman (1969) showed that other factors, such as the nature of the nest site will have a profound effect on mating systems. Many polygynous species nest in holes, making it possible for the brood to be cared for by one parent because

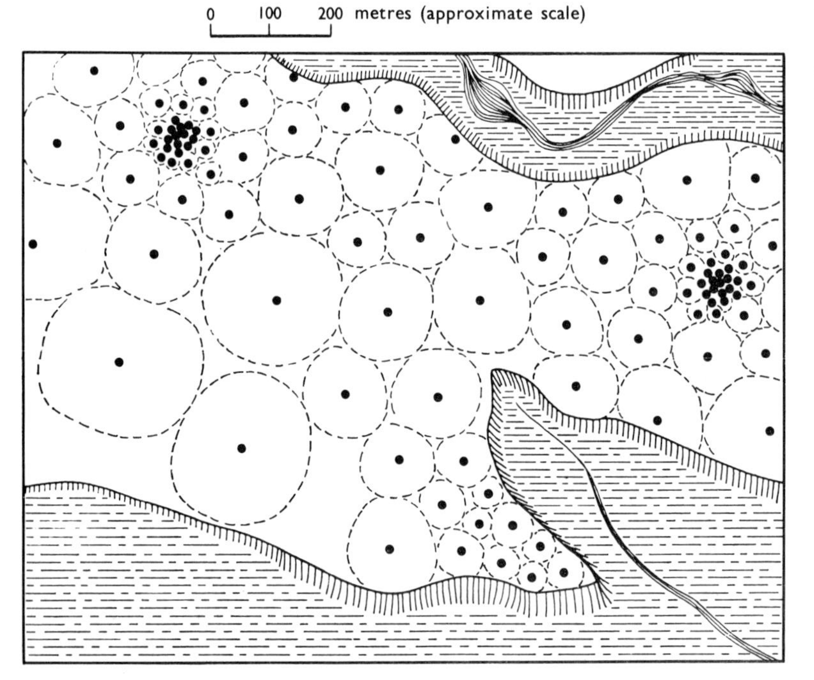

Fig. 7.10 Territorial map of males of the Uganda Kob on an area of grassland raised about two swampy river bed areas, indicated by shading. The boundaries are only very approximate, but each black dot represents the centre of a territory, close to where a male usually stands. There are two closely packed 'territorial grounds' and a third may be forming on the 'peninsula' at the lower part of the map (from Leuthold 1966).

dangers of predation are reduced. But other hole nesters, such as the great tit, are monogamous, so neither is this a sufficient explanation.

In trying to explain the diversity of mating systems, then, we have to consider the combined effects of how defendable groups of females are, plus the many different factors that can affect the survival chances of the young under different conditions. Monogamy will tend to evolve where both parents are needed to feed and care for the young, but where one parent can rear a brood on its own, the other may benefit from deserting and mating again with another individual. In mammals, the retention of young in the female's body and her feeding through lactation means that the male can often do little to improve the survival chances of his young. Consequently, it is often in his best reproductive interests not to give parental care but to seek more mates. Hence polygamy (more strictly polygyny) or promiscuity are the most common mating systems in mammals. Only where the male can contribute substantially, for example in jackals, where males feed the young through regurgitation or in marmosets, where the male carries the young through a long period of dependency, does monogamy appear in mammals. In birds, by contrast, eggs are incubated outside the female's body and the young are often fed by laborious collection of food, tasks that can as well be performed by males as by females and often very much more effectively if both parents are present. This probably explains why over 90 per cent of bird species are monogamous, and why monogamy is the exception in other groups.

As we have just seen in the case of the pied flycatcher, females may choose males because of the resources they are defending. Even where a male does not contribute to the rearing of his young by feeding them, he may nevertheless aid their survival by defending the nest against predators or intruders of the same species. However, this does not necessarily lead to strict monogamy (one male mating with one female) because, as we have seen, a male's ability to defend resources or groups of females will also affect the nature of the mating system. One male may have several females nesting in his territory (polygyny) providing resources and even paternal care for all their broods. Verner and Willson (1966) and Orians (1969) developed a useful way of looking at such a situation called the 'polygyny threshold' model. They argued that, whereas it was clearly beneficial for the male to have several females, it was not necessarily of advantage to the females, because their breeding success could be lowered by the presence of other females, for example, where the male gives less care to each brood than he would if he had only one. Nevertheless, if some males have substantially better territories than others, it may pay a female to settle in a high quality territory of an already mated male than to mate monogamously with a male with a poor territory. Success in rearing a brood may be higher for a polygynous female than for a monogamous one, up to a certain threshold (the point at which there are so many females in the good quality territory that the female would do better to go to a poorer territory).

The polygyny threshold model has been used to explain the pattern of mating in many species but, as Davies (1989) points out, it is hardly surprising to find that the model does not always fit the data. For one thing, there will be conflicts between the sexes —what is the best mating system for females may not be the best system for males, and so the result may be a compromise or what is best for one or the other. Catchpole *et al.* (1985) showed how such conflict may operate in practice. The European great reed warbler is, like the pied flycatcher, a migratory species in which the males arrive on the breeding grounds before the females and defend large territories. Female great reed warblers are attracted to characteristics of the males territories and sometimes several females settle in the particularly good territory of one male. According to the polygyny threshold model, the breeding success of polygynously mated females in good territories should be equal to or greater than the breeding success of monogamously mated females in poorer territories, but it is not: the second- or third-arriving females show reduced breeding success. Catchpole *et al.* argue that the females are deceived into choosing already mated males because the large territories in the reed beds prevent them from seeing the females that the male already has there. Alatalo *et al.* (1981) argue that pied flycatchers go one stage further and defend several different territories at once, deceiving several females into apparently 'monogamous' pairings. The male benefits but the females would seem to be worse off than if they had chosen an unmated male.

There is, then, a rough correlation between the type of mating system and the amount of care given by the two parents. In monogamous species, particularly birds, both parents usually contribute to the rearing of the young. In polygamous species, it is usually the female that does most of the parental care. The male may, if he defends a large territory and has a small number of females, contribute some care to each of his broods, obviously giving less to each one than if he had only one mate. On the other hand, there are many polygamous species where the male mates many times and contributes nothing beyond his sperm to his offspring. It is in these species, where males offer the least parental contribution, that we find the greatest sexual dimorphism in size or plumage and the most conspicuous and flamboyant mating displays. Male peacocks, lyrebirds and birds of paradise (Fig. 7.11) have spectacular plumage and courtship rituals, but do not care for their young at all. Male red deer and elephant seals (Fig. 7.12) spend much of the breeding season fighting each other in dramatic and often bloody combat but here too, the care of the young is wholly the responsibility of the female. In such species, females gain nothing in the way of paternal care or resources for their young from choosing a particular male, not even a desirable territory or nest-hole. The only strategy for a female is to choose a male who is likely to carry the best genes—that is all he contributes—and the intense male–male competition may help her to do this. The huge male elephant seals, which are three or four times the size of the females, gather harems on the breeding beaches. Only the very strongest can dominate a large area and Le Boeuf (1974; see also Le Boeuf and Peterson 1969) has found that in some seasons a mere 4 per cent of the males are responsible for 85 per cent of the matings. This shows the importance of body size and strength in the males. They usually exhaust themselves in combat and few of them manage to stay dominant for more than one or two seasons. Much the same is true of the big red deer

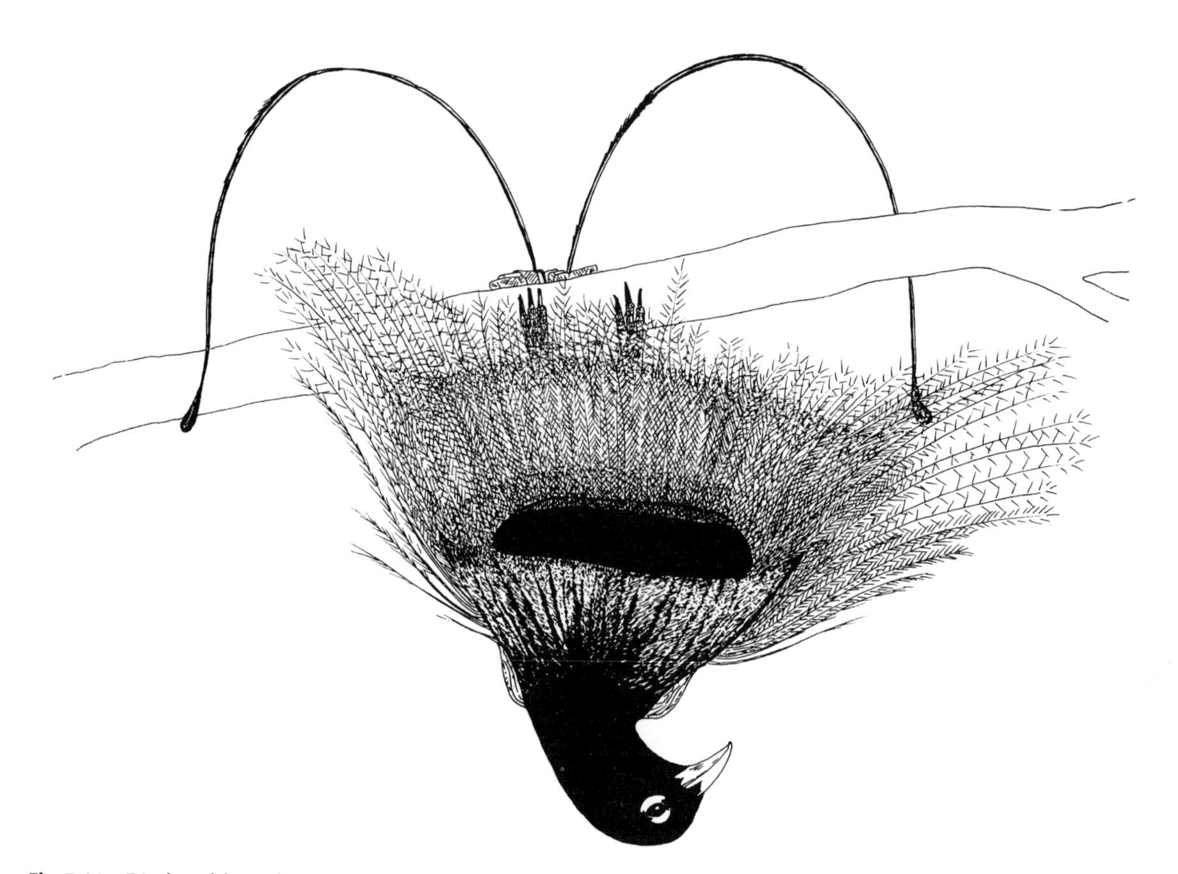

Fig. 7.11 Display of the male Prince Rubert's blue bird of paradise—a striking combination of colour, movement and sound. The bird begins by perching on a display site, calling. Then, carefully and slowly he rotates backwards and when he is hanging vertically downwards, he shakes himself and spreads out his iridescent blue plumage. With repeated movements from the hip joint, he dances—shaking his plumage up and down and uttering a soft monotonous song (drawing by Melissa Bateson).

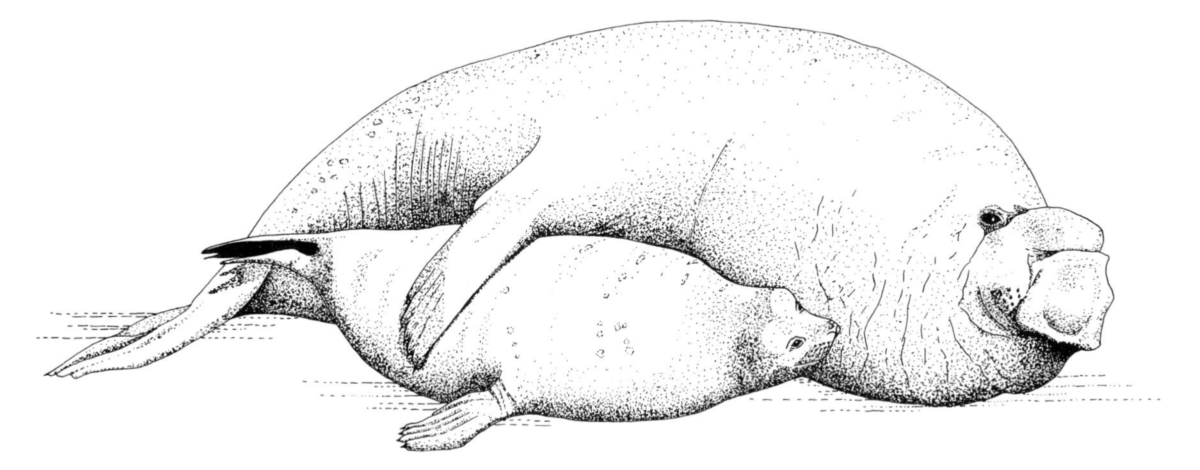

Fig. 7.12 Male (above) and female elephant seals, showing the great difference in body size between the sexes (drawing by Melissa Bateson).

stags (Clutton-Brock and Alton 1989). There is further discussion of the evolution of courtship displays and the advantages females may gain from responding to them in Chapter 3, page 60.

SOCIAL DOMINANCE

In mating systems where one sex provides most of the parental care (this will usually be the females), there will be competition between members of the opposite sex. Fighting between males is therefore a common feature of many mating systems. As we have seen, this may result in territorial defence with the males spacing themselves out from one another. However, not all species defend territories, perhaps because food or other resources are not distributed in such a way that they are easily defendable. In large ungulates such as buffalo and eland, the animals move together in large permanent herds that cover quite wide areas. The males do not exclude other males from any sort of territory but achieve mating rights through dominance (Jarman 1974). Some sort of dominance hierarchy, or peck-order, is in fact a common feature of vertebrate social organization, and is also seen in some invertebrates, such as crickets (Barnard and Burke 1979).

Schjelderup-Ebbe was one of the first to develop the concept of a social dominance hierarchy, using his work on flocks of hens (1935). He observed that a definite 'peck order' developed amongst a group of hens, one gradually emerging as the dominant in the sense that she could displace all others. Below her

	Y	B	V	R	G	YY	BB	VV	RR	GG	YB	BR
Y												
B	22											
V	8	29										
R	18	11	6									
G	11	21	11	12								
YY	30	7	6	21	8							
BB	10	12	3	8	15	30						
VV	12	17	27	6	3	19	8					
RR	17	26	12	11	10	17	3	13				
GG	6	16	7	26	8	6	12	26	6			
YB	11	7	2	17	12	13	11	18	8	21		
BR	21	6	16	3	15	8	12	20	12	6	27	

Fig. 7.13 A perfect linear hierarchy established within a group of twelve hens. Each bird is marked on its legs by colour rings, whose initials identify it. The number of times each bird pecked another flock member is given in the vertical colums (e.g. Y pecked B 22 times and pecked V 8 times) whilst the number of pecks received from another hen is given in the horizontal rows (e.g. VV received 19 pecks from YY and 9 pecks from BB). Note that no bird was ever seen to peck an individual above it in rank: hierarchies as perfect as this are probably rare in nature (from Guhl 1956).

was a second-ranking bird who could dominate all except the top bird, and so on down the group until at the bottom was a bird displaced by every other in the flock. Figure 7.13 shows one example of a peck-order of this type—a linear hierarchy as it is some-times called. The hierarchy develops as the birds dispute and it involves a good deal of fighting in the early stages as the birds test each other out. But once it is established, subordinates usually defer without question to the approach of a more dominant bird. Chickens are somewhat unusual in often having clear linear hierarchies. In other species, particularly among primates, the situation may be more complex, so that A may dominate B and B may dominate C but A does not necessarily dominate C.

It will be apparent by now that social organization in animals cannot be rigidly categorized into 'territorial' or 'hierarchical' and that mating systems and other social interactions take many different forms, and are affected by a number of factors. For many vertebrates, the changing seasons involve a change in their social organization. Male chaffinches tend to dominate females in winter flocks and displace them at feeding sites. But this situation is reversed in spring, when females tend to displace males as the flocks disperse to set up territories. Chickens, studied in a truly feral state on an island off Queensland, Australia, by McBride *et al.* (1959), alternate between a territorial system in the breeding season and a more hierarchical flock structure during the winter. The dominant male sets up a territory in the spring with a number of females in his flock. During the winter, after the young birds of the year had returned to the flocks, McBride *et al.* found that the alpha male and his harem moved about over a home range, with members of subordinate males staying at the periphery of the group, often moving between the home ranges of different alpha males. The alpha male led his flock in every sense. It was he who initiated all movements of the group, particularly across open ground, and his posture was normally more alert than those of his females. He was the first to give alarm calls and even approached predators —feral cats—while the others took cover.

Seasonal changes in behaviour are just as marked in some mammals. The classic study *A Herd of Red Deer* by Fraser Darling (1935) and the more recent studies on the Hebridean island of Rum by Clutton-Brock and Albon (1989) give us a very full picture of the social life of this species. Outside the breeding season, males and females live apart. The males live in loose bachelor herds in which there are consistent linear dominance hierarchies related to body size. High ranking stags are able to displace lower ranked stags from good food patches and often do so by lowering their antlers. The females, on the other hand, live in herds that include young animals of both sexes. Threats and displacing other animals from food are much less common among the hinds, even in winter. In early April, the antlers of the stags are shed and immediately male aggression takes on a new form. The males rise on their hind legs and 'box' with their hooves. New antlers begin growing but remain soft and tender in 'velvet' that is not shed until August. Now the stags become increasingly aggressive. Fighting with the new antlers increases until about mid-September, when the male herds break up. The males then go off singly to favoured display areas and begin roaring and gathering hinds who are coming into oestrous, and the 'rut' begins. Oestrus females are attracted to displaying stags, who defend their group of females rather than any particular area. The rutting season lasts only a few weeks and soon both sexes return to their respective herds for the winter.

In many species, of course, social behaviour does not change as dramatically from season to season as this. Animals that cooperate to hunt for prey and to defend themselves often stay together throughout the year. Wolves, for example, hunt in packs and maintain a very stable structure based on an extended family unit. There is usually one dominant male leader, but several other adult males may be included. Many primates also tend to retain a uniform social structure throughout the year. Since their social organization is also characterized by great complexity and subtlety of relationships between the individuals, we will conclude this chapter by considering their social behaviour in more detail.

PRIMATE SOCIAL ORGANIZATION

Studies on the primates have accelerated more rapidly than those on the social behaviour of any other type of animal. Workers from a number of disciplines, zoologists, psychologists, anthropologists and sociologists have converged on this group, both for their intrinsic interest and in the hope that they will provide information that is relevant to speculations on the origins of human societies. The great emphasis in the recent primate work has been on studies of natural communities in the field. Certain primates, notably the rhesus monkey, have been familiar laboratory animals for some time, but it is generally accepted that studies on captive communities are inadequate by themselves.

Here we can attempt only a short survey of the primate work. There are two journals that include many behavioural studies, *Primates* and *Folia Primatologia*, and a large number of books and symposium volumes. For an introduction to this range the reader is referred to Napier and Napier (1970); Rowell (1972); Deag (1980); Hinde (1983); Jolly (1985); Smuts *et al.* (1987); and Dunbar (1988).

Primates live in a wide variety of habitats. We tend to think of them as tropical animals but they were more widely distributed in the recent past and two of the macaques, the Barbary 'ape' of the Atlas mountains and the Japanese macaque live in areas where snow and frost are regular every winter. The majority of primates are arboreal, some of them, like the spider monkeys of South America and the colobus monkeys of Africa, exclusively so. However, a number have returned to ground living, such as the baboons and the patas monkey of Africa, whilst chimpanzees and gorillas also spend a lot of time on the ground. In general, the more arboreal primates are fruit and leaf eaters, the ground dwellers tend to be more omnivorous and include insects and perhaps small vertebrates in their diet, although a very high proportion of their food remains grass, seeds, bulbs, etc. Although the gorilla is almost exclusively vegetarian, chimpanzees and baboons will eat meat if they get the opportunity; the former have been seen to hunt and kill monkeys for food.

The living primates exhibit a tremendous range of morphological types, from the primitive lemurs that retain a long muzzle, a moist nose and claws on one of their hind toes—little modified from the ancestral primate types—to the monkeys, the great apes and humans (Fig. 7.14). Throughout this series we can observe certain trends; the enlargement of the brain, the development of the grasping hand and, in contrast to many other mammals, the great reliance on full colour vision as a dominant sense for exploration and communication.

It seems certain that from very early in their history the majority of primates were social animals, moving around in groups whose organization was stable. Jolly (1966) makes this point from her study of lemurs; here in these primitive primates we already find small mixed troops (12–20 individuals), which include several adult males and several breeding females—a very typical primate grouping. There is a dominance hierarchy within the troop (and in some lemurs females rank more highly than males) but it remains as a permanent, cohesive unit. Lemurs have group territories within their mixed woodland habitat the boundaries of which are often remarkably stable. They are marked by scent in some species and are defended by calling, which is usually sufficient to cause the neighbouring troop to retreat without further threat or fighting.

Within the troop there are frequent minor disputes but serious fights are rare outside the breeding season. This is very brief in most lemurs—at most 2 weeks—and it is at this time that subordinate males seriously challenge the older dominant ones. Apart from this short period of strife, much of lemur social life is characterized by non-aggressive interactions between individuals—indeed it is pedantic to avoid use of the term 'friendly'. There is always close contact between a mother and her infant, who clings continuously to her at first and is carried around everywhere. As it grows older other adults approach and play with the infant, as they also play with each other. Lemurs have thick, dense fur and groom frequently. Mothers groom their infants and adults frequently groom each other—this being one of the most common types of friendly contact between individuals.

In the behaviour of lemurs we can detect most of the elements that characterize all primate societies, although there are many variations on the theme. One of the most obvious types of variation concerns group size, from the almost solitary orang-utan, through the single family groups of gibbons, to the more common multimale, multifemale groups typical of howler monkeys, vervets, macaques and most baboons up to the veritable herds of gelada baboons and mandrills, which may number several hundred. Other obvious variations in primate social life concern the social structure within the groups, the type of mating system, the extent of territoriality and the nature of interchange between groups as animals become sexually mature.

A good deal of modern primatological work has been truly sociobiological in its approach. Several workers have tried to develop a general theory to explain the wide variation in primate social structure along the lines of Jarman's (1974) study on antelope that we have already discussed.

The primate counterparts of the territorial antelopes such as the dik-diks, are the marmosets and

Fig. 7.14 Some diverse representatives of the Primates. (a) A tree shrew—probably best regarded as a specialized insectivore rather than a primate, but certainly resembling the stock from which the early primates arose; (b) the tarsius of the Philippines; (c) the ring-tailed lemur of Madagascar; (d) a South American capuchin monkey; (e) the pig-tailed macaque of S.E. Asia; and two of the apes, closest relatives to humans (f) the gibbon and (g) the chimpanzee (after Le Gros Clark 1965). Not drawn to scale.

tamarins. In many ways, these have a very similar social structure—usually one, or at the most, three adults of each sex in the group together with their young. Females give birth to twins and males help with the care of infants (Goldizen 1987). The primate equivalents of the large antelopes are the baboons, but here we find a considerable diversity of social organization even among closely related species and the general theory begins to break down. Plains-living baboons (e.g. *Papio cynocephalus*) have multimale groups and no persistent male–female bonding, whereas drier country hamadryas baboons (*P. hamadryas*) have several females that are more or less permanently bonded to one male and bands that are made up of a number of such 'harem groups', which move and forage together. Evidently we have to look rather further for factors in the environment and life history of the different species to account for their diverse organization.

The usual approach has been to gather data from as many species as possible and to compare them for a wide range of ecological, morphological and behavioural factors. For example, the nature of the food supply, the degree of pressure from predators, body size, the extent of sexual dimorphism, the number of mature males in the group and group size—all of these and more—may be analysed to look for regular associations.

As food supply is unlikely ever to be permanently abundant, we have to try to explain why almost all the primates live in groups where the food in an area must be shared among several mouths. We have already discussed how living in groups may compensate individuals for having to share food by increasing the efficiency of searching. Groups can also cooperate to defend good food patches. Thus, increased feeding efficiency might favour group life. Alternatively, since most primates are fairly small animals and, alone, are very vulnerable to large predators such as the big cats, it may be that the extra protection provided in groups is the main factor. Not only is there increased vigilance but several large males combining for defence may be more than a big cat dare take on—this is certainly the case with baboon groups defending themselves against leopards.

The arguments derived from the comparisons of the ecological and behavioural factors outlined above are quite complex and we cannot go into them here. There are challenging reviews by Van Schaik (1983), Ridley (1986) and Wrangham (1987) and no complete agreement on why primate organization is so diverse. Obviously, we must expect the balance of factors leading to particular types of social organization to vary and—as with most biological systems—there will never be one solitary factor operating in isolation. Here we are more interested in the remarkable behavioural phenomena exhibited by the primates and must discuss some key examples.

Communication and social dominance in primates

We may begin by recognizing the richness and complexity of primate social life. One cannot fail to be impressed and fascinated by the high level of inter-communication that goes on between the members of a primate group. Each individual is constantly responsive to the posture, movements, gestures and calls of others. We have already referred to some remarkable examples of such responses in Chapter 3. This constant high level of attention to others is dramatic and Chance (1967, 1976) has suggested that the social structure of a monkey group is best regarded as a structure of attention itself, particularly as it relates to social rank, which we shall discuss below.

It is necessary to be somewhat cautious in our emphasis on the complexity of primate groups. We must not automatically assume that primate societies are more organized than those of, say, ungulates or carnivores. Because we ourselves are primates we find it much easier to identify elements in their communication system, in particular the mobility of their faces and the way they watch one another's faces for information on mood and intentions. Nevertheless, there are, on the most objective criteria, good grounds for the assumption. First, primates have an extended period of infancy, to which is coupled considerable longevity. The larger primates commonly live for 20 or 30 years and this means that a young primate grows up to takes its place in a group where—literally—everybody knows everybody else from long experience in their company.

Secondly, there can be no doubt that primates are amongst the most intelligent of animals, with a high learning ability and a correspondingly high flexibility in their behavioural response to a changing social situation (see Chapter 6 for further discussion of their learning). Sustaining this is a large and highly developed brain. We may be dealing here with a close coupling between behaviour and evolution. Humphrey (1976) suggests that the complex demands of social life itself formed one of the main selective factors for the growth of brain size in primates. This evolving brain, in turn, offered still more flexibility and complexity and so a mutual evolutionary relationship was established. This relationship reaches its most extreme manifestation in the human brain, with its capacity to mediate speech and all the richness of social interaction this allows.

We have already discussed the concept of social rank in connection with other species such as chickens. In primates, too, dominance and subordination are important types of relationship to others. Dominance always involves the threat of physical displacement or attack, although this may rarely be necessary once rank is established. However, it is most important to remember that relationships within primate groups are as much characterized by positive interactions as negative ones. There are many friendly contacts between animals, as when they move and rest together, invite grooming or offer to groom another. Mutual grooming, which we first mentioned in the lemurs, is very important as a placatory gesture in primates (Fig. 7.15). A dominant animal will often 'allow' itself to be groomed by a subordinate following a brief threat to which the subordinate has deferred. Sexual presentation as an appeasement gesture (see p. 83) is very common in baboons and chimpanzees and is made by males or females towards a dominant animal who threatens, or even if the subordinate wants to pass close to the dominant one (Fig. 7.16).

The pattern of grooming relationships in a group is often a good measure of its detailed structure. It will show us which animals associate together and is a good index of the cohesive forces that maintain the group as a real social unit. Figure 7.17, taken from Sade's work (1965) on rhesus macaques, illustrates just one small subunit of a group—that which hinges around a single old female. The other individuals recorded are her children and grandchildren. A very high proportion of all their grooming is to their

Fig. 7.15 Friendly grooming in the Barbary macaque. The juvenile male on the right is about 3 years old, he is grooming a subadult male aged about 4 and a half, who ranks above him in the social hierarchy. (Photo by John Deag.)

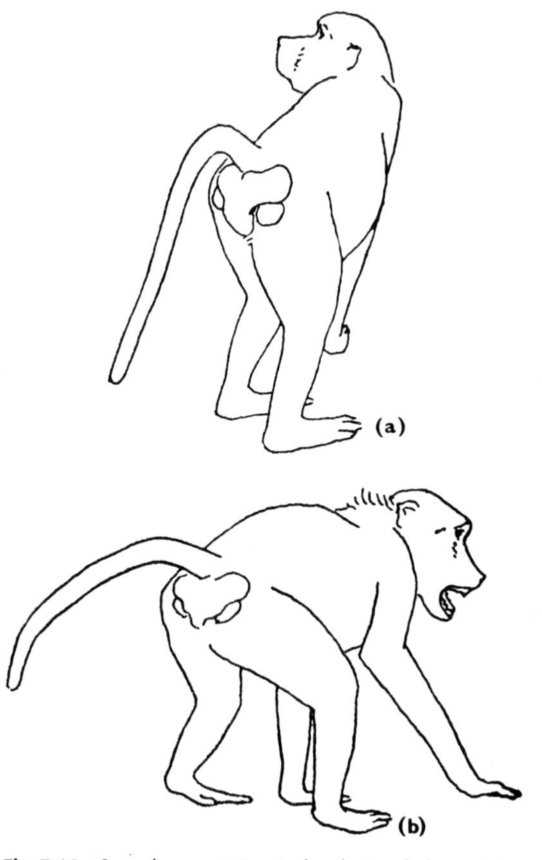

Fig. 7.16 Sexual presentation in the chacma baboon. (a) 'Genuine' presentation by an oestrus female; she characteristically looks back over her shoulder towards the male; (b) appeasement presentation towards a threatening dominant animal; the general stance is the same but the facial expression is one of fear (after Bolwig 1959).

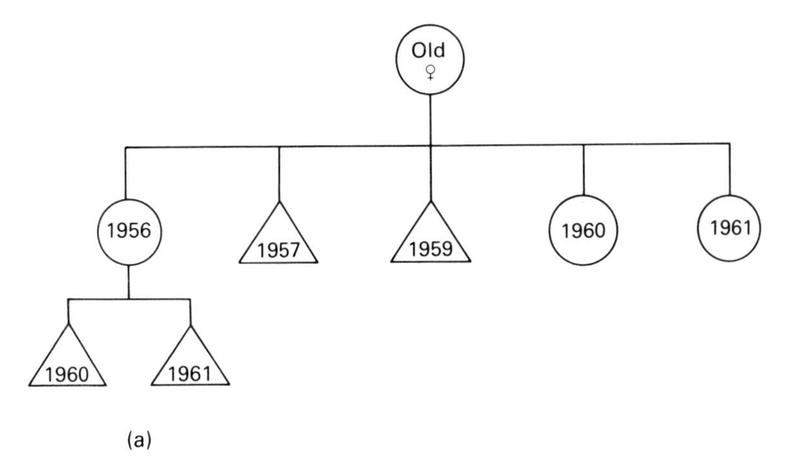

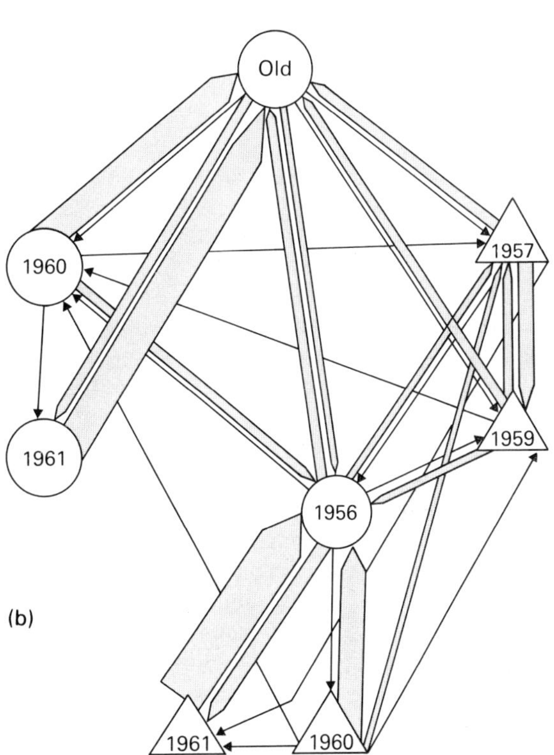

Fig. 7.17 (a) The genealogy of one matrilineal group of rhesus macaques from a free-ranging troop studied by Sade. Circles represent females, triangles males and dates of birth are inscribed. The age of the old female was not known. (b) A sociogram illustrating the grooming relationships within this group during 1961. Arrows point from groomer to groomed and their thickness represents the proportion of total grooming time spent grooming that animal. The thinnest arrows represent 1–10 per cent of grooming time, and that from the youngest male (born 1961) to his mother (1956) represents 100 per cent. Most animals within this group spent at least half their grooming time within it. The old female herself has, of course, more grooming relationships outside it. Mothers tend to groom their youngest offspring the most. Siblings, particularly the two brothers of 1957 and 1959, also groom each other a great deal (modified from Sade 1965).

mother or siblings—this genealogical unit really acts as a social unit too, and associates together. Only the old female herself spends much of her time grooming other individuals outside this group. Sade found that, as animals matured or changed their status, the grooming patterns also changed. Yet it remains true that in many primate societies the matriarchal group is much the most constant.

Figure 7.17b represents one kind of 'sociogram' —a diagram that illustrates relationships. Many different types of sociogram can be constructed (they are well reviewed in Hinde 1974, Chapter 23) based upon various behavioural measures, depending on the type of relationship one wishes to study. A dominance hierarchy, such as that shown in Figure 7.18, is one familiar example based upon agonistic interactions. Certainly the role of aggression in primate societies has attracted a great deal of attention. As may be imagined, the non-human primates have been used as evidence by both sides in the

AM1 > AM2 > AM3 > AF1 > SM1 > SM2 > AF2

SF1 > JM1

> JM1 > AF3 > JM2 > AF4 > SF1 > SF2 > AF5

SF3 JF1 IF2

> = > = > JF2 > JF3 > IM1 > = > BF1 > BM2

AF6 JM3 IF3

Fig. 7.18 The agonistic hierarchy recorded by Deag in a troop of wild Barbary macaques in the Middle Atlas Mountains of Morocco. It is based upon the distribution of threat and avoidance between pairs of individuals. The hierarchy is reasonably linear, especially at the upper end, where adult males are dominant over all other animals. Amongst the younger animals there is less strict hierarchy and this reflects changing status as animals grow up. Baby animals are not much involved in the hierarchical system and rarely give or receive threats. There is one clear exception to linearity higher up. Subadult female 1 threatened juvenile male 1 and he avoided her. A, adult; J, juvenile; I, infant; B, baby; M, male; F, female (from Deag 1977).

dispute about the nature of human aggression, which was discussed in Chapter 4. Primates vary greatly in the degree of fighting that is observed both within and between groups. Even within a single species there may be considerable variations; langur (*Presbytis entellus*) populations in northern India are far more peaceful than those in the south. The population densities are much greater in the areas where the aggressive groups live and this is probably one factor involved. Fighting may occur in crowded zoo colonies, which rarely give sufficient space for subordinate animals to keep out of the way of more dominant ones. Density, and the accompanying stress are obviously crucial factors, but it is unlikely that all primate species will respond to them in the same way, some may be inherently more aggressive than others.

However it is achieved, the stable situation in almost all primate groups can be interpreted in terms of a ranked hierarchy that determines, to a greater or lesser extent, how the individuals behave. Perhaps the best and most neutral definition of rank is that the behaviour of a high ranking or dominant animal is not limited by other individuals, whilst that of a subordinate is so limited. The limitations imposed by rank are diverse. For instance, having high rank in a group might determine access to food, preferred resting places and to females. Thus, when they come into oestrus, female baboons may move close to a dominant male and form a temporary 'consort relationship' with him.

A number of primate workers have criticized the way in which the concept of dominance has been used. Rowell (1974) quotes examples of the imprecise definition of dominance and points out that the different types of supposed advantages conferred by high rank—access to food, to oestrus females, etc. —may not always be correlated. She suggests that hierarchies, where they exist, are as much hierarchies of submission as of aggression. Certainly, it has often been observed that while rank is initially determined by threat and fighting, once established it is maintained as much by the deference of subordinate animals as by any display of threat by the dominants. Rowell suggests that dominance hierarchies are not common in wild primate groups. The fact that they arc almost invariably described from captive groups is, she claims, a reflection of the crowding and unnatural stress to which the latter are always subjected.

There can be little doubt that many of Rowell's criticisms are justified. As mentioned above, even in the largest enclosures, primate colonies in zoos are overcrowded by natural standards. When obvious dominance hierarchies have been observed in the wild they could sometimes result from human interference. For example, the rigid hierarchies described for Japanese monkey troops are probably influenced by their particular situation. The detailed work on these monkeys has been greatly helped by the success of the Japanese workers in getting them to visit artificial stations where food is regularly provided (Frisch 1968). Some troops stay close to these stations and providing food in one spot is likely to reinforce dominance structure. Goodall (1968) found just the same thing when she provided caches of bananas for her chimpanzees. The food acts as a focus that all animals are trying to approach, the dominant males sit at the best places and the rest of the group tends to space out according to rank. The hierarchy is less obtrusive when food has to be searched for individually.

Despite these valid criticisms, it remains the case that dominance hierarchies can be observed in undisturbed wild primates. Deag's (1977) study of a troop of Barbary macaques revealed the hierarchy illustrated in Figure 7.18, which is remarkably linear for the adults. He discusses how such a regular system of rank might arise from all individuals behaving to their own best advantage. The cohesion of the troop is vital for everyone's survival and some predictability about social interactions will be advantageous not only to the top animals but to low ranking ones too. Fighting is reduced, and probably stress also, because subordinate animals can keep clear of others to whom they will predictably lose in competitive encounters. Ranks change with time and subordinates will come to take over higher ranking positions as previously dominant animals become older.

Some of the most interesting complexities of primate social behaviour are revealed by studies of how ranks are acquired and changed. Linear hierarchies are by no means universal and, particularly at the top end where the adult males of a group are usually competing, we may find shifting triangular relationships. One male may be dominant over either of the next two individually, but these latter may frequently gang up to form an alliance and displace him. Subsequently he, in turn, may try to break up this pair by making conciliatory approaches to one of them and so on.

Such relationships are nowhere better described than in der Waal's fascinating account of his long term studies on chimpanzees in the Arnhem Zoo, which he called *Chimpanzee Politics* (1982). There have also been detailed studies of alliances in rhesus monkeys of the free-ranging colony on Cayo Santiago Island (off Puerto Rico) and in African vervet monkeys (Colvin 1983; Cheney 1983a). Alliances that form, break up and reform produce a shifting, dynamic pattern of social life that greatly modifies the nature and the effects of ranking.

The acquisition of rank is also a complex process, differing between the sexes and relating to their very different life histories. In many species of monkey, males never breed in the group into which they were born. The stability of primate groups might lead to a potential problem with inbreeding but this is avoided with the transfer of newly mature males between groups. By contrast, in these same species the females invariably remain in their natal troop and in close association with their mother and sisters. The sociogram of Figure 7.17b illustrates the close relationship between a mother and her children. As the males grow up they become independent of her and their rank amongst the other members of the group is largely determined by their size. There is a rise in disputing as young males become sexually mature and this is the most common age for them to leave (Cheney 1983b).

With females, on the other hand, it is the rank of their mother that often determines their position in the group. With remarkable regularity, in many cases, daughters slot in below their mothers in reverse order of age, i.e. the youngest daughter, once beyond infancy, ranks just below her mother and above her sisters. This is often because the mother supports her the most persistently (Gouzoules and Gouzoules 1987; Datta 1988). It makes good evolutionary sense to act this way because, if she survives the dangers of infancy, a young daughter has a longer potential reproductive life ahead of her than an older one. It will pay mothers, on average, to apportion their support accordingly (see Gadgil 1982, for a full discussion of this idea).

Most of the examples we have been discussing have been drawn from those primates whose social organization involves several adult males and adult females living together with no permanent male–female bonds. This is perhaps the most common type of primate society but we have already mentioned others. There are societies in which male–female bonds are more permanent, based upon 'one-male units', which consist of a single adult male, several females and their offspring. This is bound to leave a surplus of adult males and these usually form all-male bands whose members occasionally challenge those with harems. This type of organization is found in the plains-living patas monkey, the gelada and the hamadryas baboon. In the latter two, numbers of one-male units are associated together, particularly in the gelada, where hundreds of animals move around together in a loose herd structure (Dunbar 1979). The hamadryas baboon lives in semidesert regions of north east Africa, and the units move together during the day, although they spread out to comb the area for food. They come together as a troop at night to sleep on cliffs, which provide protection from predators.

The hamadryas baboon has been the subject of some remarkable field experiments by Kummer and his colleagues. Their results provide an excellent demonstration of the subtlety and flexibility of primate behaviour. Within the one-male units the male closely herds his three or four females; they move with him at all times and are not allowed to stray. Unattached females are rapidly acquired by males, so it is the more striking that there is scarcely any poaching of females between units. This remains the case even if one male is completely dominant over another in terms of access to food or resting places—the subordinate's females remain his own.

Kummer and his colleagues (Kummer *et al.* 1974; Bachmann and Kummer 1980) set up some cages of wild-caught baboons. If two males were kept together and an unattached female was introduced, they would threaten and sometimes fight each other but eventually the female became attached to one male—usually, of course, the dominant one. However, it was possible first to cage the subordinate male alone along with the female, although in full sight of and very close to the dominant one. This gave the former two a chance to interact and, depending on his attractiveness, the female would pair, more or less strongly with the subordinate male. Now the three animals would be put all together. The dominant would attempt to attract the female if she was not strongly paired to the subordinate, but if she was paired he made no attempt whatsoever. Further, such a male introduced to an established pair appears ill at ease and tends to sit with his head turned away from them. There is complete social inhibition of poaching between males of a troop.

The herding of females into a close unit is accomplished by threat. If, as she forages, a female strays too far from the male he first threatens her and, if she does not return to his side, chases and bites her. Customarily females return to the male at once when he 'stares' at them—a low intensity threat gesture. The yellow baboon is a close relative and in fact hybridizes with hamadryas baboons. Females in yellow baboon troops do not associate with particular males save at their brief oestrus period when they form a temporary consort relationship with a dominant male. For most of the time they move freely and associated with many different members of the troop.

Kummer (1968) transplanted a few yellow baboon females into a troop of hamadryas. Very quickly males began to herd them into their own units. At first the females wandered away and, if they responded at all to a male's threat, it was to flee. This of course produced the opposite effect to that which they intended. The male instantly pursued them and drove them back to his harem. Kummer found that within a few hours the yellow baboon females had learnt their harsh lesson and stayed close to their male.

This result forces us to recognize as a real possibility that some of the variation in social organization that we observe between primate species is of cultural and not genetic original. The young hamadryas baboon grows up in a one-male unit, the young yellow baboon in the less restricted social climate of the larger group. They may have similar potential and yet develop quite differently. In Chapter 5 we discussed the cultural transmission of minor items of behaviour, such as food-washing habits, but clearly with the extended infancy of primates, their long life and great learning capacities, cultural effects can extend much beyond this.

Because it is possible to study hybrids between the two baboon species—natural hybridization occurs in some areas of Ethiopia—we know that the behavioural differences are likely, in part, to be genetically based (Sugawara 1979). This is not surprising; in the primates above all other groups, we would expect both selection acting on inherited variation, and cultural transmission of acquired adaptations in behaviour, to be operating together. Cultural factors will play a major part in primate evolution, not just directly but because they will alter the background upon which genetic factors will operate.

From birth through its long infancy, both sorts of factor will be interacting to shape the developing behaviour of a young monkey or ape. Certainly, quite apart from any light they may throw on the origins of human social life (Hinde 1983, 1987) the primates offer marvellous material for the study of behavioural development. The experimental approach, which might include cross-fostering infants and cultural transplantation, has scarcely begun. If we allow the primates room to live in the natural habitats to which they have adapted, a combination of captive studies and field work is certain to yield fascinating and important results.

REFERENCES

Alatalo, R.V., Carlson, A., Lundberg, A. and Ulfstrand, S. (1981) The conflict between male polygamy and female monogamy; the case of the pied flycatcher, *Ficedula hypoleuca*. *Am. Nat.* **117**: 738–53.

Alatalo, R.V., Lundberg, A. and Glynn, C. (1986) Female pied flycatchers choose territory quality not male characteristics. *Nature (Lond.)* **323**: 152–3.

Alcock, J. (1984) *Animal Behavior. An Evolutionary Approach* (3rd ed.). Sinauer, Sunderland, MA.

Allee, W.C. (1938) *The Social Life of Animals*. Norton, New York.

Altmann, J. (1980) *Baboon Mothers and Infants*. Harvard University Press, Cambridge, MA.

Altmann, S.A. (1962) A field study of the sociobiology of rhesus monkeys, *Macaca mulatta*. *Ann. N.Y. Acad. Sci.* **102**: 338–435.

Andersson, M. (1982a) Sexual selection, natural selection and quality advertisement. *Biol. J. Linn. Soc.* **17**: 375–93.

Andersson, M. (1982b) Female choice selects for extreme tail length in a widowbird. *Nature (Lond.)* **299**: 818–20.

Andjus, R.K., Knöpfelmacher, F., Russell, R.W. and Smith, A.U. (1955) Effects of hypothermia on behaviour. *Nature (Lond)* **176**: 1015–16.

Andrew, R.J. (1985) The temporal structure of memory formation. *Perspectives in Ethology* **6**: 219–59. Plenum Press, New York.

Andrew, R. J. (ed.) (1991) *Neural and Behavioural Plasticity: the Use of The Domestic Chick as a Model*. Oxford University Press, Oxford and London.

Archer, J. (1973) Tests for emotionality in rats and mice: a review. *Anim. Behav.* **21**: 205–35.

Archer, J. (1979) *Animals Under Stress*. Studies in Biology **108**. Edward Arnold, London.

Archer, J. (1988) *The Behavioural Biology of Aggression*. Cambridge University Press, Cambridge.

Aschoff, J. (1981) Biological Rhythms. *Handbook of Behavioural Neurobiology* **4**. Plenum, New York.

Aschoff, J. (1989) Temporal orientation: circadian clocks in animals and humans. *Anim. Behav.* **37**: 881–96.

Bachmann, C. and Kummer, H. (1980) Male assessment of female choice in Hamadryas baboons. *Behav. Ecol. Sociobiol.* **6**: 315–21.

Baerends, G.P. (1976) An evaluation of the conflict hypothesis as an explanatory principle for the evolution of displays. In Baerends, G.P., Beer, C. and Manning, A. (eds), *Function and Evolution in Behaviour*, pp. 187–227. Clarendon Press, Oxford.

Baerends, G.P. (1976) The functional organisation of behaviour. *Anim. Behav.* **24**: 726–38.

Baker, M.C. and Cunningham, M.A. (1985) The biology of bird song dialects. *Behav. Brain Sci.* **8**: 85–133.

Baldwin, J.D. and Baldwin, J.L. (1973) The role of play in social organisation: comparative observations of squirrel monkeys (*Saimiri*). *Primates* **14**: 369–81.

Bakker, T.C.M. (1986) Aggressiveness in sticklebacks (*Gasterosteus aculeatus*): a behaviour genetic study. *Behaviour* **98**: 1–44.

Baptista, L.F. and Morton, M.L. (1988) Song learning in montane white-crowned sparrows: from whom and when. *Anim. Behav.* **36**: 1753–64.

Baptista, L.F. and Petrinovitch, L. (1984) Social interaction, sensitive phases and the song template hypothesis in the white-crowned sparrow. *Anim. Behav.* **32**: 172–81.

Baptista, L.F. and Petrinovitch, L. (1986) Song development in the white-crowned sparrow: social factors and sex differences. *Anim. Behav.* **35**: 1359–71.

Barfield, R.J. (1971) Gonadotrophic hormone secretion in the female ring dove in response to visual and auditory stimulation in the male. *J. Endocrinol.* **49**: 305–10.

Barnard, C.J. and Burk, T.E. (1979) Dominance hierarchies and the evolution of 'individual recognition'. *J. Theor. Biol.* **81**: 65–73.

Barnett, S.A. (1963) *A Study in Behaviour*. Methuen, London.

Barnett, S.A. (1964) The concept of stress. *Viewpoints in Biology* **3**: pp. 170–218. Butterworth, London.

Bastock, M. and Manning, A. (1955) The courtship of *Drosophila melanogaster*. *Behaviour* **8**: 85–111.

Bateson, P.P.G. (1976) Specificity and the origins of behaviour. *Advances in the Study of Behavior* **6**: 1–20.

Bateson, P.P.G. (1978) Early experience and sexual preferences. In Hutchison, J.B. (ed.), *Biological Determinants of Sexual Behaviour*, pp. 29–53. John Wiley, London.

Bateson, P.P.G. (1979) How do sensitive periods arise and what are they for? *Anim. Behav.* **27**: 470–86.

Bateson, P.P.G. (1980) Optimal outbreeding and the development of sexual preferences in Japanese quail. *Z. Tierpsychol.* **53**: 231–44.

Bateson, P.P.G. and Reese, E.P. (1969) Reinforcing properties of conspicuous objects before imprinting has occurred. *Psychon. Sci.* **10**: 379–80.

Beach, F.A. (1942) Analysis of the stimuli adequate to elicit mating behavior in the sexually inexperienced male rat. *J. Comp. Psychol.* **33**: 163–207.

Beach, F.A. and Levinson, G. (1950) Effects of androgen on the glans penis and mating behavior of castrated male rats. *J. Exp. Zool.* **114**: 159–71.

Bekoff, A. and Kauer, J.A. (1984) Neural control of hatching: role of neck position in turning on hatching leg movements in post-hatching chicks. *J. Comp. Physiol.* **145**: 497–504.

Bennet-Clark, H.C. (1970) The mechanism and efficiency of sound production in mole crickets. *J. Exp. Biol.* **52**: 619–52.

Bennet-Clark, H.C. and Ewing, A. (1969) Pulse interval as a critical parameter in the courtship song of *Drosophila melanogaster*. *Anim. Behav.* **17**: 755–9.

Bentley, D.R. and Hoy, R.R. (1970) Postembryonic development of adult motor patterns in crickets: a neural analysis. *Science (NY)* **170**: 1409–11.

Bentley, D.R. and Hoy, R.R. (1972) Genetic control of the neural network generating cricket (*Teleogryllus*) calls. *Anim. Behav.* **20**: 478–92.

Bentley, D. and Keshishian, H. (1982) Pathfinding by peripheral pioneer neurons in grasshoppers. *Science (NY)* **218**: 1082–8.

Berthold, P. and Querner, U. (1981) Genetic basis of migratory behaviour in European warblers. *Science (NY)* **212**: 77–9.

Birke, L.I.A. (1989) How do gender differences in behavior develop? A reanalysis of the role of early experience. *Perspectives in Ethology* **8**: 215–42.

Birkhead, T.R., Atkin, L. and Møller, A.P. (1987) Copulation behaviour of birds. *Behaviour* **101**: 101–133.

Blair, W.F. (1955) Mating call and stage of speciation in the *Microhyla olivacea–M. carolinensis* complex. *Evolution* **9**: 469–80.

Blakemore, C. and Cooper, G.F. (1970) Development of the brain depends on the visual environment. *Nature (Lond.)* **228**: 477–8.

Blakemore, R.P. (1975) Magnetotactic bacteria. *Science (NY)* **190**: 377–9.

Blass, E.M. and Epstein, A.N. (1971) A lateral preoptic osmosensitive zone for thirst in the rat. *J. Comp. Physiol. Psychol.* **76**: 378–94.

Blüm, V. and Fiedler, K. (1965) Hormonal control of reproductive behavior in some cichlid fish. *Gen. Comp. Endocrinol.* **5**: 186–96.

Blurton-Jones, N. (1959) Experiments on the causation of the threat postures of Canada geese. *Rep. Severn Wildfowl Trust 1960*. 46–52.

Boakes, R. (1984) *From Darwin to Behaviorism*. Cambridge University Press, Cambridge.

Bolles, R.C. (1970) Species-specific defense reactions and avoidance learning. *Psychol. Rev.* **77**: 32–48.

Bolles, R.C. (1975) *Theory of Motivation*. Harper and Row, New York.

Bonner, J.T. (1980) *The Evolution of Culture in Animals*. Princeton University Press. Princeton.

Bossema, I. and Burger, R.R. (1980) Communication during monocular and binocular looking in European jays *Garrulus g. glandarius*. *Behaviour* **74**: 274–83.

Bottjer, S.W. and Arnold, A.P. (1984) Hormones and structural plasticity in the adult brain. *Trends in Neurosci.* **7(5)**: 168–71.

Bowlby, J. (1969) *Attachment and Loss. I: Attachment*. The Hogarth Press, London.

Bowlby, J. (1973) *Attachment and Loss. II: Separation*. The Hogarth Press, London.

Boyd, H. and Fabricius, E. (1965) Observations on the incidence of following of visual and auditory stimuli in naïve mallard ducklings (*Anas platyrhynchos*). *Behaviour* **25**: 1–15.

Brain, P.F. (1975) What does individual housing mean to a mouse? *Life Sci.* **16**: 187–200.

Brain, P.F. and Jones, S.E. (1982) Assessing the unitary nature of 'aggression' by using strains of laboratory mice. *Aggressive Behavior* **8**: 108–11.

Brain, P.F., Benton, D. and Boulton, J.C. (1978) A comparison of agonistic behavior in individually-housed male mice and those co-habiting with females. *Aggressive Behavior* **4**: 201–6.

Breland, K. and Breland, M. (1961) The misbehavior of organisms. *Am. Psychol.* **16**: 681–4.

Brockmann, H.J. (1980) The control of nest depth in a digger wasp (*Sphex ichneumoneus L.*). *Anim. Behav.* **28**: 426–45.

Brockmann, H.J., Grafen, A. and Dawkins, R. (1979) Evolutionarily stable nesting strategy in a digger wasp. *J. Theor. Biol.* **77**: 473–96.

Brodsky, L.M., Ankney, C.D. and Dennis, D.G. (1988) The influence of male dominance on social interactions in black ducks and mallards. *Anim. Behav.* **36**: 1371–8.

Bronson, F.H. (1979) The reproductive biology of the house mouse. *Q. Rev. Biol.* **54**: 265–99.

Broom, D.M. (1986) Indicators of poor welfare. *Br. Vet. J.* **142**: 524–6.

Brown, C.R. (1986) Cliff swallow colonies as information centres. *Science (NY)* **234**: 83–5.

Brown, J.L. (1969) The buffer effect and productivity in tit populations. *Am. Nat.* **103**: 347–54.

Brown, R.E. and Macdonald, D.W. (eds) (1985) *Social Odours in Mammals*, Vols 1 and 2. Clarendon Press, Oxford.

Bruner, J. and Kennedy, D. (1970) Habituation: occurrence at a neuromuscular junction. *Science (NY)* **169**: 92–4.

Bruner, J. and Tauc, L. (1966) Habituation at the synaptic level in *Aplysia*. *Nature (Lond.)* **210**: 37–9.

Bruner, J.S., Jolly, A. and Sylva, K. (eds) (1976) *Play: its role in development and evolution*. Penguin, London.

Bull, H.O. (1957) Conditioned responses. In Brown, M.E. (ed.), *The Physiology of Fishes* (2 vols.), pp. 211–28. Academic Press, New York.

Bull, J.J. (1980) Sex determination in reptiles. *Q. Rev. Biol.* **55**: 3–20.

Burghardt, G.M. (1970) Defining 'communication'. In Johnston, J.W., Moulton, D.G. and Turk, A. (eds), *Advances in Chemoreception*, vol. I, pp. 5–18. Appleton Century Crofts, New York.

Burkhardt, D. (1989) UV vision: a bird's view of feathers. *J. Comp. Physiol. A.* **164**: 787–96.

Butler, C.G. (1974) *The World of the Honeybee* (3rd ed.). Collins, London.

Byrne, R. and Whiten, A. (eds) (1988) *Machiavellian Intelligence*. Clarendon Press, Oxford.

Cade, W.H. (1981) Alternative male strategies: genetic differences in crickets. *Science (NY)* **212**: 563–4.

Camhi, J.M. (1980) The escape system of the cockroach. *Sci. Am.* **243(6)**: 144–56.

Camhi, J.M. (1984) *Neuroethology: nerve cells and the natural behavior of animals*. Sinauer Associates, Sunderland, MA.

Camhi, J.M., Tom, W. and Volman, S. (1978) The escape behavior of the cockroach *Periplaneta americana*. II Detection of natural predators by air displacement. *J. Comp. Physiol.* **128**: 203–12.

Campbell, B. and Lack, E. (eds) (1985) *A Dictionary of Birds*. T. & A.D. Poyser, Berkhamstead.

Cannon, W.B. (1974) *The Wisdom of the Body*. Kegan Paul, Trench, Trubner and Co., London.

Capranica, R.R. and Moffat, A.J.M. (1975) Neurobehavioral correlates of sound communication in anurans. In Ewert, J.P., Capranica, R.R. and Ingle, D. (eds), *Advances in Vertebrate Neuroethology*, pp. 701–30. Plenum Press, New York.

Carthy, J.D. and Ebling, F.J. (eds) (1964) *The Natural History of Aggression*. Academic Press, London.

Carter, C.S. and Marr, J.H. (1970) Olfactory imprinting and age variables in the guinea-pig, *Cavia porcellus*. *Anim. Behav.* **18**: 238–44.

Catchpole, C. (1980) *Vocal Communication in Birds*. Studies in Biology No. 115. Edward Arnold, London.

Catchpole, C., Leisler, B. and Winkler, H. (1985)

The evolution of polygyny in the Great Reed Warbler, *Acrocephalus arundinaceus*: a possible case of deception. *Behav. Ecol. Sociobiol.* **16**: 285–91.

Chamove, A.S. (1980) Non-genetic induction of acquired levels of aggression. *J. Abn. Psychol.* **89**: 469–88.

Chance, E.P. (1940) *The Truth about the Cuckoo.* Country Life, London.

Chance, M.R.A. (1967) Attention structure as the basis of primate rank order. *Man* **2**: 503–18.

Chance, M.R.A. (1976) The organisation of attention in groups. In Von Cranach, M. (ed.), *Methods of Inference from Animal to Human Behavior.* Mouton/Aldine, The Hague.

Cheney, D.L. (1983a) Extrafamiliar alliances among vervet monkeys. In Hinde, R.A. (ed.), *Primate Social Relationships*, pp. 278–89. Blackwell Scientific, Oxford.

Cheney, D.L. (1983b) Proximate and ultimate factors related to the distribution of male migration. In Hinde, R.A. (ed.), *Primate Social Relationships*, pp. 241–9. Blackwell Scientific Publications, Oxford.

Cheney, D.L. and Seyfarth, R.M. (1982) How vervet monkeys perceive their grunts: field playback experiments. *Anim. Behav.* **30**: 739–51.

Clark, R.B. (1960a) Habituation of the polychaete *Nereis* to sudden stimuli. I. General properties of the habituation process. *Anim. Behav.* **8**: 82–91.

Clark, R.B. (1960b) Habituation of the polychaete *Nereis* to sudden stimuli. II. Biological significance of habituation. *Anim. Behav.* **8**: 92–103.

Clarke, A.M. and Clarke, A.D.B. (1976) *Early Experience: myth and evidence.* Open Books, London.

Clutton-Brock, T.H. and Albon, S.D. (1979) The roaring of red deer and the evolution of honest advertisement. *Behaviour* **69**: 145–70.

Clutton-Brock, T.H. and Albon, S.D. (1989) *Red Deer in the Highlands.* BSP Professional Books, Oxford.

Clutton-Brock, T.H., Guiness, F.E. and Albon, S.D. (1982) *Red Deer: behaviour and ecology of two sexes.* Edinburgh University Press, Edinburgh.

Clutton-Brock, T.H. and Harvey, P.H. (1984) Comparative approaches to investigating adaptation. In Krebs, J.R. and Davies, N.B. (eds), *Behavioural Ecology*, pp. 7–29. Blackwell Scientific Publications, Oxford.

Cohen-Salmon, C., Carlier, M., Robertoux, M., Jouhaneau, J., Semal, C. and Paillette, M. (1985) Differences in patterns of pup care in mice. V. Pup ultrasonic emissions and pup care behavior. *Physiol. and Behav.* **35**: 167–74.

Colgan, P. (1989) *Animal Motivation.* Chapman and Hall, London.

Collias, N.E. and Joos, M. (1953) The spectrographic analysis of sound signals of the domestic fowl. *Behaviour* **5**: 175–88.

Colvin, J. (1983) Description of sitting and peer relationships among immature male rhesus monkeys. In: Hinde, R.A. (ed.), *Primate Social Relationships*, pp. 20–7. Blackwell Scientific Publications, Oxford.

Cooper, K.W. (1957) Biology of eumenine wasps: V. Digital communication in wasps. *J. Exp. Zool.* **134**: 469–514.

Cowie, R.J. (1977) Optimal foraging in great tits, *Parus major. Nature (Lond.)* **268**: 137–9.

Cowie, R.J., Krebs, J.R. and Sherry, D.F. (1981) Food storage by marsh tits. *Anim. Behav.* **29**: 1252–9.

Crane, J. (1957) Basic patterns of display in fiddler crabs (*Ocypodidae* genus *Uca*). *Zoologica* **42**: 69–82.

Crook, J.H. (1965) The adaptive significance of avian social organisations. *Symp. Zool. Soc. Lond.* **14**: 181–218.

Cruze, W.W. (1935) Maturation and learning in chicks. *J. Comp. Psychol.* **19**: 371–409.

Cullen, E. (1957) Adaptations in the kittiwake to cliff-nesting. *Ibis* **99**: 275–302.

Cullen, E. (1960) Experiment of the effect of social isolation on reproductive behaviour in the three-spined stickleback. *Anim. Behav.* **8**: 235.

Curio, E., Ernst, V. and Vieth, W. (1978) Cultural transmission of enemy recognition. *Science (NY)* **202**: 899–901.

Dagan, D. and Volman, S. (1982) Sensory basis for directional wind detection in first instar cockroaches, *Periplaneta americana. J. Comp. Physiol.* **147**: 471–8.

Dane, B., Walcott, C. and Drury, W.H. (1959) The form and duration of the display actions of the 'goldeneye' *Bucephala clangula. Behaviour* **14**: 265–81.

Darling, F.F. (1935) *A Herd of Red Deer.* Oxford University Press, London.

Darwin, C. (1871) *The Descent of Man and Selection in Relation to Sex.* Murray, London.

Darwin, C. (1965) *The Expression of the Emotions in*

Man and Animals. The University of Chicago Press, London.

Datta, S. (1988) The acquisition of dominance among free-ranging rhesus monkey siblings. *Anim. Behav.* **36**: 754–72.

Davies, N.B. (1989) Sexual conflict and the polygyny threshold. *Anim. Behav.* **38**: 226–34.

Davies, N.B. and Brooke, M. de L. (1989a) An experimental study of co-evolution between the cuckoo *Cuculus canorus* and its hosts. I. Host egg discrimination. *J. Anim. Ecol.* **58**: 207–24.

Davies, N.B. and Brooke, M. de L. (1989b) An experimental study of co-evolution between the cuckoo *Cuculus canorus* and its hosts. II. Host egg markings, chick discrimination and general discussion. *J. Anim. Ecol.* **58**: 225–36.

Davies, N.B. and Houston, A.I. (1981) Owners and satellites: the economics of territory defence in the pied wagtail, *Motacilla alba*. *J. Anim. Ecol.* **50**: 157–80.

Davis, W.J., Mpitsos, G.J., Pinneo, J.M. and Rau, J.L. (1977) Modification of the behavioural hierarchy of *Pleurobranchaea*. I. Satiation and feeding mechanisms. *J. Comp. Physiol.* **117**: 99–125.

Dawkins, M.S. (1980) *Animal Suffering: the science of animal welfare*. Chapman and Hall, London.

Dawkins, M.S. and Guilford, T. (1991) The corruption of honest signalling. *Anim. Behav.* **41**: 865–74.

Dawkins, R. (1980) Good strategy or evolutionarily stable strategy? In Barlow, G.W. and Silverberg, J. (eds), *Sociobiology: beyond nature/nurture*, pp. 331–67. Westview Press, Boulder, Colorado.

Dawkins, R. (1986) *The Blind Watchmaker*. Longmans, Harlow.

Dawkins, R. (1989) *The Selfish Gene* (2nd ed.). Oxford University Press, Oxford.

Dawkins, R. and Dawkins, M. (1974) Decisions and the uncertainty of behaviour. *Behaviour* **45**: 83–103.

Dawkins, R. and Krebs, J.R. (1978) Animal signals: information or manipulation? In Krebs, J.R. and Davies, N.B. (eds), *Behavioural Ecology: an evolutionary approach* (1st ed.), pp. 282–309. Blackwell Scientific Publications, Oxford.

Deag, J.M. (1977) Aggression and submission in monkey societies. *Anim. Behav.* **25**: 465–74.

Deag, J.M. (1980) *Social Behaviour of Animals*. Studies in Biology No. 118. Edward Arnold, London.

Deag, J. and Crook, J.H. (1971) Social behaviour and agonistic buffering in the wild barbary macaque *Macaca sylvana* L. *Folia primat.* **15**: 183–200.

Delius, J.D. and Hollard, V.D. (1987) Orientation invariance of shape recognition in forebrain-lesioned pigeons. *Behav. Brain Res.* **23**: 251–9.

Denenberg, V.H. (1964) Critical periods, stimulus input and emotional reactivity: a theory of infantile stimulation. *Psychol. Rev.* **71**: 335–51.

Dethier, V.G. (1953) Summation and inhibition following contra-lateral stimulation of the tarsal chemoreceptors of the blowfly. *Biol. Bull.* **105**: 257–68.

Dethier, V.G. (1957) Communication by insects: physiology of dancing. *Science (NY)* **125**: 331–6.

Dethier, V. and Stellar, E. (1970) *Animal Behavior* (3rd ed.). Prentice-Hall, Englewood Cliffs, NJ.

Dewsbury, D.A. (1984) *Comparative Psychology in the Twentieth Century*. Hutchinson Ross, Stroudsburg, PA.

Diamond, J.N., Karasov, W.H., Phan, D. and Carpenter, F.L. (1986) Hummingbird digestive physiology: a determinant of foraging bout frequency. *Nature (Lond.)* **320**: 62–3.

Dickinson, A. (1980) *Contemporary Animal Learning Theory*. Cambridge University Press, Cambridge.

Dilger, W.C. (1962) The behavior of lovebirds. *Sci. Am.* **206(1)**: 88–98.

Douglas-Hamilton, I. and Douglas-Hamilton, O. (1975) *Among the Elephants*. Collins and Harvill, London.

D'Souza, F. and Martin, R.D. (1974) Maternal behaviour and the effects of stress in tree-shrews. *Nature (Lond.)* **251**: 309–11.

D'Udine, B. and Alleva, E. (1983) Early experience and sexual preferences in rodents. In Bateson, P.P.G. (ed.), *Mate Choice*, pp. 311–27. Cambridge University Press, Cambridge.

Dunbar, R.I.M. (1979) Structure of gelada baboon reproductive units. I. Stability of social relationships. *Behaviour* **69**: 72–87.

Dunbar, R. (1988) *Primate Social Systems*. Croom Helm, London.

Duncan, I.J.H. and Kite, V.G. (1987) Some investigations into motivation in the domestic fowl. *Appl. Anim. Behav. Sci.* **18**: 387–8.

Duvall, W.D., Müller-Schwarze, D. and Silverstein, R.M. (1986) *Chemical Signals in Vertebrates, 4*. Plenum Press, New York.

Ehrhardt, A.A. and Baker, S.W. (1974) Fetal

androgens, human central nervous system differentiation, and behavior sex differences. In Friedman, R.C., Richart, R.M. and Van de Wiele, R.L. (eds), *Sex Differences in Behavior*, pp. 33–57. Wiley, New York.

Eibl-Eibesfeldt, I. (1970) *Ethology. The Biology of Behavior.* Holt, Rinehart and Winston, New York.

Elgar, M. (1986a) The establishment of foraging flocks in house sparrows: risk of predation and daily temperature. *Behav. Ecol. Sociobiol.* **19**: 433–8.

Elgar, M. (1986b) House sparrows establish foraging flocks by giving chirrup calls if the resources are divisible. *Anim. Behav.* **34**: 169–74.

Elgar, M. (1987) Food intake rate and resource availability: flocking decisions in house sparrows. *Anim. Behav.* **35**: 1168–76.

Elgar, M. (1989) Predator vigilance and group size among mammals and birds: a critical review of the evidence. *Biol. Rev.* **64**: 13–34.

Elsner, N. (1981) Developmental aspects of insect neuroethology. In Immmelmann, K., Barlow, G.W., Petrinovitch, L. and Main, M. (eds), *Behavioural Development*, pp. 474–90. Cambridge University Press, Cambridge.

Elwood, R.W. (ed.), (1983) *Parental Behaviour of Rodents.* John Wiley, Chichester.

Emlen, S. (1975) The stellar orientation system of a migratory bird. *Sci. Am.* **233(2)**: 102.

Emlen, S. and Oring, L.W. (1977) Ecology, sexual selection and the evolution of mating systems. *Science (NY)* **197**: 215–23.

Emlen, S., Demong, N.J. and Emlen, D.J. (1989) Experimental induction of infanticide in female wattled jacanas. *Auk* **106**: 1–7.

Erber, J. (1975) The dynamics of learning in the honey bee. *J. Comp. Physiol.* **99**: 231–55.

Erber, J. (1981) Neural correlates of learning in the honey bee. *Trends in Neurosciences*, **4**, 270–3.

Erickson, C.J. (1985) Mrs. Harvey's parrot and some problems of socioendocrine response. In Bateson, P.P.G. and Klopfer (eds), *Perspectives in Ethology*, vol. 6, pp. 261–86.

Esch, H., Esch, I. and Kerr, W.E. (1965) Sound: an element common to communication of stingless bees and to dances of honeybees. *Science (NY)* **149**: 320–1.

Evans, S.M. (1968) *Studies in Invertebrate Behaviour.* Heinemann Education, London.

Ewert, J.P. (1980) *Neuroethology—An Introduction to the Neurophysiological Fundamentals of Behavior.* Springer-Verlag, Berlin.

Ewert, J.P. and Traud, R. (1979) Releasing stimuli for anti-predator behaviour in the common toad *Bufo bufo L. Behaviour* **68**: 170–80.

Fagen, R.M. (1981) *Animal Play Behaviour.* Oxford University Press, Oxford.

Ferguson, M.M. and Noakes, D.L. (1982) Genetics of social behavior in charrs (*Salvelinus spp*). *Anim. Behav.* **30**: 128–34.

Fitzgibbon, C.D. and Fanshawe, J.H. (1988) Stotting in Thomson's gazelles: an honest signal of condition. *Behav. Ecol. Sociobiol.* **23**: 69–74.

Fitzsimmons, J.T. (1972) Thirst. *Physiol. Rev.* **52**: 468–561.

Fitzsimmons, J.T. and Le Magnen, J. (1969) Eating as a regulatory control of drinking. *J. Comp. Physiol. Psychol.* **67**: 273–83.

Fletcher, D.J.C. and Michener, C.D. (eds) (1989) *Kin Recognition in Animals.* John Wiley, New York.

Fraser, A.F. and Broom, D.M. (1990) *Farm Animal Behaviour and Welfare.* Baillere Tindall, London and Saunders, New York.

Frisch, J.E. (1968) Individual behaviour and intertroop variability in Japanese macaques. In Jay, P. (ed.), *Primates: studies in adaptation and variability*, pp. 243–52. Holt, Rinehart and Winston, New York.

Frisch, K. von (1967) *The Dance Language and Orientation of Bees.* The Belknap Press of Harvard University Press, Cambridge, MA.

Fuller, J.L. and Thompson, W.R. (1978) *Foundations of Behavior Genetics.* C.V. Mosby, St. Louis, MO.

Gadgil, M. (1982) Changes with age in the strategy of social behavior. *Perspectives in Ethology* **5**: 489–502.

Galef, B.F. (1976) Social transmission of acquired behavior: a discussion of tradition and social learning in vertebrates. *Adv. Study Behav.* **6**: 77–100.

Garcia, J. and Koelling, R. (1966) Relation of cue to consequence in avoidance learning. *Psychonom. Sci.* **4**: 123–4.

Gardner, B.T. (1964) Hunger and sequential responses in the hunting behaviour of salticid spiders. *J. Comp. Physiol. Psychol.* **58**: 167–73.

Gellerman, L.W. (1933) Form discrimination in chimpanzees and two-year-old children: I Form (triangularity) per se. *J. Genet. Psychol.* **42**: 3–27.

Gerhardt, H.C. (1974) The significance of some spectral features in mating call recognition in the green treefrog *Hyla cinerea*. *J. Exp. Biol.* **61**: 229–41.

Gibson, R.M. and Bradbury, J.W. (1985) Sexual selection in lekking sage grouse: phenotypic correlates of male mating success. *Behav. Ecol. Sociobiol.* **18**: 117–23.

Gill, F.B. and Wolf, L.L. (1975) Economics of feeding territoriality in the golden-winged sunbird. *Ecology* **56**: 333–45.

Goddard, G.V. (1986) Learning. A step nearer a neural substrate. *Nature (Lond.)* **319**: 721–2.

Goldizen, A.W. (1987) Tamarins and marmosets: communal care of offspring. In Smuts, B.B. *et al.* (eds), *Primate Societies*, pp. 34–43. University of Chicago Press, Chicago.

Goodall, J. van Lawick (1968) The behaviour of free-living chimpanzees in the Gombe Stream Reserve. *Anim. Behav. Monogr.* **1**: 161–311.

Gottlieb, G. (1971) *Development of Species Identification in Birds*. University of Chicago Press, Chicago.

Gottlieb, G. (1983) Development of species identification in ducklings: X. Perceptual specificity in the wood duck embryo requires sib stimulation for maintenance. *Dev. Psychobiol.* **16**: 323–33.

Göttmark, F. and Andersson, M. (1984) Colonial breeding reduces nest predation in the common gull. *Anim. Behav.* **32**: 485–92.

Göttmark, F., Winkler, D.W. and Andersson, M. (1986) Flock-feeding on fish schools increases individual success in gulls. *Nature (Lond.)* **319**: 589–91.

Gould, S.J. and Lewontin, R.C. (1979) The spandrels of San Marco and the Panglossian paradigm: a critique of the adaptationist programme. *Proc. Roy. Soc. Lond. B* **205**: 581–98.

Gould, J.L., Dyer, F.C. and Towne, W.F. (1985) Recent progress in understanding the honey bee dance language. *Fort. der Zool.* **31**: 141–61.

Gould, J.L. and Marler, P. (1987) Learning by instinct. *Sci. Am.* **256(1)**: 62–73.

Gouzoules, S. and Gouzoules, H. (1987) Kinship. In Smuts, B.B., *et al.* (eds), pp. 299–305.

Goy, R.W. and McEwen, B.S. (1980) *Sexual Differentiation of the Brain*. MIT Press, Cambridge, MA.

Grafen, A. (1984) Natural selection, kin selection and group selection. In Krebs, J.R. and Davies, N.B. (eds), *Behavioural Ecology: an evolutionary approach*, pp. 62–84. Blackwell Scientific Publications, Oxford.

Grafen, A. (1990) Biological signals as handicaps. *J. Theor. Biol.* **144**: 517–5.

Graves, H.B., Hable, C.P. and Jenkins, T.H. (1985) Sexual selection in *Gallus*. Effects of morphology and dominance on female spatial behaviour. *Behav. Proc.* **11**: 189–97.

Greenberg, L. (1979) Genetic component of kin recognition in primitively social bees. *Science (NY)* **206**: 1095–7.

Greenough, W. T. and Bailey, C. H. (1988). The anatomy of memory: convergence of results across a diversity of tests. *Trends in Neurosciences* **11**: 142–7.

Griffin, D.R. (1958) *Listening in the Dark*. Yale University Press, New Haven, CT.

Griffin, D.R. (1984) *Animal Thinking*. Harvard University Press, Cambridge, MA.

Groebel, J. and Hinde, R.A. (eds) (1989) *Aggression and War*. Cambridge University Press, Cambridge.

Grossman, S.P. (1967) *A Textbook of Physiological Psychology*. John Wiley, New York.

Guiton, P.E. (1959) Socialization and imprinting in brown leghorn chicks. *Anim. Behav.* **7**: 26–34.

Gwadz, R. (1970) Monofactorial inheritance of early sexual receptivity in the mosquito, *Aëdes atropalpus*. *Anim. Behav.* **18**: 358–61.

Haartman, L. von (1969) Nest-site and evolution of polygamy in European passerine birds. *Orn. Fenn.* **46**: 1–12.

Hagedorn, M. and Heiligenberg, W. (1985) Court and spark: electric signals in the courtship of mating of gymnotid fish. *Anim. Behav.* **33**: 254–65.

Halliday, T.R. and Sweatman, H.P.A. (1976) To breathe or not to breathe: the newt's problem. *Anim. Behav.* **24**: 551–61.

Hamilton, W.D. (1964) The genetical evolution of social behaviour. I and II. *J. Theor. Biol.* **7**: 1–52.

Hamilton, W.D. (1971) Geometry for the selfish herd. *J. Theor. Biol.* **31**: 295–311.

Hamilton, W.D. and Zuk, M. (1984) Heritable true fitness and bright birds: a role for parasites? *Science (NY)* **218**: 384–7.

Hardy, A.C. (1965) *The Living Stream*. Collins, London.

Harlow, H.F. (1949) The formation of learning sets. *Psychol. Rev.* **56**: 51–65.

Harlow, H.F. and Harlow, M.K. (1965) The affectional systems. In Schrier, A.M., Harlow, H.F. and Stollnitz, F. (eds), *Behavior of Non-human Primates*, pp. 287–334. Academic Press, New York.

Harris, G.W. and Michael, R.P. (1964) The activation of sexual behaviour by hypothalamic implants of oestrogen. *J. Physiol. (Lond.)* **171**: 275–301.

Hay, D.A. (1985) *Essentials of Behaviour Genetics*. Blackwell, Oxford.

Heiligenberg, W. (1977) *Principles of Electrolocation and Jamming Avoidance in Electric Fish*. Springer Verlag, New York.

Heiligenberg, W. and Kramer, U. (1972) Aggressiveness as a function of external stimulation. *J. Comp. Physiol.* **77**: 332–40.

Herbers, J.M. (1981) Time resources and laziness in animals. *Oceologia* **49**: 252–62.

Herrnstein, R.J., Loveland, D.H. and Cable, C. (1976) Natural concepts in pigeons. *J. Exp. Psychol. Anim. Behav. Proc.* **2**: 285–302.

Hinde, R.A. (1954) Factors governing the changes in strength of a partially inborn response, as shown by the mobbing behaviour of the chaffinch (*Fringilla coelebs*), II. The waning of the response. *Proc. Roy. Soc. B.* **142**: 331–58.

Hinde, R.A. (1960) Factors governing the changes in strength of a partially inborn response, as shown by the mobbing behaviour of the chaffinch (*Fringilla coelebs*), III. The interaction of short-term and long-term incremental and decremental effects. *Proc. Roy. Soc. B* **153**: 398–420.

Hinde, R.A. (1970) *Animal Behaviour* (2nd ed.). McGraw-Hill, New York.

Hinde, R.A. (1974) *Biological Bases of Human Social Behaviour*. McGraw-Hill, New York.

Hinde, R.A. (1977) Mother-infant separation and the nature of inter-individual relationships. Experiments with rhesus monkeys. *Proc. Roy. Soc. Lond. B.* **196**: 29–50.

Hinde, R.A. (1983) The human species. In Hinde, R.A. (ed.), *Primate Social Relationships*, pp. 334–9. Blackwell, Oxford.

Hinde, R.A. (1987) Can nonhuman primates help us to understand human behavior? In Smuts, B.B. *et al.* (eds), *Primate Societies*, pp. 413–20. Chicago University Press, Chicago, IL.

Hinde, R.A. and Fisher, J. (1952) Further observations on the opening of milk bottles by birds. *Brit. Birds* **44**: 393–6.

Hinde, R.A. and Rowell, T.E. (1962) Communication by postures and facial expressions in the rhesus monkey (*Macaca mulatta*). *Proc. Zool. Soc. Lond.* **138**: 1–21.

Hinde, R.A. and Steel, E. (1966) Integration of the reproductive behaviour of female canaries. *Symp. Soc. Exp. Biol.* **20**: 401–26.

Hinde, R.A. and Stevenson-Hinde, J. (eds) (1973) *Constraints on Learning*. Academic Press, London.

Hodos, W. and Campbell, C.B.G. (1969) Scala Naturae: why there is no theory in comparative psychology. *Psychol. Rev.* **76**: 337–50.

Hogan, J.A. (1967) Fighting and reinforcement in Siamese fighting fish *Betta splendens*. *J. Comp. Physiol. Psychol.* **64**: 356–9.

Hollard, V.D. and Delius, J.D. (1982) Rotational invariance in visual pattern recognition by pigeons and humans. *Science (NY)* **218**: 804–6.

Holldobler, B. (1971) Communication between ants and their guests. *Sci. Am.* **224(3)**: 86–93.

Holmes, W.G. and Sherman, P.W. (1982) The ontogeny of kin recognition in two species of ground squirrels. *Am. Zool.* **22**: 491–517.

Holst, E. von and von St Paul, U. (1963) On the functional organisation of drives. *Anim. Behav.* **11**: 1–20.

Hoogland, W.G. and Sherman, P.W. (1976) Advantages and disadvantages of Bank Swallow coloniality. *Ecol. Monogr.* **46**: 33–58.

Horn, G. (1985) *Memory, Imprinting and the Brain*. Clarendon Press, Oxford.

Horn, G. (1990) Neural bases of recognition memory investigated through an analysis of imprinting. *Phil. Trans. R. Soc. Lond. B.* **329**: 133–42.

Hotta, Y. and Benzer, S. (1976) Mapping of behaviour in *Drosophila* mosaics. *Nature (Lond.)* **240**: 527–35.

Houston, A.I. and McNamara, J. (1988) A framework for the functional analysis of behavior. *Behav. Brain Sci.* **11**: 117–63.

Humphrey, N.K. (1976) The social function of intellect. In Bateson, P.P.G. and Hinde, R.A. (eds), *Growing Points in Ethology*, pp. 303–17. Cambridge University Press, Cambridge.

Hunsaker, D. (1962) Ethological isolating mechanisms in the *Sceloporus torquatus* group of lizards. *Evolution* **16**: 62–74.

Hunter, M.L. (1980) Microhabitat selection for singing and other behavior in great tits, *Parus major*: some visual and acoustical considerations. *Anim. Behav.* **28**: 468–75.

Huntingford, F.A. (1976) The relationship between anti-predator behaviour and aggression among conspecifics in the three-spined stickleback, *Gasterosteus aculeatus*. *Anim. Behav.* **24**: 245–60.

Huntingford, F.A. and Turner, A. (1987) *Animal Conflict*. Chapman and Hall, London.

Hurst, J. (1989) The complex network of olfactory communication in populations of wild house mice *Mus domesticus*. Rutty: Markings and investigation within family groups. *Anim. Behav.* **37**: 705–25.

Hutchison, J.B. (1976) Hypothalamic mechanisms of sexual behaviour with special reference to birds. *Adv. Study Behav.* **6**: 159–200.

Huxley, J.S. (1914) The courtship habits of the great crested grebe *Podiceps cristatus*. *Proc. Zool. Soc. Lond.* **1914(2)**: 491–562.

Iersel, J.J.A. van (1953) An analysis of the parental behaviour of the male three-spined stickleback. *Behav. Suppl.* **3**: 1–159.

Immelmann, K. (1972) Sexual and other long-term aspects of imprinting in birds and other species. *Adv. Study Behav.* **4**: 147–74.

Immelmann, K. (1975) The evolutionary significance of early experience. In Baerends, G.P., Beer, C. and Manning, A. (eds), *Function and Evolution of Behaviour*, pp. 243–53. Clarendon Press, Oxford.

Janowitz, H.D. and Grossman, M.I. (1949) Some factors affecting the food intake of normal dogs and dogs with esophagotomy and gastric fistulas. *Am. J. Physiol.* **159**: 143–8.

Jarman, P.J. (1974) The social organisation of antelope in relation to ecology. *Behaviour* **48**: 215–55.

Jarvis, J.U.M. (1981) Eusociality in a mammal: cooperative breeding in naked mole rat colonies. *Science (NY)* **212**: 571–3.

Jerison, H.J. (1985) Animal intelligence and encephalization. *Phil. Trans. Roy. Soc. Lond. B* **308**: 21–35.

Joffe, J.M. (1965) Genotype and prenatal and premating stress interact to affect adult behavior in rats. *Science (NY)* **150**: 1844–5.

Jolly, A. (1966) *Lemur Behavior*. Chicago University Press, Chicago, IL.

Jolly, A. (1985) *The Evolution of Primate Behavior* (2nd ed.). Macmillan Publishing Co., New York.

Kammer, A. and Rheuben, E. (1976) Adult motor pattern produced by moth pupae during development. *J. Exp. Biol.* **65**: 65–84.

Kandel, E.R. and Schwartz, J.H. (1982) Molecular biology of learning: Modulation of transmitter release. *Science (NY)* **218**: 433–43.

Kater, S.B. and Rowell, C.H.F. (1973) Integration of sensory and centrally programmed components in generation of cyclical feeding activity of *Helisoma trivolvis*. *J. Neurophysiol.* **36**: 142–55.

Kennedy, J.S. (1965) Coordination of successive activities in an aphid. Reciprocal effects of settling on flight. *J. Exp. Biol.* **43**: 489–509.

Kennedy, J.S. (1987) Animal motivation: the beginning of the end? In Chapman, R.F., Bernays, E.A. and Stoffolano J.G. Jr (eds), *Perspectives in Chemoreceptors and Behaviour*, pp. 17–31. Springer Verlag, New York.

Klopfer, P.H. (1959) An analysis of learning in young Anatidae. *Ecology* **40**: 90–102.

Koehler, O. (1951) The ability of birds to 'count'. *Bull. Anim. Behav.* **9**: 41–5.

Köhler, W. (1927) *The Mentality of Apes* (2nd ed.). Kegan Paul, Trench, Trubner and Co., London.

Komisaruk, B.R. (1967) Effects of local brain implants of progesterone on reproductive behavior in ring doves. *J. Comp. Physiol. Psychol.* **64**: 219–24.

Komisaruk, B.R., Adler, N.T. and Hutchison, J. (1972) Genital sensory field: enlargement by oestrogen treatment in female rats. *Science (NY)* **178**: 1295–8.

Kommers, P.E. and Dhindsa, M.S. (1989) Influence of dominance and age on mate choice in black-billed magpies: an experimental study. *Anim. Behav.* **37**: 645–55.

Konishi, M. (1965) The rôle of auditory feedback on the control of vocalisation in the white-crowned sparrow. *Z. Tierpsychol.* **22**: 770–83.

Konorski, J. (1948) *Conditioned Reflexes and the Nervous System*. Cambridge University Press, Cambridge.

Kortlandt, A. (1940) Ein Übersicht der angeborenen Verhaltungsweisen des Mittel-Europäischen Kormorans (*Phalacrocorax carbo sinensis*), ihre Funktion, ontogenetische Entwicklung und phylogenetische Herkunft. *Arch. Néerl. Zool.* **4**: 401–42.

Krebs, J.R. and Davies, N.B. (1987) *An Introduction to Behavioural Ecology* (2nd ed.). Blackwell Scientific Publications, Oxford.

Krebs, J.R. and Dawkins, R. (1984) Animal signals: mind reading and manipulation. In Krebs, J.R. and Davies, N.B. (eds), *Behavioural Ecology: an evolutionary approach* (2nd ed.), pp. 380–402. Blackwell Scientific Publications, Oxford.

Krebs, J.R., MacRoberts, M.H. and Cullen, J.M. (1972) Flocking and feeding in the great tit, *Parus major*—an experimental study. *Ibis* **114**: 507–30.

Krebs, J.R., Sherry, D.F., Healy, S.D., Perry, V.H. and Vaccarino, A.L. (1989) Hippocampal specialization of food-storing birds. *Proc. Natl. Acad. Sci. USA* **86**: 1388–92.

Kroodsma, D.E. (1982) Learning and the ontogeny of sound signals in birds. In Kroodsma, D.E. and Miller, D.H. (eds), *Acoustic Communication in Birds*, pp. 1–23. Academic Press, New York.

Kruijt, J.P. (1964) Ontogeny of social behaviour in Burmese Red Junglefowl (*Gallus gallus spadiceus*). *Behaviour* **12**: 1–201.

Kruuk, H. (1972) *The Spotted Hyena*. University of Chicago Press, Chicago, IL.

Kummer, H. (1968) Two variations in the social organization of baboons. In Jay, P. (ed.), *Primates: studies in adaptation and variability*, pp. 293–312. Holt, Rinehart and Winston, New York and London.

Kummer, H., Gotz, W. and Angst, W. (1974) Triadic differentiation: an inhibitory process protecting pair bands in baboons. *Behaviour* **49**: 62–87.

Lack, D. (1943) *The Life of the Robin*. H.F. & G. Witherby, Ltd, London.

Lack, D. (1968) *Ecological Adaptations for Breeding in Birds*. Methuen, London.

Lagerspetz, K.M.J. and Lagerspetz, K.Y.H. (1971) Changes in the aggressiveness of mice resulting from selective breeding, learning and social isolation. *Scand. J. Psychol.* **12**: 241–8.

Lall, A.B., Seligar, B.H., Biggley, W. and Lloyd, J.E. (1980) Ecology of colors of firefly bioluminescence. *Science (NY)* **210**: 560–2.

Lashley, K.S. (1950) In search of the engram. *Symp. Soc. Exp. Biol.* **4**: 454–82.

Lazarus, J. (1979) The early warning function of flocking in birds: an experimental study with captive Quelea. *Anim. Behav.* **27**: 855–65.

Le Boeuf, B.J. (1974) Male–male competition and reproductive success in elephant seals. *Am. Zool.* **14**: 163–76.

Le Boeuf, B.J. and Peterson, R.S. (1969) Social status and mating activities in elephant seals. *Science (NY)* **163**: 91–3.

Lee, P. (1983) Play as a means for developing relationships. In Hinde, R.A. (ed.), *Primate Social Relationships*, pp. 82–9. Blackwell, Oxford.

Leger, D.W. and Owings, D.H. (1978) Responses to alarm calls by California ground squirrels: effect of call structure and maternal status. *Behav. Ecol. Sociobiol.* **3**: 177–86.

Le Gros Clark, W.E. (1960) *The Antecedents of Man*. Quadrangle Books, Chicago, IL.

Lehrman, D.S. (1964) The reproductive behavior of ring doves. *Sci. Am.* **211**: 48–54.

Lenington, S. and Egid, K. (1989) Environmental influences on the preferences of wild female house mice for males of differing t-complex genotypes. *Behavior Genetics* **19**: 257–66.

Lettvin, J.Y., Maturana, H.R., McCulloch, W.S. and Pitts, W.H. (1959) What the frog's eye tells the frog's brain. *Proc. Inst. Radio Engrs.* **47**: 1940–51.

Leuthold, W. (1966) Variations in territorial behaviour of Uganda Kob, *Adenota kob thomasi*. *Behaviour* **27**: 215–58.

Levine, R.B. (1986) Reorganization of the insect nervous system during metamorphosis. *Trends in Neurosci.* **9**: 315–19.

Levine, R.B. and Truman, J.W. (1985) Dendritic reorganisation of abdominal motoneurons during metamorphosis of the moth *Manduca sexta*. *J. Neurosci.* **5**: 2424–31.

Lindauer, M. (1961) *Communication Among Social Bees*. Harvard University Press, Cambridge, MA.

Lindauer, M. (1975) Evolutionary aspects of orientation and learning. In Baerends, G.P., Beer, C. and Manning, A. (eds), *Function and Evolution in Behaviour*, pp. 228–42. Clarendon Press, Oxford.

Lissman, H.W. (1963) Electric location by fishes. *Sci. Am.* **207**: 50–9.

Lloyd, J.E. (1965) Aggressive mimicry in *Photuris* firefly *femmes fatales*. *Science (NY)* **149**: 653–4.

Lloyd, J.E. (1975) Aggressive mimicry in *Photuris* fireflies: signal repertoires by *femmes fatales*. *Science (NY)* **187**: 452–3.

Lorenz, K.Z. (1937) The companion in the bird's world. *Auk* **54**: 245–73.

Lorenz, K.Z. (1941) Vergleichende Bewegungsstudien an Anatinen. *J. Orn. Lpz.* **89**: 194–293. (An English translation appeared in several parts, in Vols 57–59 of Avicultural Magazine.)

Lorenz, K. (1950) The comparative method in studying innate behaviour patterns. *Symp. Soc. Exp. Biol.* **4**: 221–68.

Lorenz, K.Z. (1952) *King Solomon's Ring*. Methuen, London.

Lorenz, K.Z. (1966) *Evolution and Modification of Behaviour*. Methuen, London.

Lorenz, K.Z. (1966) *On Aggression*. Methuen, London.

Luria, A.R. (1975) *The Mind of a Mnemonist*. Penguin, London.

McBride, G., Parer, I.P. and Foenander, F. (1959) The social organisation and behaviour of the feral domestic fowl. *Anim. Behav. Mongr.* **2**: 127–81.

Macdonald, D.W. (1986) A meerkat volunteers for guard duty so its comrades can live in peace. *Smithsonian.* April: 55–64.

McFarland, D.J. (1971) *Feedback Mechanisms in Animal Behaviour*. Academic Press, New York.

McFarland, D.J. and Houston, A.I. (1981) *Quantitative Ethology*. Pitman, Oxford.

McGaugh, J.L. (1989) Involvement of hormonal and neuromodulatory systems in the regulation of memory storage. *Ann. Rev. Neurosci.* **12**: 255–87.

McGill, T.E. (1965) *Readings in Animal Behavior.* Holt, Rinehart and Winston, New York.

McGrew, W.C. and Tutin, C.E.G. (1978) Evidence for a social custom in wild chimpanzees? *Man* **13**: 234–51.

McGrew, W.C., Tutin, C.E.G. and Baldwin, P.J. (1979) Chimpanzees, tools and termites: cross-cultural comparisons of Senegal, Tanzania and Rio Muni. *Man* **14**: 185–215.

Mackintosh, N.J. (1965) Discrimination learning in the octopus. *Anim. Behav. Suppl.* **1**: 129–34.

Mackintosh, N.J. (1983) General principles of learning. In Halliday, T.R. and Slater, P.J.B. (eds), *Animal Behaviour, Vol. 3 Genes, Development and Learning*, pp. 149–77. Blackwell Scientific, Oxford.

Mackintosh, N.J., Wilson, B. and Boakes, R.A. (1985) Differences in mechanisms of intelligence among vertebrates. *Phil. Trans. Roy. Soc. Lond. B* **308**: 53–65.

Macphail, E. (1985) Vertebrate intelligence: the null hypothesis. *Phil. Trans. Roy. Soc. Lond. B.* **308**: 37–51.

Macphail, E.M. (1987) The comparative psychology of intelligence. *Behav. Brain Sci.* **10**: 645–95.

Magnus, D.B.E. (1958) Experimental analysis of some overoptimal sign stimuli in the mating behaviour of the fritillary butterfly *Argynnis paphia* L. Lepidoptera Nymphalidae) *Proc. 10th Int. Congr. Ent. Montreal* **2**: 405–18.

Maier, N.R.F. and Schneirla, T.C. (1935) *Principles of Animal Psychology*. McGraw-Hill, New York.

Maier, N.R.F. (1932) Cortical destruction of the posterior part of the brain and its effects on reasoning in rats. *J. Comp. Neurol.* **56**: 179–214.

Manning, A. (1961) The effects of artificial selection for mating speed in *Drosophila melanogaster. Anim. Behav.* **9**: 82–92.

Marler, P. (1984) Song learning: innate species differences in the learning process. In Marler, P. and Terrace, H.S. (eds), *The Biology of Learning.* Springer-Verlag, New York.

Marler, P., Dufty, A. and Pickert, R.A. (1986) Vocal communication in the domestic chicken. I. Does a sender communicate information about the quality of a food referent to a receiver? *Anim. Behav.* **34**: 188–93.

Marler, P. and Tamura, M. (1964) Culturally transmitted patterns of vocal behavior in sparrows. *Science (NY)* **146**: 1483–6.

Martin, G.M. and Lett, B.T. (1985) Formation of associations of colored and flavoured food with induced sickness in five main species. *Behav. and Neural Biol.* **43**: 223–37.

Martin, P. and Caro, T.M. (1985) On the functions of play and its role in behavioural development. *Adv. Study Behav.* **15**: 59–103.

Martin, R.D. (1968) Reproduction and ontogeny in tree-shrews (*Tupaia belangeri*) with reference to their general behaviour and taxnoomic relationships. *Z. Tierpsychol.* **25**: 409–95.

Masserman, J.H. (1950) Experimental neuroses. *Sci. Am.* **182**(3): 38–43.

May, D.J. (1949) Studies on a community of willow warblers. *Ibis* **91**: 24–54.

Maynard Smith, J. (1964) Group selection and kin selection. *Nature (Lond.)* **201**: 1145–7.

Maynard Smith, J. (1982) *Evolution and the Theory of Games*. Cambridge University Press. Cambridge.

Maynard Smith, J. and Harper, D. (1988) The evolution of aggression: can selection generate variability? *Phil. Trans. Roy. Soc. Lond. B.* **319**: 557–70.

Maynard Smith, J. and Riechert, S.E. (1984) A conflicting-tendency model of spider agonistic behaviour: hybrid–pure population line comparisons. *Anim. Behav.* **32**: 564.

Meddis, R. (1983) The evolution of sleep. In Mayes, A. (ed.), *Sleep Mechanisms and Functions*, pp. 57–106. Van Nostrand, London.

Menzel, R. (1985) Learning in honey bees in an

ecological and behavioral context. In Holldobler, B. and Lindauer, M. (eds), *Experimental Behavioural Ecology and Sociobiology*, pp. 53–74. Gustav Fischer Verlag, Stuttgart and New York.

Menzel, R. and Erber, J. (1978) Learning and memory in bees. *Sci. Am.* **239**(1): 80–7.

Michelsen, A. (1989) Ein mechanisches Modell der tanzenden Honigbiene. *Biologie in unserer Zeit.* **19(4)**: 121–6.

Michelsen, A., Andersen, B.B., Kirchner, W.H. and Lindauer, M. (1989) Honeybees can be recruited by a mechanical model of a dancing bee. *Naturwissenschaften* **76**: 277–80.

Milinski, M. (1984) A predator's cost of overcoming the confusion effect of swarming prey. *Anim. Behav.* **32**: 1157–62.

Milinski, M. and Heller, R. (1978) Influence of a predator on the optimal foraging behaviour of sticklebacks (*Gasterosteus aculeatus*). *Nature (Lond.)* **275**: 642–4.

Miller, N.E. (1941) The frustration–aggression hypothesis. *Psychol. Rev.* **48**: 337–42.

Miller, N.E. (1956) Effects of drugs on motivation: the value of using a variety of measures. *Ann. N.Y. Acad. Sci.* **65**: 318–33.

Miller, N.E. (1957) Experiments on motivation. Studies combining psychological, physiological and pharmacological techniques. *Science (NY)* **216**: 1271–8.

Miller, N.E., Sampliner, R.I. and Woodrow, P. (1957) Thirst reducing effects of water by stomach fistula versus water by mouth, measured by both a consummatory and an instrumental response. *J. Comp. Physiol. Psychol.* **50**: 1–5.

Mishkin, M. and Appenzeller, T. (1987) The anatomy of memory. *Sci. Am.* **256(6)**: 62–71.

Mitchell, P. and Thompson, N. (eds) (1985) *Deception: perspectives on human and nonhuman deceit.* SUNY Press, New York.

Mock, D.W. and Fujioka, M. (1990) Monogamy and long-term pair bond in vertebrates. *Trends in Ecol. and Evol.* **5**: 39–43.

Mock, D.W., Lamey, T.C., Williams, C.F. and Pelletier, A. (1987) Flexibility in the development of heron sibling aggression: an intraspecific test of the prey-size hypothesis. *Anim. Behav.* **35**: 1386–93.

Moehlman, P.D. (1979) Jackal helpers and pup survival. *Nature (Lond.)* **277**: 382–3.

Møller, A. (1987) Social control of deception among status signalling House Sparrows. *Behav. Ecol. Sociobiol.* **20**: 307–11.

Møller, A. (1988) Female choice selects for male sexual tail ornaments in the monogamous swallow. *Nature (Lond.)* **332**: 640–2.

Moltz, H. and Stettner, L.J. (1961) The influence of patterned-light deprivation on the critical period for imprinting. *J. Comp. Physiol. Psychol.* **54**: 279–83.

Money, J. and Ehrhardt, A.A. (1973) *Man and Woman, Boy and Girl.* John Hopkins University Press, Baltimore.

Montagu, M.F.A. (1968) *Man and Aggression*. Oxford University Press, London.

Moore, B.R. (1973) The role of directed Pavlovian reactions in simple instrumental learning in the pigeon. In Hinde, R.A. and Stevenson-Hinde, J. (eds), *Constraints on Learning*, pp. 159–88. Academic Press, London.

Morgan, L. (1894) *An Introduction to Comparative Psychology*. Scott, London.

Morris, D. (1959) The comparative ethology of grassfinches (*Erythrurae*) and mannikins (*Amadinae*). *Proc. Zool. Soc. Lond.* **131**: 389–439.

Morris, R.G.M. (1983) Modelling amnesia and the study of memory in animals. *Trends in Neurosciences* **6**: 479–83.

Morris, R.G.M. (1989) Synaptic plasticity, neural architecture and forms of memory. In McGaugh, J.L., Weinberger, N.M. and Lynch, G. (eds), *Brain Organisation and Memory: cells, systems and circuits*. Oxford University Press, New York.

Morris, R.G.M., Garrud, P., Rawlins, J.N.P. and O'Keefe, J. (1982) Place navigation impaired in rats with hippocampal lesions. *Nature (Lond.)* **297**: 681–3.

Morton, D.S. (1975) Ecological sources of selection on avian sounds. *Am. Nat.* **109**: 17–34.

Moynihan, M. (1967) Comparative aspects of communication in New World primates. In Morris, D. (ed.), *Primate Ethology*, pp. 236–66. Weidenfeld and Nicholson, London.

Munn, C.A. (1986) Birds that 'cry wolf'. *Nature (Lond.)* **319**: 143–5.

Munn, N.L. (1950) *Handbook of Psychological Research on the Rat*. Houghton Mifflin, Boston.

Murphey, R.K. (1985) Competition and chemoaffinity in insect sensory systems. *Trends in Neurosci.* **8**: 120–5.

Murphey, R.K. (1986) The myth of the inflexible invertebrate: competition and synaptic remodel-

ling in the development of invertebrate nervous systems. *J. Neurobiol.* **17**: 585–91.
Namikas, J. and Wehmer, F. (1978) Gender composition of the litter affects behaviour of male mice. *Behav. Biol.* **23**: 219–24.
Napier, J.R. and Napier, P.H. (1967) *A Handbook of Living Primates*. Academic Press, London.
Napier, J.R. and Napier, P.H. (eds) (1970) *Old World Monkeys, Evolution, Systematics and Behavior*. Academic Press, London.
Narins, P.M. and Capranica, R.R. (1976) Sexual differences in the auditory system of the tree frog, *Eleutherodactylus coqui*. *Science* **192**: 378–80.
Nelson, B. (1980) *Seabirds, their Biology and Ecology*. Hamlyn, London.
Nicol, C. (1987) Behavioral responses of laying hens following a period of spatial restriction. *Anim. Behav.* **35**: 1709–19.
Noble, G.K. (1936) Courtship and sexual selection of the flicker (*Colaptes auratus lutens*). *Auk* **53**: 269–82.
Nottebohm, F. (1980) A brain for all seasons: cyclic anatomical changes in song control nuclei of the canary brain. *Science* **214**: 1368–70.
Nottebohm, F. (1989) From bird song to neurogenesis. *Sci. Am.* **260**(2): 74–9.
Olson, G.C. and Krasne, F.B. (1981) The crayfish lateral giants are command neurons for escape behaviour. *Brain Res.* **214**: 89–100.
Oppenheim, R.W. (1974) The ontogeny of behavior in the chick embryo. *Adv. Study Behav.* **5**: 133–72.
Oppenheim, R.W. (1981) Ontogenetic adaptations and retroprogressive processes in the development of the nervous system and behaviour. In Connolly, K.J. and Prechtl, H.F.R. (eds), *Maturation and Development: biological and psychological perspectives*, pp. 73–109. J.P. Lippincott, Philadelphia.
Orians, G. (1969) On the evolution of mating systems in birds and mammals. *Am. Nat.* **103**: 589–603.
Owen-Smith, N. (1971) Territoriality in the white rhinoceros (Ceratomerium simum) Burchell. *Nature (Lond.)* **231**: 294–6.
Packer, C. (1977) Reciprocal altruism in *Papio anubis*. *Nature (Lond.)* **265**: 441–3.
Packer, C. (1986) The ecology of sociality in felids. In Rubenstein, D.I. and Wrangham, R.W. (eds), *Ecological Aspects of Social Evolution*, pp. 429–51. Princeton University Press, N.J.
Parker, G. (1984) Evolutionarily stable strategies. In Krebs, J.R. and Davies, N.B. (eds), *Behavioural Ecology* (2nd ed.), pp. 30–61. Blackwell Scientific, Oxford.
Partridge, B.L. and Pitcher, T.J. (1979) Evidence against a hydrodynamic function for fish schools. *Nature (Lond.)* **279**: 418–19.
Paul, R.C. and Walker, T.J. (1979) Arboreal singing in a burrowing cricket, *Anurogryllus arboreus*. *J. Comp. Physiol.* **132**: 217–23.
Pavlov, I.P. (1941) *Lectures on Conditioned Reflexes* (2 vols). International Publishers, New York.
Payne, R.S. and McVay, S. (1971) Songs of humpback whales. *Science* **173**: 585–97.
Payne, T.L., Birch, M.C. and Kennedy, C.E.J. (eds) (1986) *Mechanisms in Insect Olfaction*. Clarendon Press, Oxford.
Pepperberg, I.M. (1990) Some cognitive abilities of an African grey parrot (*Psittacus erithacus*). *Adv. Study Behav.* **19**: 357–409.
Plotkin, H.C. and Odling-Smee, F.J. (1979) Learning, change and evolution. *Adv. Study Behav.* **10**: 1–41.
Pollock, G.S. and Hoy, R.R. (1979) Temporal pattern as a cue for species-specific calling song recognition in crickets. *Science (NY)* **204**: 429–32.
Powell, G.V.N. (1974) Experimental analysis of the social value of flocking by starlings (*Sturnus vulgaris*) in relation to predation and foraging. *Anim. Behav.* **22**: 501–5.
Premack, D. and Woodruff, G. (1978) Does the chimpanzee have a theory of mind? *Behav. Brain Sci.* **1**: 515–26.
Provine, R.R. (1976) Eclosion and hatching in cockroach first instar larvae: a stereotyped pattern of behaviour. *J. Insect Physiol.* **22**: 127–31.
Ralls, K. (1971) Mammalian scent marking. *Science (NY)* **171**: 443–9.
Ramsay, A.O. and Hess, E.H. (1954) A laboratory approach to the study of imprinting. *Wilson Bull.* **66**: 196–206.
Rand, A.S. and Rand, W.M. (1978) Display and dispute settlement in nesting iguanas. In Greenberg, N. and Maclean, P.O. (eds), *Behaviour and Neurology of Lizards*, pp. 245–52. National Institute of Mental Health, Rockville, MD.
Read, A.F. and Harvey, P.H. (1989) Reassessment of comparative evidence for Hamilton and Zuk's hypothesis on parasites and sexual selection. *Nature (Lond.)* **328**: 68–70.
Rescorla, R.A. (1988) Pavlovian conditioning: it's

not what you think it is. *Am. Psychol.* **43**: 151–60.
Rhijn, J.G. van (1973) Behavioural dimorphism in male ruffs *Philomachus pugnax* (L.). *Behaviour* **47**: 153–229.
Rhijn, J.G. van (1980) Communication by agonistic displays: a discussion. *Behaviour* **74**: 288–93.
Ridley, M. (1986) The number of males in a primate troop. *Anim. Behav.* **34**: 1848–58.
Riechert, S. (1984) Games spiders play III Cues underlying context-associated changes in agonistic behaviour. *Anim. Behav.* **32**: 1–15.
Riechert, S. and Maynard Smith, J. (1989) Genetic analysis of two behavioral traits linked to individual fitness in the desert spider *Agelenopsis aperta*. *Anim. Behav.* **37**: 624–37.
Rolls, B.J. and Rolls, E.T. (1982) *Thirst*. Cambridge University Press, Cambridge.
Romer, A.S. (1962) *The Vertebrate Body*. W.B. Saunders, Philadelphia.
Roper, T.J. (1983) Learning as a biological phenomenon. In Slater, P.J.B. and Halliday, T.R. (eds), *Animal Behaviour, Vol 3. Genes, Development and Learning*, pp. 178–212. Blackwell Scientific, Oxford.
Roper, T.J. (1984) Response of thirsty rats to absence of water: frustration disinhibition or compensation? *Anim. Behav.* **32**: 1225–35.
Roper, T.J. (1986) Cultural evolution of feeding behaviour in animals. *Sci. Progress* **70**: 571–83.
Rosenblatt, J.S. and Siegel, H.I. (1983) Physiological and behavioural changes during pregnancy and parturition underlying the onset of maternal behaviour in rodents. In Elwood, R.W. (ed.), *Parental Behaviour of Rodents*, pp. 23–66. Wiley, Chichester.
Rosenzweig, M.R. (1984) Experience, memory and the brain. *Am. Psychol.* **39**: 365–76.
Roth, L.M. (1948) An experimental laboratory study of the sexual behavior of *Aëdes aegypti* (L.). *Am. Midl. Nat.* **40**: 265–352.
Rowell, C.H.F. (1961) Displacement grooming in the chaffinch. *Anim. Behav.* **9**: 38–63.
Rowell, T. (1972) *The Social Behaviour of Monkeys*. Penguin Books, Harmondsworth.
Rowell, T. (1974) The concept of social dominance. *Behav. Biol.* **2**: 131–54.
Rowland, W.J. (1989) Mate choice and the supernormality effect in female sticklebacks (*Gasterosteus aculeatus*). *Behav. Ecol. Sociobiol.* **24**: 433–8.
Rozin, P. and Kalat, J.W. (1971) Specific hungers and poison avoidance as adaptive specialisations of learning. *Psychol. Rev.* **78**: 459–86.
Rushen, J. (1985) Stereotypies, aggression and the feeding schedules of tethered sows. *Appl. Anim. Behav. Sci.* **14**: 137–47.
Rutter, M. (1979) Maternal deprivation 1972–1978: new findings, new concepts, new approaches. *Child Dev.* **50**: 283–305.
Ryan, M.J., Fox, J.H., Wilczynski, W. and Rand, A.S. (1990) Sexual selection for sensory exploitation in the frog *Physalaemus pustulosus*. *Nature (Lond.)* **343**: 66–7.
Ryan, M.J., Tuttle, M.D. and Rand, A.S. (1982) Bat predation and sexual advertisement in a neotropical anuran. *Am. Natur.* **119**: 136–9.
Sacks, O. (1985) *The Man Who Mistook His Wife for a Hat*. Picador, Pan Books, London.
Sade, D.S. (1965) Some aspects of parent–offspring and sibling relations in a group of rhesus monkeys with a discussion of grooming. *Am. J. Phys. Anthropol.* **23**: 1–18.
Saunders, D.S. (1977) *An Introduction to Biological Rhythms*. Blackie, Glasgow.
Savage-Rumbaugh, S., Raumbaugh, D. M. and Boysen, S. (1978) Linguistically mediated tool use and exchange by chimpanzees *Pan troglodytes*. *Behav. Brain Sci.* **1**: 539–54.
Schaller, G. (1972) *The Serengeti Lion: a study of predator prey relations*. University of Chicago Press, Chicago, IL.
Schjelderup-Ebbe, T. (1935) Social behaviour of birds. In Murchison, C. (ed.), *Handbook of Social Psychology*, pp. 947–72. Clark University Press, Worcester, MA.
Schleidt, W.M. (1964) Uber die Spontaneität von Erbkoordinationen. *Z. Tierpsychol.* **21**: 235–56.
Schleidt, W.M., Schleidt, M. and Magg, M. (1960) Störung der Mutter-Kind-Beziehung bei Truthühnern durch Gehörverlust. *Behaviour* **16**: 254–60.
Schneider, D. (1966) Chemical sense communication in insects. *Symp. Soc. Exp. Biol.* **20**: 273–97.
Schneider, D. (1969) Insect olfaction: deciphering system for chemical messages. *Science (NY)* **163**: 1031–7.
Schneirla, T.C. (1966) Behavioral development and comparative psychology. *Q. Rev. Biol.* **41**: 283–302.
Scott, J.P. (1958) *Aggression*. University of Chicago Press, Chicago, IL.

Scott, J.P. (1962) Critical periods in behavioral development. *Science (NY)* **138**: 949–58.

Scott, J.P. and Fuller, J.L. (1965) *Genetics and the Social Behavior of the Dog*. University of Chicago Press, Chicago, IL.

Searcy, W.A. and Brenowitz, E.A. (1988) Sexual differences in species recognition of avian song. *Nature (Lond.)* **332**: 152–4.

Searcy, W.A. and Marler, P. (1981) A test for responsiveness to song structure and programming in female sparrows. *Science (NY)* **213**: 926–8.

Seely, T. (1985) *Honeybee Ecology*. Princeton University Press, N.J.

Seligman, M.E.P. and Hager, J.L. (eds) (1972) *Biological Boundaries of Learning*. Prentice Hall, Englewood Cliffs, N.J.

Sevenster, P. (1961) A causal analysis of a displacement activity (fanning in *Gasterosteus aculeatus L.*). *Behav. Suppl.* **9**: 1–170.

Sevenster-Bol, A.C.A. (1962) On the causation of drive reduction after a consummatory act (in *Gasterosteus aculeatus L.*). *Archs. néerl. Zool.* **15**: 175–236.

Seyfarth, R.M. and Cheney, D.L. (1984) The natural vocalisations of non-human primates. *Trends in Neurosci.* **7**: 66–73.

Seyfarth, R.M., Cheney, D.L. and Marler, P. (1980) Vervet monkey alarm calls: evidence of predator classification and semantic communication. *Science (NY)* **210**: 801–3.

Shapiro, D.Y. (1979) Social behavior, group structure and the control of sex reversal in hermaphroditic fish. *Adv. Stud. Behav.* **10**: 43–102.

Sherrington, C.S. (1906) *The Integrative Action of the Nervous System*. Scribner's, New York.

Sherrington, C.S. (1917) Reflexes elicitable in the cat from pinna, vibrissae and jaws. *J. Physiol.* **51**: 404–31.

Sherry, D. (1985) Food storage by birds and mammals. *Adv. Study Behav.* **15**: 153–88.

Sherry, D. and Galef, B.G. (1984) Cultural transmission without imitation: milk bottle opening by birds. *Anim. Behav.* **32**: 937–8.

Shettleworth, S.J. (1983) Memory in food-hoarding birds. *Sci. Am.* **248**(3): 86–94.

Short, R.V. (1982) Sex determination and differentiation. In Austin, C.R. and Short, R.V. (eds), *Reproduction in Mammals, 2. Embryonic and Fetal Development*, pp. 70–113. Cambridge University Press, Cambridge.

Skinner, B.F. (1938) *The Behavior of Organisms*. Appleton-Century-Crofts, New York.

Slater, P.J.B. (1983) Bird song learning: theme and variations. In Brush, A.H. and Clark, G.A. Jr (eds), *Perspectives in Ornithology*, pp. 475–511. Cambridge University Press, Cambridge.

Slater, P.J.B. (1989) Bird song learning: causes and consequences. *Ethology, Ecology and Evolution* **1**: 19–46.

Slater, P.J.B. and Ince, S.A. (1979) Cultural evolution in chaffinch song. *Behaviour* **71**: 146–66.

Slater, P.J.B., Ince, S.A. and Colgan, P.W. (1980) Chaffinch song types: their frequencies in the population and distribution between repertoires of different individuals. *Behaviour* **75**: 207–18.

Smuts, B.B., Cheney, D.L., Seyfarth, R.M., Wrangham, R.W. and Struhsaker, T.T. (eds) (1987) *Primate Societies*. University of Chicago Press, Chicago.

Sonnemann, P. and Sjölander, S. (1977) Effects of cross fostering on the sexual imprinting of the female zebra finch *Taeniopygia guttata*. *Z. Tierpsychol.* **45**: 337–48.

Spalding, D. (1873) Instinct: with original observations on young animals. *Macmillan's Mag.* **27**: 282–93. (Reprinted in *Br. J. Anim. Behav.* (1954) **2**: 1–11.)

Sparks, J. (1982) *The Discovery of Animal Behaviour*. Collins/BBC, London.

Stent, G.S. (1980) The genetic approach to developmental neurobiology. *Trends in Neurosci.* **3**: 49–51.

Stephens, D.W. and Krebs, J.R. (1986) *Foraging Theory*. Princeton University Press, N.J.

Struhsaker, T.T. and Leland, L. (1987) Colobines: infanticide by adult males. In Smuts, B. *et al.* (eds) *Primate Societies*, pp. 38–97.

Sugawara, K. (1979) Sociobiological study of a wild group of hybrid baboons between *Papio anubis* and *P. hamadryas* in the Awash Valley, Ethiopia. *Primates* **20**: 21–56.

Taylor, A., Sluckin, W. and Hewitt, R. (1969) Changing colour preference of chicks. *Anim. Behav.* **17**: 3–8.

Teitelbaum, P. and Epstein, A.N. (1962) The lateral hypothalamic syndrome: recovery of feeding and drinking after lateral hypothalamic lesions. *Psychol. Rev.* **69**: 74–90.

Ten Cate, C. (1982) Behavioural differences between zebra finches and Bengalese finch (foster) parents raising zebra finch offspring. *Behaviour* **81**: 152–72.

Ten Cate, C., Los, L. and Schilperood, L. (1984) The influence of differences in social experience on the development of species recognition in zebra finch males. *Anim. Behav.* **32**: 852–60.

Thorpe, W.H. (1961) *Bird Song*. Cambridge University Press, Cambridge.

Thorpe, W.H. (1963) *Learning and Instinct in Animals* (2nd ed.). Methuen, London.

Tiefer, L. (1978) The context and consequences of contemporary sex research: a feminist perspective. In McGill, T.E., Dewsbury, D.A. and Sachs, B. (eds), *Sex and Behavior: status and prospectus*, pp. 363–85. Plenum Press, New York.

Tinbergen, N. (1951) *The Study of Instinct*. Oxford University Press, Oxford.

Tinbergen, N. (1952) 'Derived' activities, their causation, biological significance, origin and emancipation during evolution. *Q. Rev. Biol.* **27**: 1–32.

Tinbergen, N. (1959) Comparative studies of the behaviour of gulls (Laridae): a progress report. *Behaviour* **15**: 1–70.

Tinbergen, N. (1963) On aims and methods of ethology. *Z. Tierpsychol.* **20**: 410–33.

Tinbergen, N. (1968) On war and peace in man and animals. *Science (NY)* **160**: 1411–18.

Tinbergen, N. and Perdeck, A.C. (1950) On the stimulus situation releasing the begging response in the newly-hatched herring gull chick (*Larus a. argentatus Pont*). *Behaviour* **3**: 1–38.

Tinbergen, N., Kruuk, H., Paillette, M. and Stamm, R. (1962a) How do black-headed gulls distinguish between eggs and egg-shells. *Br. Birds.* **55**: 120–9.

Tinbergen, N., Broekhuysen, G.J., Feekes, F., Houghton, J.C., Kruuk, H. and Szuk, E. (1962b) Eggshell removal by the black-headed gull *Larus ridibundus L.*: a behavioural component of camouflage. *Behaviour* **19**: 74–117.

Toates, F. (1986) *Motivational Systems*. Cambridge University Press, Cambridge.

Tolman, E.C. (1932) *Purposive Behaviour in Animals and Men*. The Century Co., New York.

Trivers, R.L. (1974) Parent–offspring conflict. *Am. Zool.* **14**: 249–64.

Trivers, R.L. (1985) *Social Evolution*. Benjamin Cummings, Menlo Park, CA.

Tyndale-Biscoe, C.H. (1973) *The Life of Marsupials*. Edward Arnold, London.

Valenstein, E.S., Cox, V.C. and Kakolewski, J.W. (1970) Reexamination of the role of the hypothalamus in motivation. *Psychol. Rev.* **77**: 16–31.

Van Schaik, C.P. (1983) Why are diurnal primates living in groups? *Behaviour* **87**: 120–44.

Vaysse, G. and Médioni, J. (1982) *L'Emprise des Gènes*. Privat, Toulouse.

Verner, J. and Willson, M.F. (1966) The influence of habitats on mating systems of North American passerine birds. *Ecology* **47**: 143–7.

Vestergaard, K. (1980) The regulation of dustbathing and other patterns in the laying hen: a Lorenzian approach. In Moss, R. (ed.), *The Laying Hen and its Environment*, pp. 101–20. Martinus Nijhoff, The Hague.

Vidal, J.M. (1980) The relations between filial and sexual imprinting in the domestic fowl: effects of age and social experience. *Anim. Behav.* **28**: 880–91.

Vince, M.A. (1969) Embryonic communication, respiration and the synchronisation of hatching. In Hinde, R.A. (ed.), *Bird Vocalisations*, pp. 233–60. Cambridge University Press, Oxford.

Vines, G. (1981) Wolves in dog's clothing. *New Sci.* **91**: 648–52.

Vom Saal, F.S. and Bronson, F. (1980) Sexual characteristics of adult females correlates with their blood testosterone levels during development in mice. *Science (NY)* **208**: 597–9.

Vowles, D.M. (1965) Maze learning and visual discrimination in the wood ant (*Formica rufa*). *Br. J. Psychol.* **56**: 15–31.

Waal, F. van der (1989) *Chimpanzee Politics*. The Johns Hopkins University Press, Baltimore, MD.

Waldman, B., Frumhoff, P.C. and Sherman, P.W. (1988) Problems of kin recognition. *Trends in Ecol. and Evol.* **3**: 8–13.

Walker, S. (1983) *Animal Thought*. Routledge and Kegan Paul, London.

Walsh, E.G. (1964) *Physiology of the Nervous System* (2nd ed.). Longmans Green, London.

Warren, J.M. (1965) Primate learning in comparative perspective. In Schrier, A.M., Harlow, H.F. and Stollnitz, F. (eds), *Behavior of Non-human Primates*, Vol. I, pp. 249–81. Academic Press, New York.

Watson, J.B. (1924) *Behaviorism*. University of Chicago Press, Chicago, IL.

Wells, M.J. (1958) Factors affecting reactions to *Mysis* by newly hatched *Sepia*. *Behaviour* **13**: 96–111.

Wells, M.J. (1962) *Brain and Behaviour in Cephalopods*. Heinemann Educational, London.

Wiersma, C.A.G. (1947) Giant nerve fibre system of the crayfish. A contribution to comparative physiology of synapse. *J. Neurophysiol.* **10**: 23–38.

Wilcox, R.S. (1979) Sex discrimination in *Gerris remigris*: role of a surface wave signal. *Science (NY)* **206**: 1325.

Wiley, R.H. (1973) The strut display of male sage grouse: a 'fixed' action pattern. *Behaviour* **47**: 129–52.

Wiley, R.H. and Richards, D.G. (1978) Physical constraints on acoustic communication in the atmosphere: implications for the evolution of animal vocalisations. *Behav. Ecol. Sociobiol.* **3**: 69–94.

Wilkinson, G.S. (1984) Reciprocal food sharing in the vampire bat. *Nature (Lond.)* **308**: 181–4.

Wilson, E.O. (1965) Chemical communication in the social insects. *Science (NY)* **149**: 1064–71.

Wilson, E.O. (1971) *The Insect Societies*. Belknap Press of Harvard University Press, Cambridge, MA.

Wilson, E.O. (1975) *Sociobiology*. Harvard, The Belknap Press, Cambridge, MA.

Wilson, E.O. and Bossert, W.H. (1963) Chemical communication among animals. *Rec. Progr. Hormone. Res.* 673–716.

Wiltschko, W. and Wiltschko, R. (1988) Magnetic versus celestial orientation in migrating birds. *Trends in Ecol. and Evol.* **3**: 13–15.

Wilz, K.J. (1970) Causal and functional analysis of dorsal pricking and nest activity in the courtship of the three-spined stickleback *Gasterosteus aculeatus*. *Anim. Behav.* **18**: 115–24.

Wine, J.J. and Krasne, F.B. (1982) The cellular organisation of crayfish escape behavior. In Bliss, D.E., Atwood, H. and Sandeman, D. (eds), *The Biology of Crustacea. Vol IV, Neural Integration*. Academic Press, New York.

Wirtshafter, D. and Davis, J.D. (1977) Set points, settling points and the control of body weight. *Physiol. and Behav.* **19**: 75–8.

Woolfenden, G.E. and Fitzpatrick, J.W. (1984) *The Florida Scrub Jay*. Princeton University Press, NJ.

Wrangham, R.W. (1987) Evolution of social structure. In Smuts, B.B. *et al.* (eds), *Primate Societies*, pp. 282–96.

Wyles, J.S. Kunkel, J.G. and Wilson, A.C. (1983) Birds, behavior and anatomical evolution. *Proc. Natl. Acad. Sci.* **80**: 4394–7.

Young, D. (1989) *Nerve Cells and Animal Behaviour*. University of Cambridge Press, Cambridge.

Zahavi, A. (1975) Mate selection—a selection for a handicap. *J. Theor. Biol.* **53**: 205–14.

Zahavi, A. (1987) The theory of signal selection and some of its implications. In Delfino, V.P. (ed.), *International Symposium on Biological Evolution*, pp. 305–27. Adriatica Editrice.

Zarrow, M.X., Denenberg, V.H. and Anderson, C.O. (1965) Rabbit: frequency of suckling in the pup. *Science (NY)* **150**: 1835–6.

Latin Names of Species where not given in text

INVERTEBRATES

Mollusca

sea slug	*Tritonia* sp., *Pleurobranchaea* sp.
octopus	*Octopus vulgaris*
cuttle-fish	*Sepia officinalis*
water snail	*Helisoma trivolvis*

Arthropoda

water flea	*Daphnia* sp.
fiddler crab	*Uca* spp.
crayfish	*Procambarus clarkii*
jumping spiders	*Salticidae*
cockroach, American	*Periplaneta americana*
cricket	*Gryllus* sp., *Teleogryllus*
katydids	*Tettigonidae*
praying mantis	*Mantis religiosa*
termites	Isoptera
earwigs	*Forficula* sp.
pond-skater (water-strider)	*Gerris remigris*
aphid	*Aphis* sp.
silver-washed fritillary	*Dryas paphia*
cinnabar moth	*Hypocrita jacobaeae*
hawk moth	*Manduca sexta*
silk moth	*Bombyx mori*
blowfly	*Phormia regina*
mosquito	*Aëdes* sp.
ant	*Formica* sp.
weaver ant	*Oecophylla* sp.
mason wasp	*Manobia quadridens*
digger wasp	*Ammophila* sp.
great golden digger wasp	*Sphex ichneumoneus*
sweat bee	*Lasioglossum zephyrum*
wasp	*Vespa* sp.
bumble-bee	*Bombus* sp.
honey-bee	*Apis mellifera*

VERTEBRATES

Fish

minnow	*Phoxinus phoxinus*
three-spined stickleback	*Gasterosteus aculeatus*
Siamese fighting-fish	*Betta splendens*
goldfish	*Carassius auratus*
pike	*Esox lucius*

Amphibia

green tree frogs	*Hyla cinerea, H. arborea*
cricket frog	*Acris crepitans*
Tungara frog	*Physalaemus pustulosus*
toads	*Bufo bufo*
newt	*Triturus vulgaris*

Reptiles

turtle	*Chrysemys picta*
garter snake	*Thamnophis sirtalis*
water snake	*Natrix sipedon*
rattlesnake	*Crotalus* sp.

Birds

great crested grebe	*Podiceps cristatus*
cormorant	*Phalacrocorax carbo*
emperor penguin	*Aptenodytes forsteri*
gannet	*Sula bassana*
great blue heron	*Ardea herodias*
purple heron	*Ardea purpura*
mallard	*Anas platyrhynchos*
garganey	*Anas querquedula*
pintail duck	*Anas acuta*
yellow-billed teal	*Nettion flavirostre*
shelduck	*Tadorna tadorna*
mandarin duck	*Aix galericulata*
wood-duck	*Aix sponsa*
goldeneye duck	*Bucephala clangula*
Canada goose	*Branta canadensis*
goose, greylag or domestic	*Anser anser*
martial eagle	*Polemaetus bellicosus*
hawks	Falconiformes
goshawk	*Accipiter gentilis*
turkey, wild or domestic	*Meleagris gallopavo*
sage grouse	*Centrocercus urophasianus*
black grouse	*Lyrurus tetrix*
red grouse	*Lagopus lagopus*
prairie chicken	*Tympanuchus cupido*
pheasant	*Phasianus* sp.
quail (Japanese)	*Coturnix coturnix*
chicken, domestic and jungle fowl	*Gallus gallus*
peafowl (blue)	*Pavo cristatus*
moor- or water-hen	*Gallinula chloropus*
coot	*Fulica atra*
jacana	*Jacana* sp.
oystercatcher	*Haematopus ostralegus*
ruff	*Philomachus pugnax*
black-headed gull	*Larus ridibundus*
lesser black-backed gull	*Larus fuscus*
common gull	*Larus canus*

herring gull	*Larus argentatus*
laughing gull	*Larus atricilla*
kittiwake	*Rissa tridactyla*
terns	*Sterna* spp.
pigeon	*Columba livia*
ring dove	*Streptopelia risoria*
cuckoo (European)	*Cuculus canorus*
owl	*Strix* sp.
love-birds	*Agapornis* spp.
African grey parrot	*Psittacus erithacus*
superb lyrebird	*Menura novaehollondiae*
yellow-shafted flicker	*Colaptes auratus*
lark	*Alauda* sp.
pied wagtail	*Motacilla alba*
meadow pipit	*Anthus pratensis*
tree pipit	*Anthus trivialis*
swallow	*Hirundo rustica*
bank swallow (sand martin)	*Riparia riparia*
cliff swallow	*Petrochelidon pyrrhonata*
dunnock (hedge sparrow)	*Prunella modularis*
red-winged blackbird	*Agelaius phoeniceus*
carib grackle	*Quiscalus lugubris*
Steller's jay	*Cyanocitta stelleri*
Florida scrub jay	*Aphelocoma coerulescens*
jay (European)	*Garrulus glandarius*
crow	*Corvus* sp.
magpie	*Pica pica*
blue titmouse	*Parus caeruleus*
great titmouse	*Parus major*
marsh titmouse	*Parus palustris*
black-capped chickadee	*Parus atricapillus*
nuthatch	*Sitta europea*
blackbird (European)	*Turdus merula*
song thrush	*Turdus philomelos*
robin, European	*Erithacus rubecula*
wheatear	*Oenanthe oenanthe*
redstart	*Phoenicurus phoenicurus*
reed warbler	*Acrocephalus scirpaceus*
great reed warbler	*Acrocephalus arundinaceus*
sedge warbler	*Acrocephalus schoenobaenus*
blackcap	*Sylvia atricapilla*
willow warbler	*Phylloscopus trochilus*
chiffchaff	*Phylloscopus collybita*
wren	*Troglodytes troglodytes*
pied flycatcher	*Ficedula hypoleuca*
golden-winged sunbird	*Nectarina reichenowi*
bou-bou shrike	*Laniarius aethiopicus*
Arabian babbler	*Turdoides squamiceps*
starling	*Sturnus vulgaris*
chaffinch	*Fringilla coelebs*
bullfinch	*Pyrrhula pyrrhula*
greenfinch	*Carduelis chloris*
linnet	*Acanthis cannabina*
canary	*Serinus canaria*
reed bunting	*Emberiza schoeniclus*
song sparrow	*Melospiza melodia*
swamp sparrow	*Melospiza georgiana*
Lincoln sparrow	*Melospiza lincolnii*
white-crowned sparrow	*Zonotrichia leucophrys*
indigo bunting	*Passerina cyanea*
zebra finch	*Taeniopygia guttata*
Bengalese finch (striated finch)	*Lonchura striata*

spice finch	*Lonchura punctulata*
cut-throat finch	*Amandina fasciata*
strawberry finch	*Amandava amandava*
house sparrow	*Passer domesticus*
red-billed weaver	*Quelea quelea*
long-tailed widow, bird	*Euplectes progne*
Prince Rupert's blue bird of paradise	*Paradisea rudolphi*

Mammals

INSECTIVORA	
water shrew	*Neomys fodiens*
tree shrew	*Tupaia belangeri*
RODENTIA	
mouse, domestic	*Mus musculus*
brown rat, *also* domestic rat, laboratory rat	*Rattus norvegicus*
flying squirrel	*Glaucomys volans*
guinea pig	*Cavia porcellus*
naked mole rat	*Heterocephalus glaber*
squirrel	*Sciurus* sp.
Belding's ground squirrel	*Spermophilus beldingi*
LAGOMORPHA	
rabbit	*Oryctolagus cuniculus*
CETACEA	
Humpback whale	*Megaptera novaeangliae*
PROBOSCIDEA	
African elephant	*Loxodonta africana*
PERISSODACTYLA	
white rhinoceros	*Ceratotherium simum*
horse	*Equus caballus*
ARTIODACTYLA	
pig	*Sus scrofa*
hippopotamus	*Hippopotamus amphibius*
pygmy hippopotamus	*Choeropsis liberiensis*
red deer	*Cervus elephas*
white-tailed deer	*Odocoileus virginianus*
moose	*Alces alces*
buffalo	*Synceros caffer*
eland	*Taurotragus oryx*
duiker	*Cephalophus* sp.
wildebeeste	*Connochaetes taurinus*
Uganda kob	*Kobus kob*
water buck	*Kobus ellipsiprymnus*
impala	*Aepyceros melampus*
gazelle	*Gazella* sp.
dik-dik	*Madoqua* sp.
muskox	*Ovibos moschatus*
sheep, domestic	*Ovis aries*
goat	*Capra* sp.
CHIROPTERA	
frog-eating bat	*Trachops cirrhosus*
vampire bat	*Desmodus rotundus*
CARNIVORA	
polecat	*Mustela putorius*
stoat	*Mustela erminea*
skunk	*Mephitis* sp.
spotted hyaena	*Crocuta crocuta*
cape hunting dog	*Lycaon pictis*
fox	*Vulpes* sp.
coyote	*Canis latrans*
black-backed jackal	*Canis mesomelas*
dog, domestic	*Canis familiaris*

wolf	*Canis lupus*
leopard	*Panthera pardus*
tiger	*Panthera tigris*
lion	*Panthera leo*
cat, domestic	*Felis domesticus*
cheetah	*Acinonyx jubatus*
polar bear	*Ursus maritimus*
meerkat	*Suricata suricatta*
elephant seal	*Mirounga angustirostris*
PRIMATES	
tarsier	*Tarsius* sp.
lemurs	Lemuridae
ring-tailed lemur	*Lemur catta*
marmoset	*Callithrix* sp.
tamarin	*Saguinus* sp.
dusky titi	*Callicebus moloch*
capuchin monkey	*Cebus* sp.
squirrel monkey	*Saimiri sciurea*
howler monkey	*Alouatta* sp.
spider monkey	*Ateles* sp.
rhesus monkey	*Macaca mulatta*
Japanese macaque	*Macaca fuscata*
Barbary macaque (Barbary ape)	*Macaca sylvanus*
pig-tailed macaque	*Macaca nemestrina*
vervet monkey	*Cercopithecus aethiops*
patas monkey	*Erythrocebus patas*
chacma baboon	*Papio ursinus*
yellow baboon	*Papio cynocephalus*
hamadryas baboon	*Papio hamadryas*
gelada baboon	*Theropithecus gelada*
black and white colobus	*Colobus polykomos*
red colobus	*Colobus badius*
mangabey	*Cercocebus albigena*
langur	*Presbytis entellus*
gibbons	*Hylobatidae*
orangutan	*Pongo pygmaeus*
chimpanzee	*Pan troglodytes*
gorilla	*Gorilla gorilla*

INDEX